Petroleum Engineering Handbook

Petroleum Engineering Handbook

Edited by **Andy Margo and**
Michael Dedini

NY RESEARCH PRESS
New York

Published by NY Research Press,
23 West, 55th Street, Suite 816,
New York, NY 10019, USA
www.nyresearchpress.com

Petroleum Engineering Handbook
Edited by Andy Margo and Michael Dedini

International Standard Book Number: 978-1-63238-355-6 (Hardback)

Printed in China.

Contents

	Preface	**VII**
Chapter 1	**The Mechanism of Wellbore Weakening in Worn Casing-Cement-Formation System** Zheng Shen, Frederick E. Beck and Kegang Ling	**1**
Chapter 2	**Mysteries behind the Low Salinity Water Injection Technique** Emad Waleed Al-Shalabi, Kamy Sepehrnoori and Gary Pope	**9**
Chapter 3	**Effect of Polymer Adsorption on Permeability Reduction in Enhanced Oil Recovery** Saurabh Mishra, Achinta Bera and Ajay Mandal	**20**
Chapter 4	**Investigation of CO_2 Enhanced Oil Recovery Using Dimensionless Groups in Wettability Modified Chalk and Sandstone Rocks** Vahid Alipour Tabrizy	**29**
Chapter 5	**CFD Method for Predicting Annular Pressure Losses and Cuttings Concentration in Eccentric Horizontal Wells** Titus N. Ofei, Sonny Irawan and William Pao	**45**
Chapter 6	**Study and Application of Polymer Film-Forming Shielding Plugging Drilling Fluid System** Yuxue Sun, Fei Teng and Jingyuan Zhao	**61**
Chapter 7	**Optimization of Multiple Hydraulically Fractured Horizontal Wells in Unconventional Gas Reservoirs** Wei Yu and Kamy Sepehrnoori	**66**
Chapter 8	**Dispersing of Petroleum Asphaltenes by Acidic Ionic Liquid and Determination by UV-Visible Spectroscopy** Eshagh Rezaee Nezhad, Fariba Heidarizadeh, Sami Sajjadifar and Zahra Abbasi	**82**
Chapter 9	**Corrosion Inhibition of Tubing Steel during Acidization of Oil and Gas Wells** M. Yadav, Sumit Kumar and P. N. Yadav	**87**
Chapter 10	**Electromagnetic Heating of Heavy Oil and Bitumen: A Review of Experimental Studies and Field Applications** Albina Mukhametshina and Elena Martynova	**96**

Chapter 11 **Simulation of Formation Damage after Long-Term Water Flooding** 103
Liu He, Gao Yang, Li Guoxin and Li Yiliang

Chapter 12 **Rethinking Petroleum Products Certification** 110
Thiago Feital, Públio Lima, José Carlos Pinto, Maurício B. de Souza Jr., Gilberto Xavier, Mario Jorge Lima and Marcel Joly

Chapter 13 **Modeling of Partially Hydrolyzed Polyacrylamide-Hexamine-Hydroquinone Gel System Used for Profile Modification Jobs in the Oil Field** 122
Upendra Singh Yadav and Vikas Mahto

Chapter 14 **Knowledge Discovery for Classification of Three-Phase Vertical Flow Patterns of Heavy Oil from Pressure Drop and Flow Rate Data** 133
Adriane B. S. Serapião and Antonio C. Bannwart

Chapter 15 **Improving the Demulsification Process of Heavy Crude Oil Emulsion through Blending with Diluent** 141
K. K. Salam, A. O. Alade, A. O. Arinkoola and A. Opawale

Chapter 16 **A Field Study on Simulation of CO_2 Injection and ECBM Production and Prediction of CO_2 Storage Capacity in Unmineable Coal Seam** 147
Qin He, Shahab D. Mohaghegh and Vida Gholami

Chapter 17 **Well Screen and Optimal Time of Refracturing: A Barnett Shale Well** 155
Shayan Tavassoli, Wei Yu, Farzam Javadpour and Kamy Sepehrnoori

Chapter 18 **Review of Sand Production Prediction Models** 165
Hossein Rahmati, Mahshid Jafarpour, Saman Azadbakht, Alireza Nouri, Hans Vaziri, Dave Chan and Yuxing Xiao

Chapter 19 **Effect of Bed Deformation on Natural Gas Production from Hydrates** 181
Mohamed Iqbal Pallipurath

Permissions

List of Contributors

Preface

Petroleum Engineering is a field which is concerned with activities related to the production of crude oil or natural gas. The profession started around the early 1900s and has evolved significantly over the decades. With the ever increasing demand for reservoirs of oil and gas, this is a field that is highly critical in today's industry and economy. This field requires at least satisfactory knowledge of many related fields like petroleum geology, drilling, economics, reservoir engineering and geophysics, among many others. Since its inception, petroleum engineering has evolved towards solving highly difficult situations, keeping in mind the constantly depleting global oil and gas reservoirs. New methodologies are constantly being developed to break through problems in depletion, like better computer modelling, analysis, new materials, as well as new technologies such as enhanced oil recovery and horizontal drilling. These have drastically improved the chances of finding new and previously untapped reservoirs. Those recruited to this industry have usually been from the fields of chemical engineering, mining and physics, while further training in specialised fields is usually done with respective companies. Engineers in this industry broadly divide themselves into several types which include reservoir engineers, production engineers and drilling engineers. A base career in petroleum engineering usually requires the search for natural oil and gas reservoirs or perhaps in developing the latest machinery and equipment to be used.

This book is an effort to collate all new research findings in petroleum engineering and we are thankful to those whose hard work and toil resulted in these findings.

Editor

The Mechanism of Wellbore Weakening in Worn Casing-Cement-Formation System

Zheng Shen,[1] Frederick E. Beck,[1] and Kegang Ling[2]

[1] *Texas A&M University, College Station, TX 77843, USA*
[2] *University of North Dakota, Grand Forks, ND 58202, USA*

Correspondence should be addressed to Zheng Shen; zheng.shen@weatherford.com

Academic Editor: Guillaume Galliero

Maintaining casing integrity, in terms of downhole zonal isolations and well stability, is extremely important in oil/gas wells. Casing wear occurs not only in directional drilling, but also in vertical drilling with a slight deviation angle. In most hydrocarbon wells, deteriorated casing was reported from the onset of casing wear by the presence of friction force during the rotation of drillpipe. The friction force against the casing wall causes the reduction of casing strength. Furthermore, the rotation of drillpipe combined with corrosive drilling fluids could dramatically degrade the casing strength. We used a finite element analysis to focus on the stress evolution in worn casings. Comparison study between worn casing and perfect casing was conducted. Our study showed that the thermal load significantly increases the stress concentration of the worn casing in the wellbore. Finite element solutions indicated that the radial stress of the worn casing is not affected as much as the hoop stress. Along with the increased burst pressure or the elevated temperature, the unworn portion of the casing also suffers from severe compression stress. This work is important to broadening the understanding of well engineers through addressing the true stress profile of worn casing in cemented wellbore.

1. Introduction

Casing wear in the oil and gas industry is recorded on a world basis. Rotation of drillpipe during the drilling process creates significant contact forces that result in the reduction of casing wall thickness. Thickness reduction of the casing wall weakens the burst and collapse resistances of casing, where stress concentration at the worn location is expected. Casing wear will be accelerated in the presence of corrosive fluids.

Casing wear is a serious problem that is not limited to directional or extended wells. Because the contact pressure generated on the inner surface of the casing becomes much harder to control during the drill bit penetration into deep formations, worn casing is also found in vertical wells. The strength of the casing depends on the material and geometry of the casing; the calculation of the burst and collapse pressures of casing is given in the Appendix. All criteria indicate that the burst and collapse resistances strengthen with the increase of casing wall thickness.

Within the limited information about the stress profile of worn casings, casing wear is analyzed alone in models without including cement sheath and formation. The stress profiles of worn casing developed by using the existing isothermal models [1] are insufficient to provide the whole picture regarding how worn casing behaves in a downhole field condition. The drilling process becomes more complicated in deep formations where the temperature of wellbore fluids is significantly different in conventional wells. In event of the worn casing, maintaining the stability of the cased wellbore in high-pressure and high-temperature (HPHT) wells is challenging because of large potential variations of temperature and pressure. High temperature can bring significant pressure increase in the sealed annuli, which can lead to further failures of the casing string or the production liner [2].

Casing wear can be caused by different parts of drilling tools. Casing wear typically results from contact pressure of the casing and the tool joint. As shown in Figure 1, the tool joint represented by part 2 has a relatively small diameter compared to the casing represented by part 1. The thickness of casing wall decreases because of the crescent-shaped wear by the tool joint indicted in the red area. After the reduction of casing wall thickness, stress caused by wellbore fluid pressure

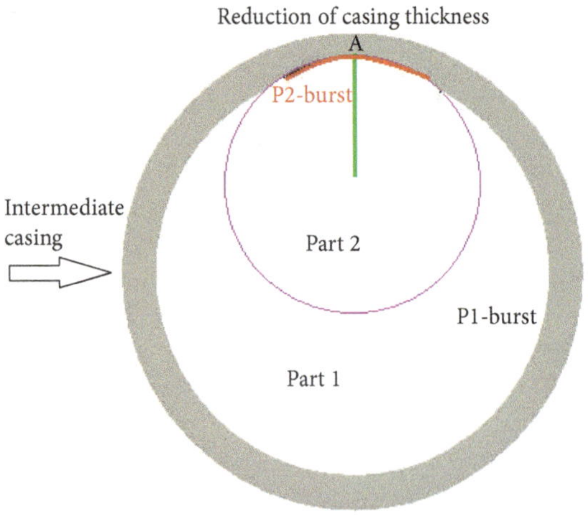

FIGURE 1: Crescent shape worn casing by the tool joint.

and formation *in situ* stresses is expected to be concentrated on the worn casing.

Thorough understanding of the true stress profile of worn casing leads to economic and safe casing design. Unfortunately, mechanical behaviors of casing from past studies were incomplete because of ignoring the formation and time-dependent temperature effect. Researchers have investigated the stress profile of worn casing, focusing on the standalone casing [3, 4]. Others found the relation of wear depth and contact force through experimental tests, with each wear depth recorded at a given contact force [5]. Researchers [6, 7] have also studied the effect of formation to casing. They concluded that the surrounding formation significantly increased the burst resistance of casing, but the work was applied to the isothermal downhole condition. None of the existing studies considered the thermal and mechanical behaviors of worn casing in a cemented well. The mechanism of wellbore weakening due to worn casing has not yet been sufficiently understood. The limitations of the existing models motivated us to determine true stress profile in and around worn casing in the cased hole.

This work aims to describe the stress concentration of casing after wear in a cemented well. The results showed that the formation *in situ* stresses and temperature largely impact the stress profile of the worn casing in the cemented wellbore. Casing uncemented with formation suffers from tension when the burst pressure is applied in the wellbore. Without considering the temperature effect in the worn casing, the stress concentration of the worn part is apparently underestimated. The traditional Lame equations for stress calculation in the wellbore are weakened by not including the temperature effect, cement sheath, and formation. The results also showed that the elevated temperature has expanded the worn casing and resulted in further compression in the worn casing. This work established the mechanical responses of worn casing through addressing the true stress profile of worn casing in cemented wellbore. The findings of the present work will be beneficial to well engineers.

2. Previous Studies

Uncemented casing suffers from serious tension when the burst pressure is applied on the inner wall of casing. For a perfect casing without wear in an isothermal condition, the radial stress and hoop stress can be evaluated using (1). The equations do not take into account the effects of cement sheath and formation:

$$\begin{aligned}\sigma_r &= \frac{p_w r_w^2}{r_o^2 - r_w^2}\left(1 - \frac{r_o^2}{r^2}\right) - \frac{p_o r_o^2}{r_o^2 - r_w^2}\left(1 - \frac{r_w^2}{r^2}\right),\\ \sigma_\theta &= \frac{p_w r_w^2}{r_o^2 - r_w^2}\left(1 + \frac{r_o^2}{r^2}\right) - \frac{p_o r_o^2}{r_o^2 - r_w^2}\left(1 + \frac{r_w^2}{r^2}\right).\end{aligned} \tag{1}$$

The reduction of casing wall thickness mainly results from casing corrosion and casing wear. It is well known that drilling through sand formation zones causes serious erosion randomly along the casing pipe. Casing corrosion is categorized into several types. Decarburization is a hydrogen-carbon attack of a component of any alloy under high temperature. Galvanic corrosion occurs when two dissimilar metals are in contact. This type of corrosion, most likely occurring in the surface casing, is normally seen as a result of the active macrocorrosion cells when the surface casing is not well cemented. Biological corrosion happens as a result of activities of living organisms. This happens in the environment of temperature ranging from 30°F to 180°F, pH measurement of 0 to 11, and pressures up to 15,000 psi [8].

In drilling, the rotation of drillpipes could uniformly reduce the casing thickness from the inner surface of casing or create a crescent-shaped wear pattern. Researchers have focused on the wear volume of casing through laboratory measurements [9]. Casing wear can be caused by tool joints, drillpipe, and wireline. Casing wear by the drillpipes and wirelines is not as significant as that by the tool joints. Casing wear by tool joints is determined by the factors such as drill string rotation time and the speed, drilling mud properties, casing strength, and well dogleg severity.

Stress concentration around worn casing has been studied by some researchers. A casing wear model is largely weakened by not considering the temperature and formation. Researchers [1] focused on the rupture capacity of casing after wear. The analytical solution for the hoop stress at the surfaces of worn casing was constructed by dividing the entire worn casing into three superimposable shapes. To obtain the induced hoop stress of the worn casing, the superposition principle was adopted in a bipolar coordinate. Others argued that casing wear models can be simplified by assuming a slotted ring in the casing inner wall [3]. In their method, the resistance of the hoop stress is decreased because of the reduction of casing wall plus the extra burst pressure acting on the surfaces of the slotted ring. The stress under the worn surface was also investigated. The experimental work indicates that the area below the wear surface experiences unusual strain [10].

Researchers investigated the maximum wear groove depth resulting from the contact pressure applied to the inner wall of casing [4]. Various-size drill strings were used to find a relation for the groove depth of crescent-shaped casing wear as a function of time. The casing wear was tested in single, sharp, and blunt groove forms. Another study emphasized the alleviation and prevention of casing wear by using optimized well operations, such as dogleg severity control and sealant application in the worn sections [11].

3. Numerical Modeling

Casing wear causes significant reduction of the casing strength. Worn casing is under the risk of tangential collapse and radial cracking in the wellbore system. Force equilibrium can be expressed in terms of the normal force, shear force, and body force. The general relation of the forces in a cylindrical coordinate system is given in (2) [12]. Ideally, an analytical solution for the hoop stress in the worn casing is required. Typical Lame equations used in casing integrity analysis do not qualify for the effects of worn shape and formation. It is difficult to develop a reasonable and accurate analytical model investigating worn casing. Because the existing force balance equations are not suitable for addressing the effects of cement sheath, formation, and temperature in worn casing system, it is difficult to develop a reasonable and accurate analytical model investigating worn casing:

$$\frac{\partial \sigma_r}{\partial r} + \frac{1}{r}\frac{\partial \tau_{r\theta}}{\partial_\theta} + \frac{\sigma_r - \sigma_\theta}{r} + \sigma f_r = 0. \quad (2)$$

Alternatively, a finite element model can be used to overcome the limitations of the analytical models. Finite element analysis (FEA) is a computer-based numerical technique applied to many engineering problems. This method is popular in performing prejob designs for primary jobs or postmodified jobs for remedial actions. By using a finite element method, a structure, such as the casing-cement-formation system, can be broken down into small elements composed of nodes in a global coordinate system. The deflections and stresses of all nodes must be calculated using computer programs because of the complexity of the global system. The computation time of solving for the finite element model depends on the number of total elements and computer performance.

The numerical simulator ANSYS was used. Numerical simulations were performed to analyze thermal and mechanical behaviors of worn casing in the cemented wellbore. In the present model, the worn casing was located in a vertical well and was perfectly cemented with the surrounding formation. Plane strain assumes that one direction of the investigated body is much larger than the other two directions. In our case, the vertical section of the wellbore is much larger than the radial direction. It is reasonable to construct a 2D numerical model based on the concept of plane strain, which leads to the following relations in (3). The strain of normal to the x-y plane and the shear strains γ_{xz} and γ_{yz} are zero:

$$\varepsilon_z = \gamma_{xz} = \gamma_{yz} = 0. \quad (3)$$

Drilling fluids have significant effect on the stability of the wellbore system. Because of the geometry of the casing cement formation, the thermal energy exchange in the near-wellbore system reaches equilibrium in a short time. A past study [13] has shown that wellbore fluids induce major stress change to the casing in the wellbore at the first hour. In this model, the minimum horizontal stress is equal to maximum stress. Fluids are pumping into the wellbore and circulating back to the surface through the wellbore annulus. The duration of the circulation is one hour. Wellbore fluid temperature is different from the formation temperature. The pressure and temperature at the boundaries are specified in the following:

$$\begin{aligned} r &= r_w, & \sigma &= p_w, \\ r &= r_o, & \sigma &= p_o, \\ r &= r_w, & T &= T_w, \\ r &= r_o, & T &= T_o. \end{aligned} \quad (4)$$

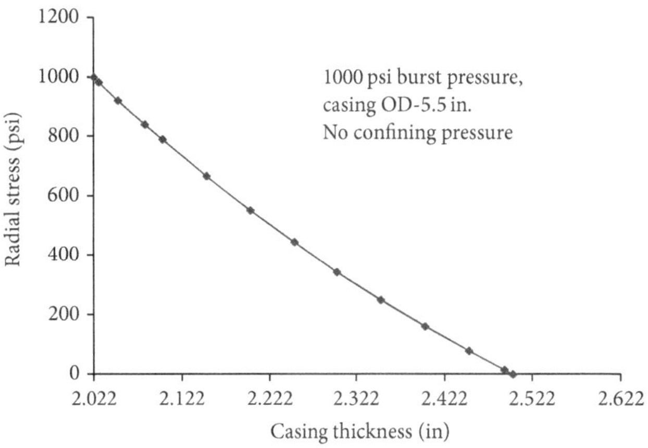

Figure 2: Casing radial stress in the radial direction.

Accuracy of a finite element scheme measures the closeness between the analytical solution and the numerical solution. We first simplified the boundary conditions of the numerical model we used; thus the analytical solutions to the same problem can be found to verify the numerical results. Figures 2 and 3 show the radial stress and tangential stress of casing subject to the internal burst pressure of 1000 psi while ignoring the formation effect. The numerical solution demonstrated a good match compared with the analytical result.

In this following study, the cased wellbore diameter of 6 in. is used, and the outer diameter of the formation is 10 times as large as the wellbore diameter. Cohesion force and friction angle of cement and rock are included in the simulations. The cohesion force is the force of attraction between the molecules of the same substance, and the friction angle measures the ability of rock or cement to resist the shear stresses. Other material properties of the worn casing cement formation used in the numerical model are listed in Tables 1 and 2.

A crescent shape of worn casing is assumed as shown in Figure 3. The reduction factor of worn casing is defined as the biggest reduction of casing thickness over the original casing thickness in (5). Field observations have shown that worn casing in many wells lost approximately 20% of the casing thickness [1]. The original casing thickness is 0.5 in. In the

TABLE 1: Geometric and geomechanical properties.

	ID (in.)	OD (in.)	Elastic modulus (psi)	Poisson ratio	Cohesion force (psi)	Friction angle (°)
Casing	6	7	2.80E + 07	0.3	Not applicable	Not applicable
Cement sheath	7	9	3.00E + 06	0.24	8.00E + 03	30
Formation	9	60	6.00E + 06	0.2	6.00E + 03	30

TABLE 2: Thermal properties.

	Density (lb/in^3)	Conductivity (btu/in-hr-F)	Heat capacity (Btu/lb-F)	Thermal expansion (F^{-1})
Casing	0.253	0.6	0.12	1.8E − 05
Cement sheath	0.065	0.05	0.5	1.2E − 05
Formation	0.072	0.04	0.4	1.4E − 05

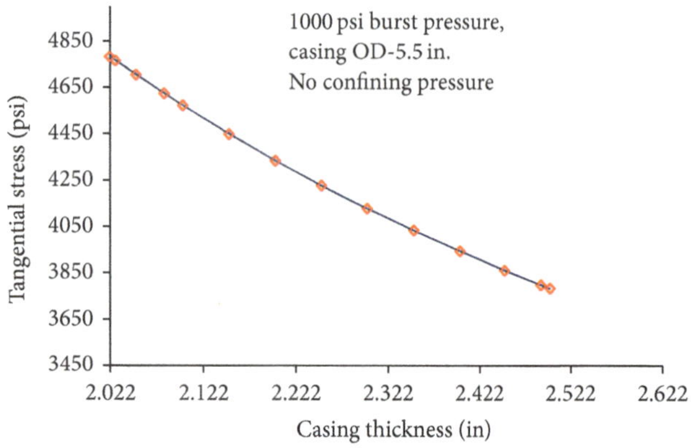

FIGURE 3: Casing tangential stress in the radial direction.

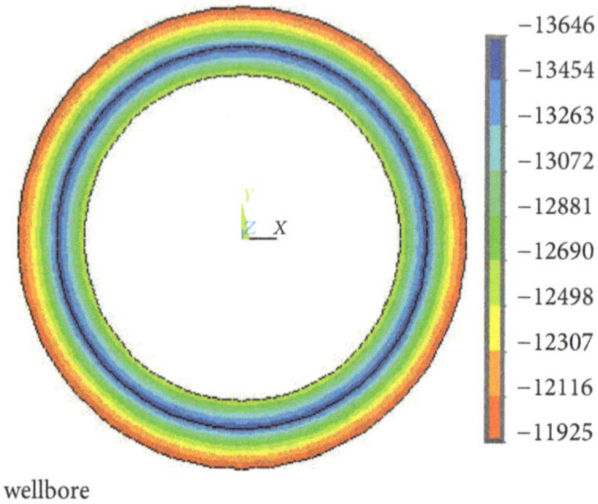

FIGURE 4: Radial stress inside the perfect casing cement for Case 1.

deepest worn point, 0.1 in. thickness of casing is lost after casing wear:

$$\mathrm{RF} = \frac{h_{\text{initial}} - h_{\text{current}}}{h_{\text{initial}}}. \quad (5)$$

Table 3 lists the fluid temperatures and pressures under different scenarios. To evaluate the stress concentration in the worn casing, six base models were built with either different temperatures or wellbore pressures in the model. Case 1 simulates the perfect casing-cement-formation system without casing wear. Case 2 through 6 are used for the worn casing-cement-formation system in different wellbore fluid pressure and temperature. The results analysis is presented in the next section.

4. Results

The results presented are near-wellbore mechanical behaviors when the wellbore fluids flow in the worn casing-cement-formation system. Although stresses and displacements at any location can be found using this model, our discussions focus on the casing-cement case. Contour figures are used to show the stress distributions in different downhole conditions. In the model, a negative sign represents the compression and a positive sign is for tension.

Case 1. The aim of this case is to validate this model. The results presented in Figures 4 and 5 are consistent with the existing solutions. In previous studies, different researchers predict the maximum radial stress at the casing-cement interface in a perfect casing-cement system [6]. In Case 1, the perfect casing without thermal effects and wear is analyzed. The figures reveal the contour profiles of radial stress and hoop stress in the casing-cement part. The maximum radial stress reaches approximately 13,600 psi, which is uniformly distributed at the casing-cement interface. The inner wall of the casing suffers from the highest hoop stress.

Case 2. The reduction of casing thickness decreases the casing strength. No heat transfer occurs in Case 2 because the temperatures of the fluid and rock are identical. Although the casing is worn, the maximum radial stress still occurs at the interface of casing and cement. There is little difference between the radial stresses in the worn part and those in the unworn part, as shown in Figure 6.

The tension failure of worn casing will be overestimated without considering the confining effect of cement and rock. A previous study shows that the wellbore burst pressure causes severe tangential tension at the worn casing [3].

TABLE 3: Fluid temperature and burst pressure for six cases.

	Wellbore fluid temperature (°F)	Formation temperature (°F)	Initial temperature (°F)	Burst pressure (psi)	Rock *in situ* stress (psi)	Casing
Case 1	350	350	350	12,500	11,000	Perfect
Case 2	350	350	350	12,500	11,000	Worn
Case 3	450	350	350	12,500	11,000	Worn
Case 4	450	350	350	13,500	11,000	Worn
Case 5	450	350	350	14,500	11,000	Worn
Case 6	450	350	350	13,500	0	Worn

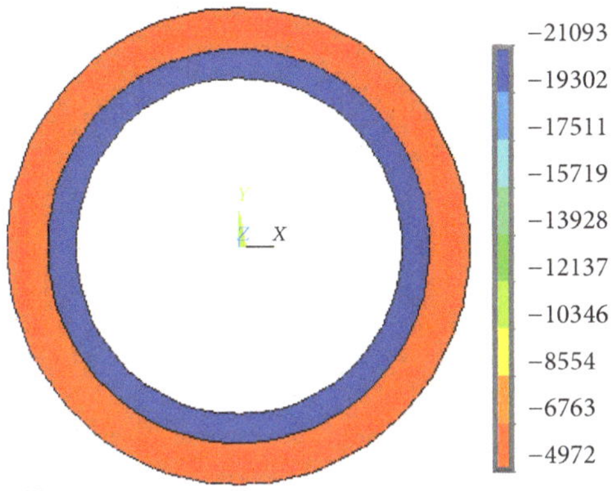

FIGURE 5: Hoop stress inside the perfect casing cement for Case 1.

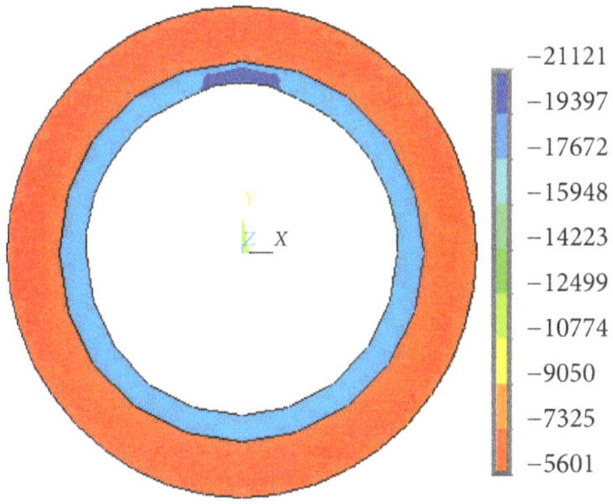

FIGURE 7: Hoop stress inside the worn casing cement for Case 2.

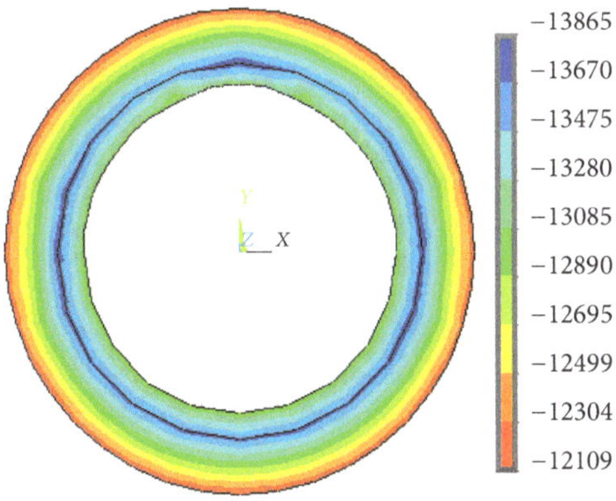

FIGURE 6: Radial stress inside the worn casing cement for Case 2.

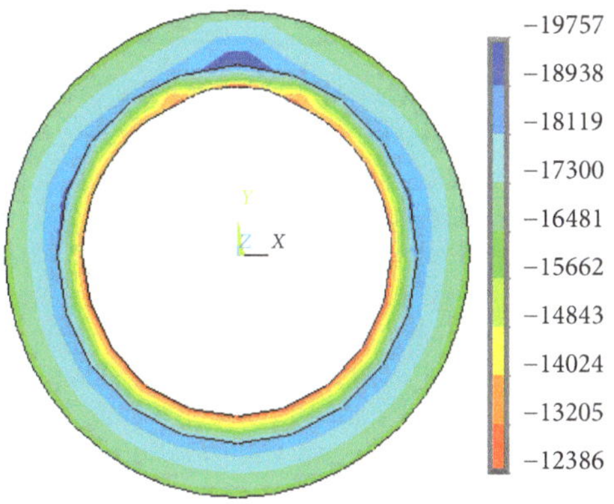

FIGURE 8: Radial stress inside the worn casing cement for Case 3.

This occurs because the burst pressure in the uncemented wellbore produces the tension only when the casing stands alone. Figure 7 shows that the large compression hoop stress occurring around the worn area likely causes the yield of worn casing.

Case 3. The fluid temperature is 450°F, which is much higher than the rock temperature of 350°F. Thermal stress acts on the casing-cement part until the heat transfer reaches equilibrium. Compared with the maximum radial stress of 13,866 psi in Case 2, maximum radial stress in the casing-cement part is increased by one-third because of the thermal stress (Figure 8).

The casing is still constrained by the cement and formation, but the elevated temperature in the casing-cement-formation system resulting from the wellbore fluids tends to cause expansion of the worn casing. The results shown in Figures 8 and 9 indicate that the consequence of the casing expansion is to produce further compression on the casing-cement part. The maximum hoop stress of 73,791 psi in Case 3 is close to three times larger than the 21,121 psi in Case 2.

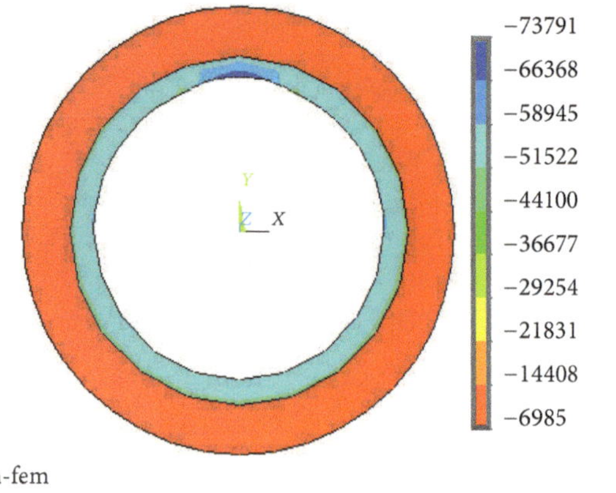

FIGURE 9: Hoop stress inside the worn casing cement for Case 3.

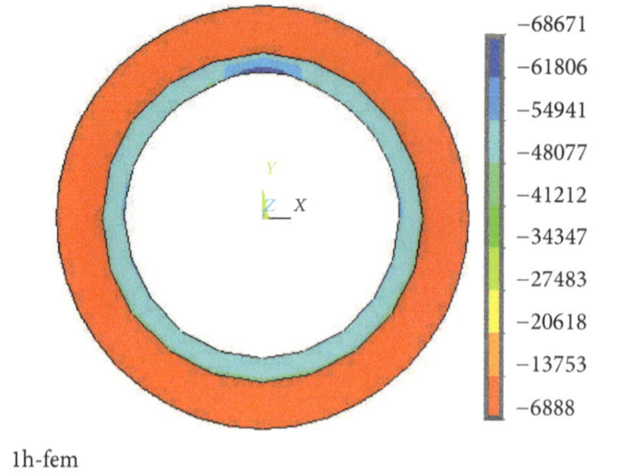

FIGURE 11: Hoop stress inside the worn casing cement for Case 4.

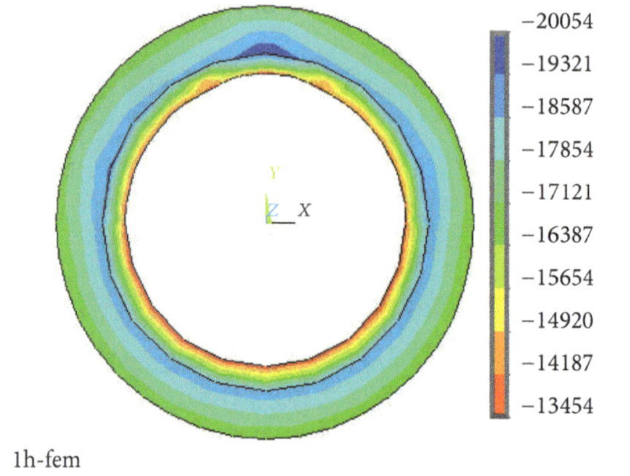

FIGURE 10: Radial stress inside the worn casing cement for Case 4.

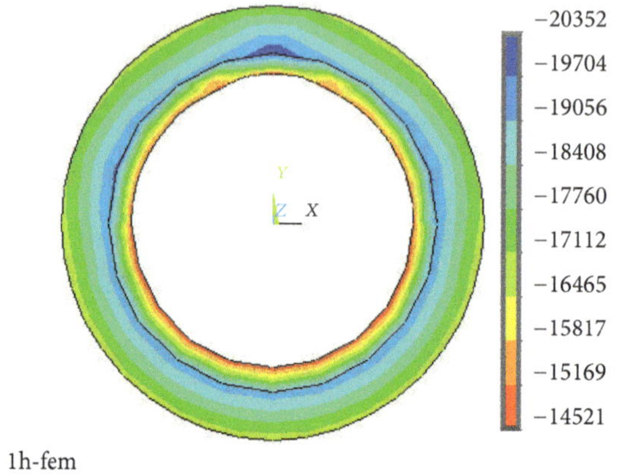

FIGURE 12: Radial stress inside the worn casing cement for Case 5.

It is concluded that the increase of hoop stress is much more than the increase of radial stress in the worn casing cement due to the thermal effect. Several types of casings will fail under this circumstance. The minimum yield strength of casing strings such as J55 and K55 is 55,000 psi. The high temperature difference in the wellbore could put the worn casing-cement-formation system under high risk of failure.

Case 4. The temperature profile is unchanged. The burst pressure of 13,500 psi is used. The compression radial stress in the worn casing cement is slightly larger than that in Case 3, as shown in Figure 10. This indicates that the larger the wellbore burst pressure, the greater the induced casing radial stress resulting from the higher fluid pressure acting on the inner wall of the wellbore.

It is interesting to note that the hoop stress inside the worn casing cement caused by the burst pressure of 13,500 psi is lower than that by the burst pressure of 12,500 psi, as shown in Figure 11. This hoop stress is decreased from 73,791 psi to 68,671 psi. The thermal expansion of casing in the constrained environment due to the cement and formation causes extremely large compression tangential stress, whereas the increased burst pressure tends to expand the worn casing in the opposite direction. It is concluded that a higher burst pressure in the wellbore could reduce the maximum hoop stress in the worn casing-cement part.

Case 5. Figures 12 and 13 reveal the radial stress and hoop stress profile in the worn casing-cement system. The temperature profile in Case 5 is identical with that used in Cases 2 and 3. The burst pressure increases to 14,500 psi. The results show that the radial stress in the system slightly increases with a bigger burst pressure. As expected, the maximum hoop stress occurs on the worn part, and it decreases by 5,000 psi along with the increase of burst pressure.

Cases 2 to 5 present stress profiles of the worn casing under different burst pressures. The stress concentration on the worn part in the casing-cement-formation is predicted in all cases. The risk of tangential compression failure in the worn casing decreases along with the increase of burst pressure.

Case 6. Figures 14 and 15 show the radial stress and hoop stress of the worn casing cement, respectively. The results reveal that the formation is important to prevent tension failure on the worn casing. To better illustrate mechanical behaviors of the worn casing caused by the cement and formation, zero formation *in situ* stress is used in this case.

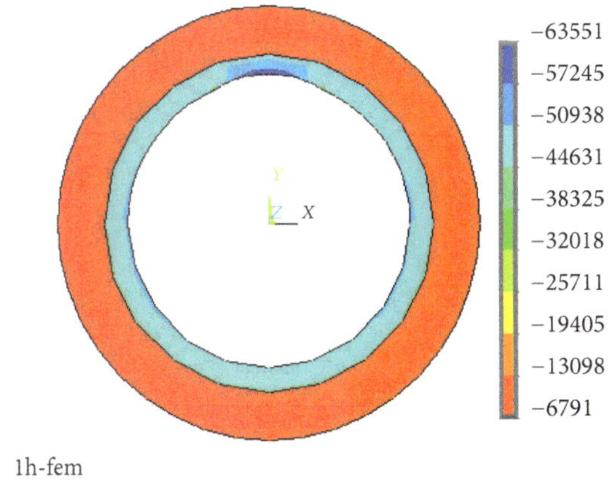

FIGURE 13: Hoop stress inside the worn casing cement for Case 5.

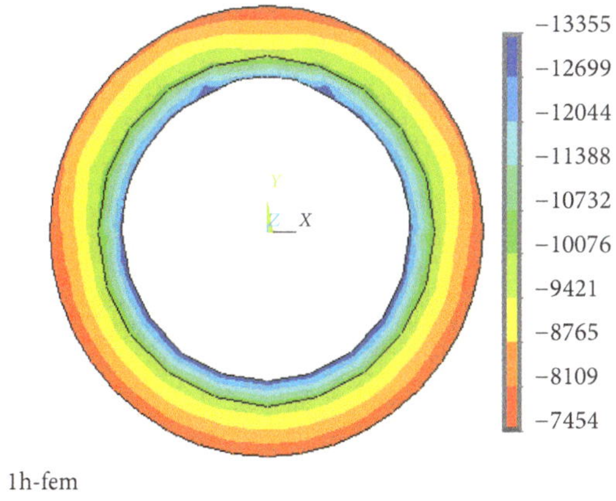

FIGURE 14: Radial stress inside the worn casing cement for Case 6.

The maximum radial stress occurs on the inner wall of the worn casing, which differs from Cases 1 to 4. The maximum hoop stress, as shown in Figure 15, occurs on the worn part. The worn casing suffers from extreme tension without the formation *in situ* stress.

5. Conclusions

The present numerical model has addressed the complete wellbore system consisting of wellbore fluid, casing, cement, and formation. Two principal stresses are analyzed in different downhole conditions. A worn casing-cement-formation model is built using finite element analysis. Ignoring the effect of temperature, the induced stresses in the worn casing-cement-formation system were slightly higher than that in a perfect casing-cement-formation system. In this study, casing expansion due to the temperature difference between wellbore and formation produces severe compression on the worn casing-cement part. The results show that the induced stress could exceed the minimum yield strength of casing strings such as J55 and K55.

Although the casing thickness is partially reduced after wear, worn casing in the downhole system suffers from apparent stress concentration. It is concluded that the increase of the hoop stress is much more than the increase of radial stress in the worn casing cement due to the thermal effect. The burst pressure tends to cause the tension of worn casing. The effects of formation and temperature dominate the stress distribution of the worn casing. It is important to note that high wellbore burst pressures facilitate lowering the compression tangential failure in the worn casing-cement part in the cemented wellbore.

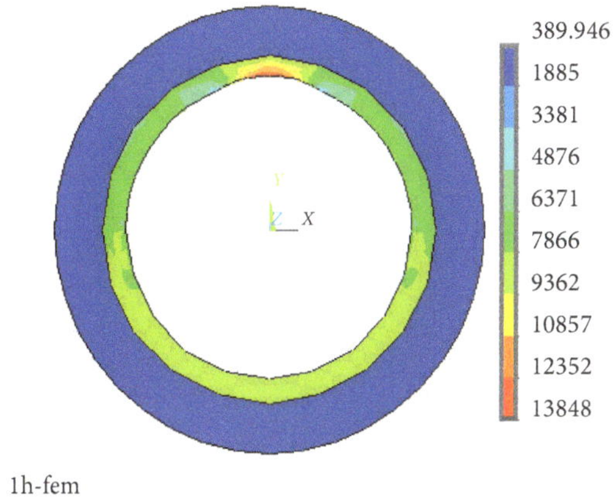

FIGURE 15: Hoop stress inside the worn casing cement for Case 6.

Appendix

Burst Pressure. The burst criteria for casing design are used to guarantee the integrity of casing burst resistance. Besides the API standard shown in (A.1), other researchers propose the initial yield burst standard, the full yield burst standard, and the rupture burst standard [3]. The initial stress criteria in (A.2) emphasize the casing initial yield at the inner wall of casing. Equation (A.3) describes the casing yield across the entire casing wall under the full yield criteria. The rupture criterion is based on casing ductile and tensile failure, as shown in (A.4). Consider the following:

$$p_{\text{API}} = \frac{1.75Y_m t}{D}, \tag{A.1}$$

$$p_{\text{IY}} = \frac{1.75Y_m}{\sqrt{3}} \frac{2t}{D}\left(1 - \frac{t}{D}\right), \tag{A.2}$$

$$p_{\text{FY}} = \frac{1.75Y_m}{\sqrt{3}} \frac{2t}{D}\left(1 + \frac{t}{D}\right), \tag{A.3}$$

$$p_{\text{DR}} = \frac{1.75Y_m t}{D - t}. \tag{A.4}$$

Collapse Pressure. API presents four types of collapse criteria for casing collapse design, which are indicated in (A.5) to (A.8). They comprise yield, plastic, transition, and elastic criteria. The proper selection of each standard depends on the casing wall thickness and diameter. To overcome the limitation of the elastic-plastic boundary under the API yield criteria, Tamano [14] proposes a general yield pressure equation, shown in (A.9). Consider the following:

$$p_Y = 2Y_m \frac{D/t - 1}{(D/t)^2}, \quad \text{(A.5)}$$

$$p_p = Y_m \left(\frac{F_1}{d/t} - F_2 \right) - F_3, \quad \text{(A.6)}$$

$$p_E = 2\frac{E}{1-\nu^2} \frac{1}{D/t(D/t-1)^2}, \quad \text{(A.7)}$$

$$p_T = Y_m \left(\frac{F_4}{d/t} - F_5 \right), \quad \text{(A.8)}$$

$$p_{\text{Ta}} = 2Y_m \left(\frac{D/t - 1}{(D/t)^2} \right) \left(1 + \frac{1.47}{D/t - 1} \right). \quad \text{(A.9)}$$

Nomenclature

σ_r: Radial stress
σ_θ: Hoop stress
σ_z: Axial stress
$\varepsilon_{i=r,\theta,z}$: Strain
r_w: Inner radius of casing
r_o: Outer radius of casing
r_2: The outer radius of tool joint
p_w: Burst pressure
p_o: Collapse pressure
Y_m: Yield stress of material
t: Casing thickness
D: Casing outer diameter
F_i ($i = 1$ to 5): American Petroleum Institute (API) coefficient for casing collapse criteria
f_r: Body force
E: Young's modulus
ν: Poisson's ratio.

Conflict of Interests

The authors declare that there is no conflict of interests regarding the publication of this paper.

References

[1] J. S. Song, J. Bowen, and F. Klementich, "Internal pressure capacity of crescent-shaped wear casing," in *Proceedings of the IADC/SPE Drilling Conference*, pp. 547–553, New Orleans, La, USA, February 1992.

[2] P. Oudeman and M. Kerem, "Transient behavior of annular pressure build-up in HP/HT wells," in *Proceedings of the 11th ADIPEC: Abu Dhabi International Petroleum Exhibition and Conference*, pp. 665–674, Abu Dhabi, UAE, October 2004.

[3] J. Wu and M. G. Zhang, "Casing burst strength after casing wear," in *Proceedings of the SPE Production and Operations Symposium 2005*, pp. 517–526, Oklahoma City, Okla, USA, April 2005.

[4] R. W. Hall Jr. and K. P. Malloy Sr., "Contact pressure threshold: an important new aspect of casing wear," in *SPE Production and Operations Symposium 2005: Anticipate the Future, Build on the Present, Celebrate the Past*, pp. 499–505, Society of Petroleum Engineers, Oklahoma City, Okla, USA, April 2005.

[5] D. Gao, L. Sun, and J. Lian, "Prediction of casing wear in extended-reach drilling," *Petroleum Science*, vol. 7, no. 4, pp. 494–501, 2010.

[6] W. W. Fleckenstein, A. W. Eustes III, and M. G. Miller, "Burst-induced stresses in cemented wellbores," *SPE Drilling and Completion*, vol. 16, no. 2, pp. 74–82, 2001.

[7] W. J. Rodriguez, W. W. Fleckenstein, and A. W. Eustes, "Simulation of collapse loads on cemented casing using finite element analysis," in *Proceedings of the SPE Annual Technical Conference and Exhibition*, pp. 5239–5247, Denver, Colo, USA, October 2003.

[8] S. Talabani, B. Atlas, M. B. Al-Khatiri, and M. R. Islam, "An alternate approach to downhole corrosion mitigation," *Journal of Petroleum Science and Engineering*, vol. 26, no. 1-4, pp. 41–48, 2000.

[9] W. B. Bradley and J. E. Fontenot, "The prediction and control of casing wear (includes associated papers 6398 and 6399)," *SPE Journal of Petroleum Technology*, vol. 27, no. 2, pp. 233–245, 1975.

[10] W. M. Rainforth, "Microstructural evolution at the worn surface: a comparison of metals and ceramics," *Wear*, vol. 245, no. 1-2, pp. 162–177, 2000.

[11] B. Calhoun, S. Langdon, J. Wu, P. Hogan, and K. Rutledge, "Casing wear prediction and management in deepwater wells," in *Proceedings of the SPE Deepwater Drilling and Completions Conference 2010*, pp. 232–241, Society of Petroleum Engineers, Galveston, Tex, USA, October 2010.

[12] E. Fjar, R. M. Holt, A. M. Raaen, R. Risnes, and P. Horsrud, *Petroleum Related Rock Mechanics*, Elsevier, Oxford, UK, 2004.

[13] Z. Shen and F. E. Beck, "Three-dimensional modeling of casing and cement sheath behavior in layered, nonhomogeneous formations," in *Proceedings of the IADC/SPE Asia Pacific Drilling Technology Conference 2012*, pp. 949–958, Society of Petroleum Engineers, July 2012.

[14] T. Tamano, "A new empirical formula for collapse resistance of commercial casing," in *ASME Transactions of Energy Resources Technology*, 2005.

2

Mysteries behind the Low Salinity Water Injection Technique

Emad Waleed Al-Shalabi, Kamy Sepehrnoori, and Gary Pope

Department of Petroleum and Geosystems Engineering, The University of Texas at Austin, Austin, TX 78712, USA

Correspondence should be addressed to Emad Waleed Al-Shalabi; ealshalabi@utexas.edu

Academic Editor: Alireza Bahadori

Low salinity water injection (LSWI) is gaining popularity as an improved oil recovery technique in both secondary and tertiary injection modes. The objective of this paper is to investigate the main mechanisms behind the LSWI effect on oil recovery from carbonates through history-matching of a recently published coreflood. This paper includes a description of the seawater cycle match and two proposed methods to history-match the LSWI cycles using the UTCHEM simulator. The sensitivity of residual oil saturation, capillary pressure curve, and relative permeability parameters (endpoints and Corey's exponents) on LSWI is evaluated in this work. Results showed that wettability alteration is still believed to be the main contributor to the LSWI effect on oil recovery in carbonates through successfully history matching both oil recovery and pressure drop data. Moreover, tuning residual oil saturation and relative permeability parameters including endpoints and exponents is essential for a good data match. Also, the incremental oil recovery obtained by LSWI is mainly controlled by oil relative permeability parameters rather than water relative permeability parameters. The findings of this paper help to gain more insight into this uncertain IOR technique and propose a mechanistic model for oil recovery predictions.

1. Introduction

Oil recovery from carbonate rocks is a challenge due to the high fracture density and the rock wettability state which ranges from mixed-wet to oil-wet. One of the recently recommended improved oil recovery (IOR) techniques is low salinity water injection (LSWI), which is believed to shift the wettability state of the rock towards more water-wet state. The LSWI technique has several advantages including high efficiency in displacing light to medium gravity crude oils, ease of injection into oil-bearing formations, availability and affordability of water, and lower capital and operating costs compared to other IOR methods, which leads to favorable economics. Other names proposed in the literature for the same mechanism are LoSal, Smart Waterflood, and Advanced Ion Management. The LSWI effect on oil recovery from carbonates was shown both at laboratory scale and to a limited extent at field scale. Although most researchers believe that wettability alteration is the main mechanism for the LSWI on oil recovery from carbonates, there are others who believe in the presence of other contributing mechanisms. Therefore, work is still progressing to understand the chemical interactions in crude oil-brine-rock (COBR) in the porous media.

The LSWI effect on oil recovery from carbonate is not well addressed compared to sandstone rocks due to the previous thoughts of relating wettability alteration by low salinity water to the presence of clay, which is not the case in carbonate rocks. Nevertheless, the effect of LSWI on oil recovery from carbonate rocks was investigated at laboratory scale using both spontaneous imbibition and coreflooding studies.

Høgnesen et al. [1] concluded from their spontaneous imbibition experiments on reservoir limestone cores that increasing sulfate ion concentration at high temperature leads to oil recovery increase due to the role of sulfate ion as a wettability modifying agent for carbonate rocks from mixed-wet to water-wet state. Webb et al. [2] investigated the effect of sulfate on oil recovery from North Sea carbonate core samples through spontaneous imbibition experiments. They reported that seawater has the ability to alter wettability of the carbonate system to more water-wet state compared to sulfate-free water. Zhang et al. [3] studied wettability alteration of North Sea chalk reservoirs in Ekofisk field showing the effect of adding calcium and/or magnesium ions at various temperatures. They concluded that wettability alteration occurs if the imbibing water contains either Ca^{2+} and SO_4^{2-} or Mg^{2+} and SO_4^{2-}. Moreover, Strand et al. [4] observed 15% increase in oil recovery from limestone cores when seawater was imbibed compared to seawater free of sulfate. Increased oil recovery using low salinity water injection in limestone formations was noticed by Fjelde [5].

Bagci et al. [6] reported high oil recovery of 35.5% of OOIP from their corefloods by using 2 wt% KCl on limestone cores and high pH effluent brine due to ions exchange reactions with the clay present in the rock. They considered wettability alteration as the reason behind recovering more oil without further explanation. Yousef et al. [7] investigated the applicability of low salinity water injection (Smart Waterflood) on carbonate rocks for improving oil recovery by using different dilutions of seawater. The results of coreflooding tests showed increasing oil recovery with stepwise dilution of seawater upon which 18% incremental oil recovery was achieved due to tertiary water injection. Coreflooding experiments on both dolomite cores from West Texas and limestone cores from the Middle East were performed by Gupta et al. [8]. Experiments showed incremental 5–9% OOIP recovery from both dolomite and limestone cores as a result of adding sulfate ions. For limestone cores, 7–9% OOIP was obtained due to reducing hardness of the injected water, not the total dissolved solids. Another interesting finding is the 15% and 20% OOIP by using borate (BO_3^{3-}) and phosphate (PO_4^{3-}) as modified ions, respectively. In a later work, Yousef et al. [9] demonstrated that wettability alteration is the reason behind LSWI through NMR, contact angle measurement, and zeta potential studies. The results showed that wettability alteration occurs through changing the surface charge from the zeta potential measurements and dissolution of $CaSO_4$ from NMR tests.

Al-Harrasi et al. [10] provided direct evidence of low salinity water flooding effect on oil recovery from Omani carbonate rocks through spontaneous and coreflooding experiments. Wettability alteration was referred to as the reason for low salinity water injection with negligible reduction in interfacial tension. Also, they reported that low salinity water injection by lowering the ionic strength has more pronounced effect on oil recovery from oil-bearing zone cores compared to hardening the injected water. Nevertheless, the case is just the opposite for Stevns Klint outcrop chalk cores because they are more responsive to hardening of the injected low salinity water as was reported by Romanuka et al. [11] through their spontaneous imbibition experiments.

Extensive research was performed by Austad and coworkers [12–14] which showed the possibility of wettability alteration and enhancing oil recovery from carbonate rocks by modifying the ionic composition in the injected water. Wettability alteration is the main and most acceptable mechanism for the incremental oil recovery obtained from carbonate rocks using LSWI. Wettability alteration in carbonate rocks using smart water can be achieved by injecting water containing SO_4^{2-} and either Ca^{2+} or Mg^{2+} or both of them in the presence of high temperatures (>90°C). It was proposed that, with increasing temperature, the affinity of sulfate to chalk rock surface increases and sulfate adsorption occurs. At the same time, Ca^{2+} adsorption increases as well, as the initial positive charge of the rock decreases. Hence, more excess Ca^{2+} ions are present close to the surface which reacts with the carboxylic material and releases some of them. Moreover, with increasing the temperature Mg^{2+} becomes more active, Ca^{2+} substitution by Mg^{2+} occurs, and sulfate becomes less active as it reacts with Mg^{2+}. Otherwise, $CaSO_4$ precipitation occurs which causes injection problems [13].

The first ever LoSal application in carbonate reservoirs was reported by Yousef et al. [15]. Two single-well chemical tracer tests (SWCTT) were applied in an upper jurassic carbonate reservoir using a diluted version of Qurayyah seawater. The tests resulted in about 7 saturation units' reduction in the residual oil beyond conventional seawater injection. The results obtained matched their previous experimental work which is encouraging to plan a multiwell demonstration pilot.

Only a few modeling works for carbonate rocks have been performed so far due to various reasons. These reasons include the complex chemical interaction in rock-oil-brine and the heterogeneity of carbonate rocks, which complicate oil recovery predications by LSWI. Also, the uncertainty in the controlling mechanism and the clash in some of the published experimental results shifted the focus on experimental work rather than modeling work. This paper includes history-matching of Chandrasekhar and Mohanty's [16] experimental work using the UTCHEM simulator, which is a 3D multiphase flow, transport, and chemical flooding simulator for black oil, developed at The University of Texas at Austin.

2. Experimental Data

Chandrasekhar and Mohanty [16] conducted several vertical corefloods to investigate the low salinity water injection effect on oil recovery from Middle Eastern carbonate core plugs. The coreflood of our interest is the one with different injected seawater dilutions. Heterogeneous carbonate core plug was used with average porosity and liquid permeability of 26.4% and 7.59 mD, respectively. The coreflood was conducted at reservoir temperature of 248°F and atmospheric pressure at an injection rate of 1 ft/day. At each injection cycle, the injection rate was increased to 10 ft/day to make sure that the oil maximum recovery was reached. The brines used were field water of 179,726 ppm and different diluted versions of seawater of 43,619 ppm by weight. Rock properties and fluid properties are shown in Tables 1, 2, and 3. The oil recovery and pressure drop data are shown in Figures 1 and 2. Here an increase in oil recovery is observed with stepwise dilution of the seawater by tertiary water injection. More details about the experimental work are described elsewhere [16]. The UTCHEM software was used to history-match both oil recovery and pressure drop data. The pressure drop data of each injection cycle is matched just for the low injection rate (1 ft/day). The second injection cycle is an exception, where the pressure drop is matched for both injection rates (1 and 10 ft/day) because of the incremental oil recovery obtained with increasing the rate at 10 ft/day in this cycle.

TABLE 1: Reservoir core properties (Chandrasekhar and Mohanty, 2013 [16]).

Pore Volume	15.8		cc
Porosity		0.264	
Permeability	7.59		mD
Diameter	3.79		cm
Cross Sectional Area	11.28		cm^2
Length	5.3		cm
S_{wi}		0.3181	
S_{oi}		0.6819	

TABLE 2: Field oil sample properties (Chandrasekhar and Mohanty, 2013 [16]).

Oil properties	
Viscosity (cp) at 248°F	1.05
Density (g/cc) at 248°F	0.618
API (degrees)	40

TABLE 3: Properties of various dilutions of seawater (248°F) (Chandrasekhar and Mohanty, 2013 [16]).

	Water propertiesat 248°F	
Injected water salinity (ppm)	Water density (g/cc)	Water viscosity (cp)
43619	0.9760	0.260
21809.5	0.9595	0.246
4361.9	0.9464	0.234
2180.95	0.9448	0.233

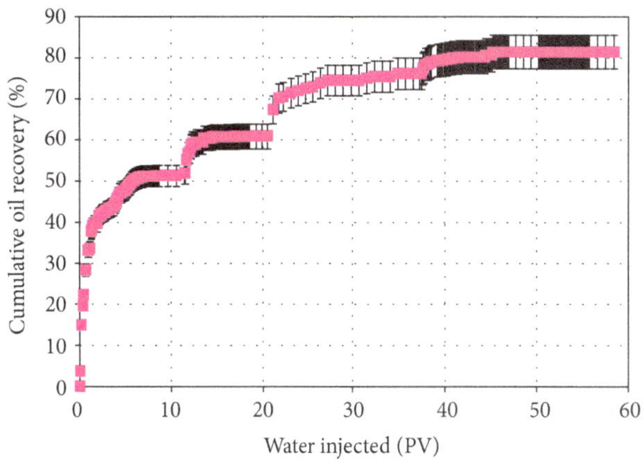

FIGURE 1: Cumulative oil recovery curve for the coreflood conducted [16].

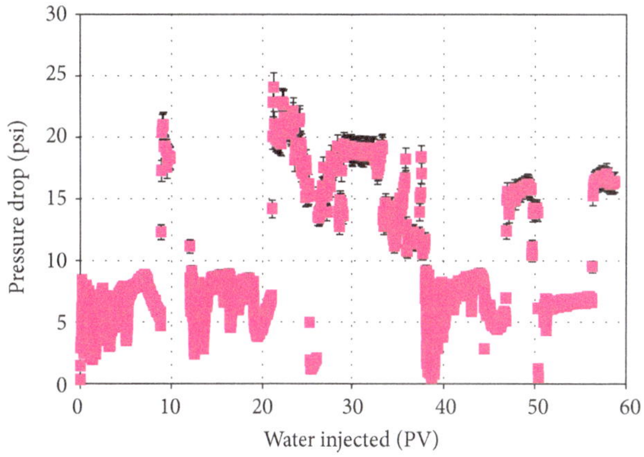

FIGURE 2: Pressure drop data for the coreflood conducted [16].

3. Simulation Data

This section includes both simulation model description and experimental data analysis to obtain the simulation inputs needed.

3.1. Simulation Model Description. A 2D Cartesian grid system was used with 10 × 1 × 10 grid blocks to simulate the heterogeneous core plug for the coreflood. The decision on using a heterogeneous model is discussed later in the seawater cycle match section. The heterogeneous model was considered by generating permeability distribution with an arithmetic mean of 7.59 mD and Dykstra Parson's coefficient of 0.85. A spherical variogram and a log normal permeability distribution were used. The y-direction correlation length was assumed to be similar to the x-direction; however, a high z-direction correlation length compared to the x-direction was chosen, which generated vertical layers of different permeabilities (Figure 3). The simulation model has two horizontal wells, injector at the bottom and producer at the top. Table 4 shows length, width, and height dimensions for the grid blocks. We assumed a negligible capillary end effect.

3.2. Experimental Data Analysis. This section includes a description of pressure drop data analysis and application of JBN method to find relative permeability curves for the seawater injected cycle.

3.2.1. Pressure Drop Data Analysis. As was previously mentioned, the analysis of the pressure drop curve (Figure 2) is performed for the 1 ft/day injection rate, except for the second cycle where the injection rate of 10 ft/day resulted in additional oil recovery. The water endpoint relative permeability for each cycle at 1 ft/day and even for the second cycle at 10 ft/day can be calculated using Darcy's law and stabilized average pressure drop value. The oil endpoint relative permeability was provided experimentally for the seawater cycle ($K_{ro}^* = 0.203$). Table 5 summarizes the obtained endpoint relative permeabilities for water and oil. Table 5 shows endpoint permeability calculation for the second cycle including both low salinity water injection and trapping number effects. We can see a slight decrease in the water endpoint relative permeability values for the LSWI effect in all injection cycles, which indicates the presence of wettability alteration by LSWI.

TABLE 4: Heterogeneous core model data.

Parameter	Value	Comments
Number of grid blocks	100	2D ($10 \times 1 \times 10$)
Grid block sizes (ΔI, ΔJ, ΔK), m	x-**direction:** 1–10, Δx is 0.0033588 m y-**direction:** 1-1, Δy is 0.033588 m z-**direction:** 1–10, Δz is 0.0053 m	Constant grid size in the x-, y-, and z-direction.
Composite core model dimensions, m	0.033588 m × 0.033588 m × 0.053 m	Length × width × height

TABLE 5: Endpoint relative permeability data analysis.

Oil viscosity	1.05	cP			
Composite core length	5.3	cm	Oil-water IFT	30	Dynes/cm
Cross-sectional area	11.28	cm^2			
Injection rate (main)	0.045	cc/min	$K_{o@Swirr}$	1.54	mD
Absolute brine permeability	7.59	mD			
Injection cycle	Water viscosity (cP)	Pressure drop (psi)	$k_{rw}{}^*$	S_{or}	$k_{ro}{}^*$
First	0.26	7.20	0.025	0.329	0.203
Second 1 (LSWI effect)	0.246	7.00	0.024	0.267	
Second 2 (trapping number effect)	0.246	18.90	0.089	0.163	
Third	0.234	7.00	0.023	0.127	
Fourth	0.233	7.20	0.022	0.127	

3.2.2. JBN Method. Johnson, Bossler, and Naumann (JBN) method was applied to find relative permeability curves for the seawater cycle. The data obtained are shown in Figure 4; Corey's model was fitted to find relative permeability endpoints and exponents for the seawater cycle. Moreover, the analysis was taken a step further to calculate the fractional flow curve, upon which both mobility ratio and gravity effects are considered. The fractional flow in this case is defined as

$$f_w = \frac{S^{n_w} M^o \left[1 - N_g^o (1 - S)^{n_o} \sin(\alpha)\right]}{S^{n_w} M^o + (1 - S)^{n_o}}. \tag{1}$$

The relative permeability data calculated from JBN method was fitted to an exponential function that was used later to calculate the fractional flow curve through mobility ratio calculations (Figure 5). The equation of relative permeability ratio as function of saturation is given by

$$\frac{K_{ro}}{K_{rw}} = ae^{-bS_w}. \tag{2}$$

The latter fractional flow curve (experimental) was matched with the analytical solution of fraction flow, where the main matching parameters are n_w of 1.3 and n_o of 3.5 (Figure 6). The water and oil endpoint relative permeability values were obtained using pressure drop curve analysis from Table 5. The coreflood was conducted vertically which makes $\sin(\alpha)$ equal to 1. The endpoint mobility ratio used is 0.47 and the endpoint gravity number is 0.01. The definitions of endpoint mobility ratio (M^o) and endpoint gravity number (N_g^o) are given by

$$M^o = \frac{k_{rw}^o \mu_o}{k_{ro}^o \mu_w},$$
$$N_g^o = \frac{k k_{ro}^o \Delta\rho g}{u \mu_o}. \tag{3}$$

Nevertheless, there is a mismatch between the Buckley Leverett analytical solution compared to the experimental data provided by Chandrasekhar and Mohanty [16] (Figure 7). The main difference between the analytical solution and the experimental data is the heterogeneity effect which is not taken into account in analytical solution added to the capillary pressure effect. Both of these effects can be considered using the UTCHEM simulator to history-match the data.

4. Results and Discussion

This section covers seawater cycle history-matching and methods used for the wettability alteration effect matching for different dilutions of seawater injected cycles.

4.1. Seawater Cycle Match. The final set of relative permeability curves for the seawater cycle as a result of pressure drop analysis and JBN method is shown in Figure 8. The figure shows a weakly oil-wet rock, where k_{ro}^* is 0.203, k_{rw}^* is 0.025, n_o is 3.5, n_w is 1.3, and the intersection point is at about 0.5 water saturation. The heterogeneity effect was taken into consideration by applying a Dykstra Parson coefficient of

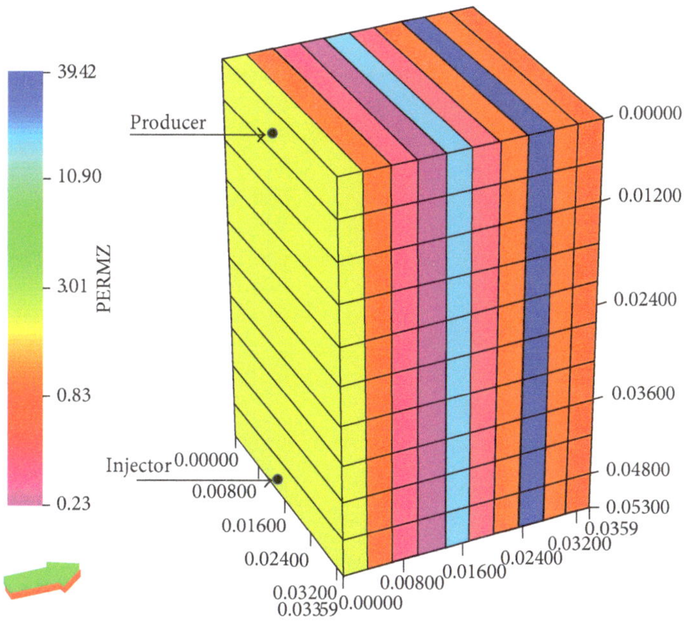

FIGURE 3: Simulation model used in different runs with heterogeneous permeability.

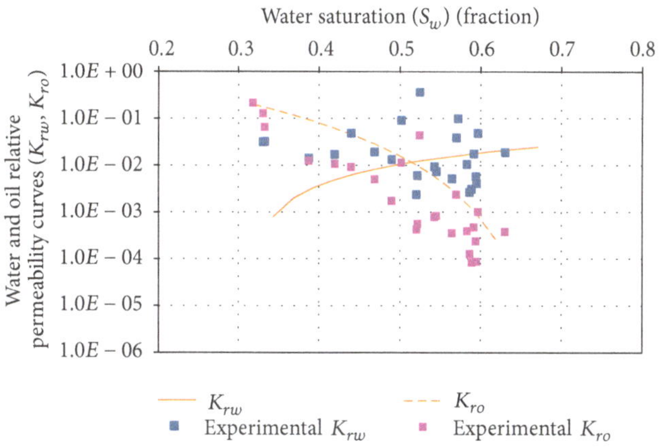

FIGURE 4: Relative permeability data analysis using the JBN method (first cycle).

0.85 along with correlation lengths which resulted in vertical layers of different permeabilities as was previously shown in Figure 3. Moreover, the capillary pressure contribution was considered by applying Brooks-Corey model for imbibition capillary pressure for mixed-wet rocks as follows.

(i) Water-wet part of capillary pressure curve ($S < S^*$) is

$$P_{c12} = \mathrm{CPC}_1 \sqrt{\frac{\phi}{k}} \left(\frac{S_w^* - S_w}{S_w^* - S_{wr}} \right)^{\mathrm{EPC}_1}; \tag{4}$$

(ii) Oil-wet part of capillary pressure curve ($S > S^*$) is

$$P_{c12} = \mathrm{CPC}_2 \sqrt{\frac{\phi}{k}} \left(\frac{S_w - S_w^*}{1 - S_{or} - S_w^*} \right)^{\mathrm{EPC}_2}, \tag{5}$$

TABLE 6: Summary of relative permeability and capillary pressure parameters (seawater cycle).

Seawater cycle match parameters			
Relative permeability parameters			
$k_{rw}{}^*$	0.025	n_w	1.3
$k_{ro}{}^*$	0.203	n_o	3.5
Capillary pressure parameters			
CPC_w	2	EPC_w	2
CPC_o	−2	EPC_o	2
S^*		0.5	

where CPC is a parameter related to the maximum capillary pressure, EPC is capillary pressure exponent, and S^* is the water saturation at zero capillary pressure value. The capillary pressure curve used in matching oil recovery and pressure drop data along with summary of capillary pressure parameters and relative permeability parameters is presented in Figure 9 and Table 6, respectively. History-matching of oil recovery and pressure drop data showed that the CPC_2 parameter controls the ultimate oil recovery value; however, CPC_1 controls the initial hump of the oil recovery and the pressure drop data match. It can be seen clearly that the capillary pressure does not contribute much to data history-matching. Hence, the capillary pressure is neglected for history-matching the successive dilutions of seawater injection.

The results of history-matching of oil recovery and pressure drop data are depicted in Figures 10 and 11, respectively. In the latter figures, two curves are presented for the homogeneous 1D model with an average permeability and heterogeneous models. The history-matching shows the importance

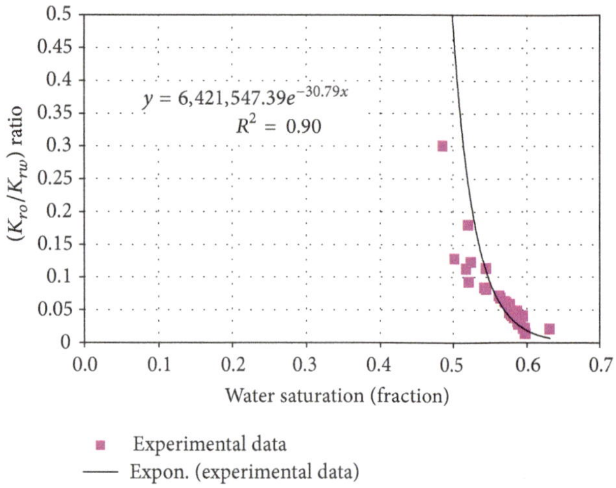

FIGURE 5: Relative permeability ratio versus water saturation (first cycle).

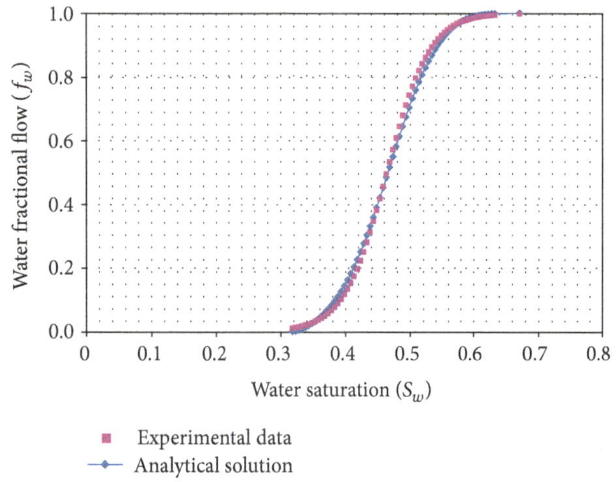

FIGURE 6: Fractional flow curve history-match (first cycle).

of heterogeneity incorporation to match reasonably the oil recovery and pressure drop curves.

4.2. Dilutions of Seawater Injected Cycles Match. This section includes history-matching of the LSWI cycles of Chandrasekhar and Mohanty [16] coreflood using two proposed methods.

4.2.1. First Method. In this method, seawater cycle's relative permeability parameters are used for the different dilution cycles, while only changing the residual oil saturation for each cycle based on the reported values. As expected, history-matching of data is not possible using this method which validates the necessity of tuning relative permeability parameters for LSWI cycles because S_{or} contribution by itself is not enough. This is supported by the findings of oil recovery and pressure drop history-matching curves using this approach (Figures 12 and 13). It is worth mentioning that the jump in the second cycle of pressure drop curve

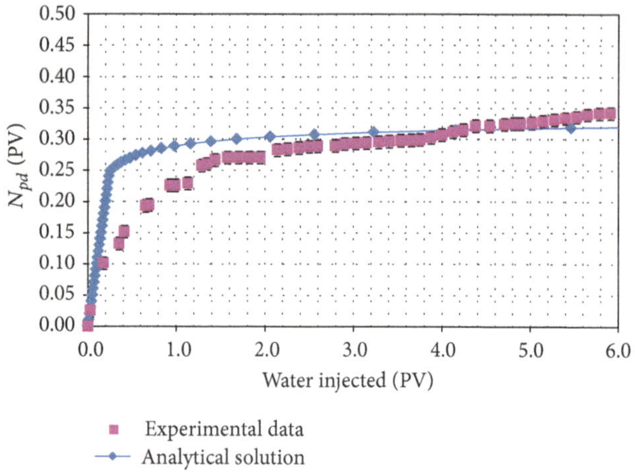

FIGURE 7: Buckley Leverett analytical solution (first cycle).

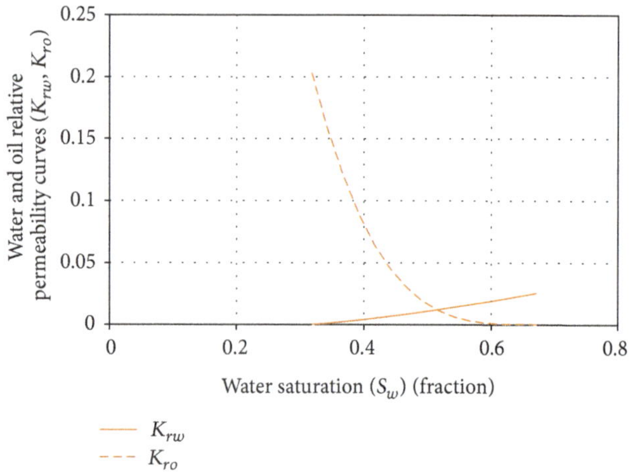

FIGURE 8: Relative permeability curves (first cycle).

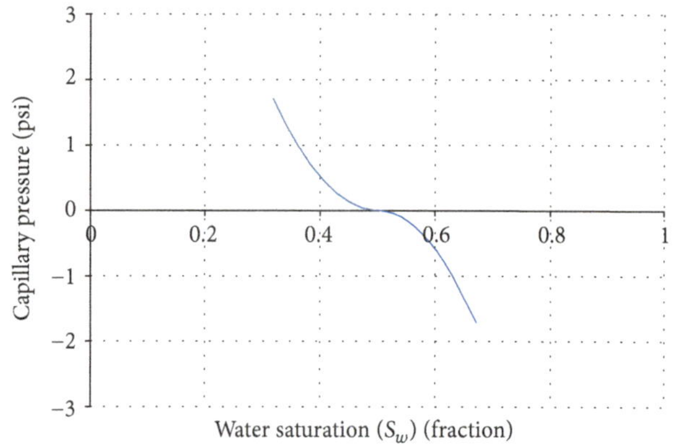

FIGURE 9: Capillary pressure curve (first cycle).

(Figure 13) is due to the trapping number effect as the injection rate was increased to 10 ft/day without changing the relative permeability parameters.

4.2.2. Second Method. This method includes three approaches: changing Corey's exponents only (first approach),

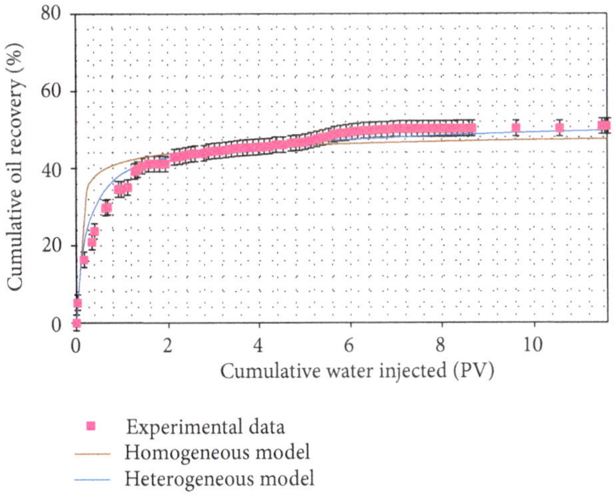

FIGURE 10: Oil recovery match for seawater cycle.

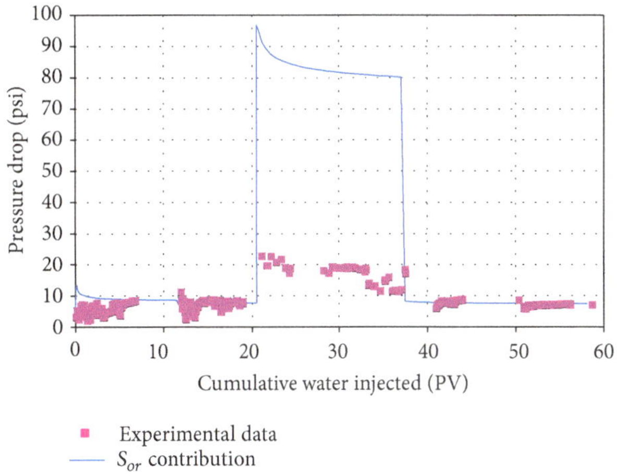

FIGURE 13: Overall pressure drop match using the first method.

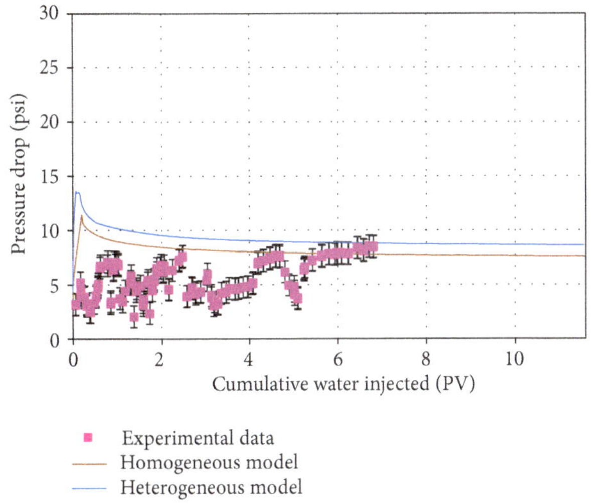

FIGURE 11: Pressure drop match for seawater cycle.

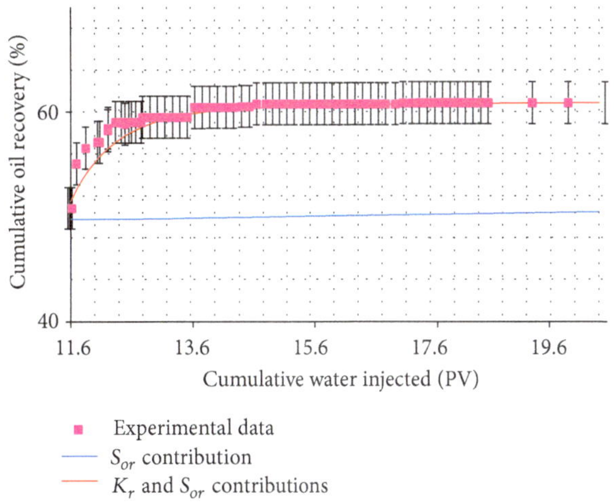

FIGURE 14: LSWI effect on second cycle oil recovery match using the second method (third approach).

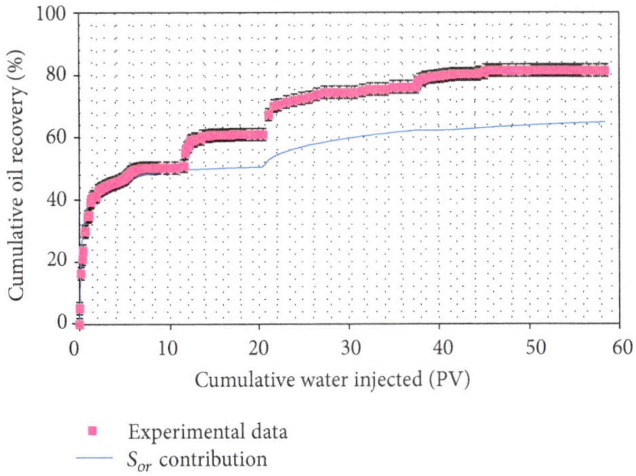

FIGURE 12: Cumulative oil recovery match using the first method.

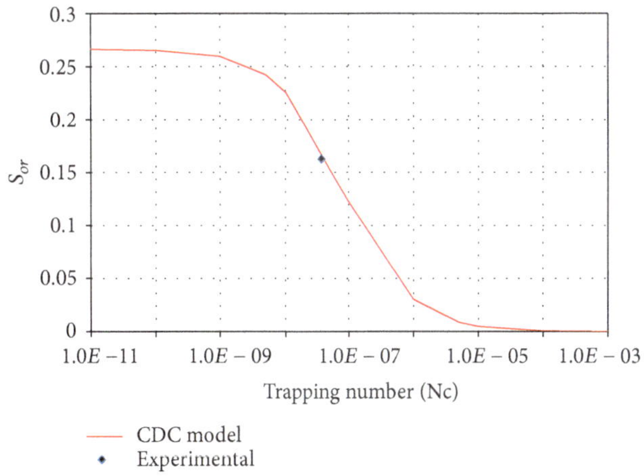

FIGURE 15: Modeled CDC curve for the coreflooding experiment.

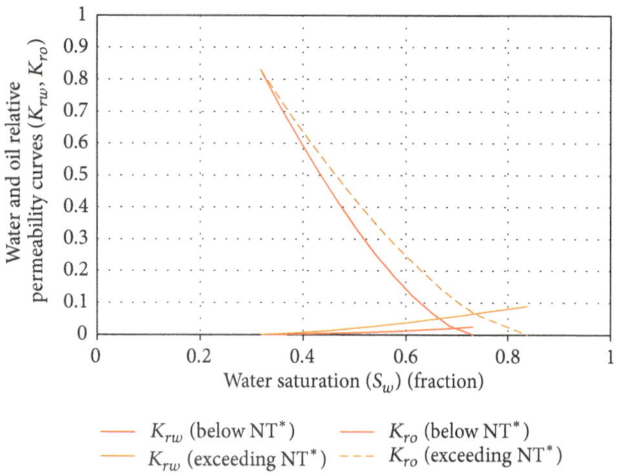

FIGURE 16: Relative permeability curves before and after exceeding critical N_T (second cycle-trapping number).

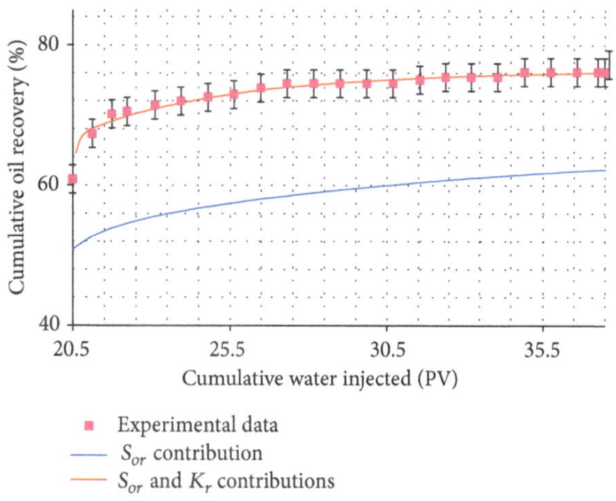

FIGURE 17: Trapping number effect on second cycle oil recovery match using the second method (third approach).

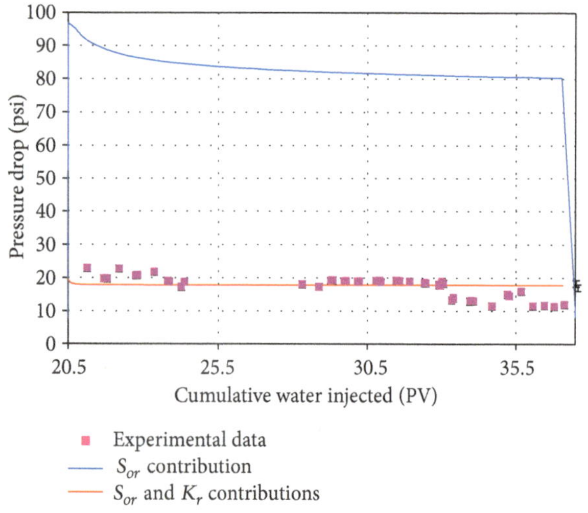

FIGURE 18: Trapping number effect on second cycle pressure drop match using the second method (third approach).

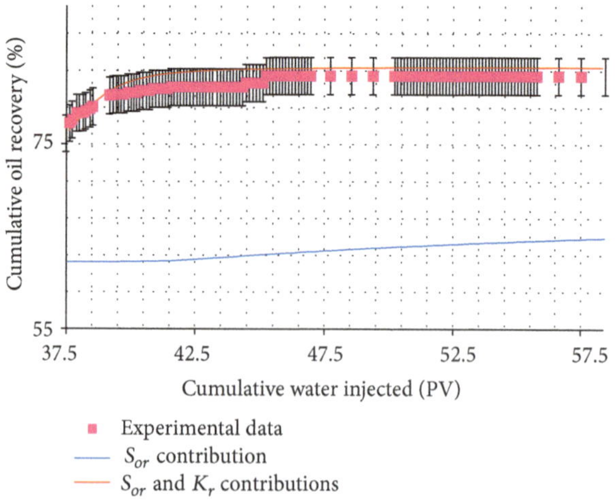

FIGURE 19: Third and fourth cycles oil recovery match using the second method (third approach).

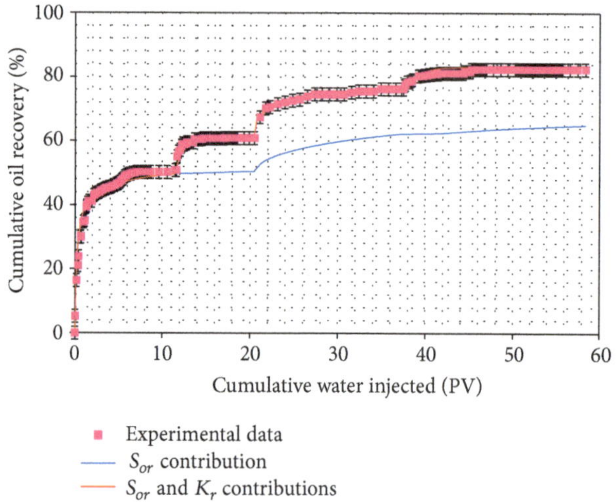

FIGURE 20: Cumulative oil recovery match using the second method (third approach).

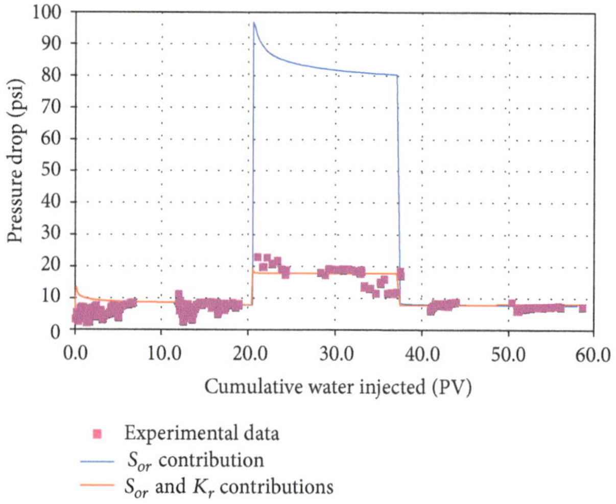

FIGURE 21: Overall pressure drop match using the second method (third approach).

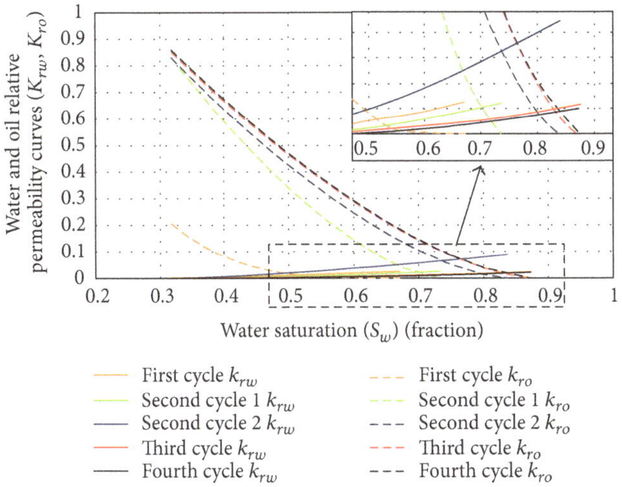

FIGURE 22: Relative permeability curves using the second method (third approach).

changing endpoint relative permeabilities only (second approach), and changing both Corey's exponents and endpoint relative permeabilities (third approach). The first two approaches were not successful in history-matching pressure drop and oil recovery data. The third approach of the second method is applied on Chandrasekhar and Mohanty [16] coreflood by tuning relative permeability parameters, including endpoints and Corey's exponents to match the data in each cycle starting with the second cycle. The k_r and S_{or} contributions curve for the LSWI effect on the second cycle is shown in Figure 14.

The trapping number effect on the second cycle is considered using the capillary desaturation curve (CDC). The relation for adjusting residual oil saturation as a function of trapping number was proposed by Pope et al. [17] as follows:

$$S_{lr} = S_{lr}^{\text{high}} + \frac{S_{lr}^{\text{low}} - S_{lr}^{\text{high}}}{1 + T_l N_{T_l}^{\tau}}, \quad \text{for } l = 1, \ldots, np. \tag{6}$$

Figure 15 shows the modeled CDC curve for the second cycle where the experimental trapping number calculated for the injection rate of 10 ft/day is matched using S_{lr}^{low} of 0.267, S_{lr}^{high} of zero, τ of 0.82, and T_l parameter of 650,000. The detailed calculations of the CDC curve are listed in Table 7. The effect of trapping number on relative permeability parameters was also considered using Delshad et al.'s [18] proposed model as follows:

$$k_{rl}^{o} = k_{rl}^{o^{\text{low}}} + \frac{S_{l'r}^{\text{low}} - S_{l'r}}{S_{l'r}^{\text{low}} - S_{l'r}^{\text{high}}} \left(k_{rl}^{o^{\text{high}}} - k_{rl}^{o^{\text{low}}} \right), \quad \text{for } l, l' = 1, \ldots, np,$$
$$n_l = n_l^{\text{low}} + \frac{S_{l'r}^{\text{low}} - S_{l'r}}{S_{l'r}^{\text{low}} - S_{l'r}^{\text{high}}} \left(n_l^{\text{high}} - n_l^{\text{low}} \right) \quad \text{for } l, l' = 1, \ldots, np. \tag{7}$$

TABLE 7: CDC curve parameters.

$S_{or\,(\text{high})}$	0.000
$S_{or\,(\text{low})}$	0.267
T_{22} (parameter)	650,000
Tau (N_T exponent)	0.82
N_T	S_{or}
$1.00E-11$	0.267
$1.00E-11$	0.267
$1.00E-10$	0.266
$1.00E-09$	0.260
$5.00E-09$	0.242
$1.00E-08$	**0.226**
$1.00E-07$	**0.122**
$1.00E-06$	0.030
$5.00E-06$	0.009
$1.00E-05$	0.005
$1.00E-04$	0.001
$1.00E-03$	0.000

TABLE 8: Relative permeability parameters before and after exceeding critical N_T.

Second cycle matching parameters trapping number effect			
Below N_T^* (critical)		Exceeding N_T^* (critical)	
n_w	1.7	n_w	1.43
n_o	1.55	n_o	1.55
k_{rw}^*	0.024	k_{rw}^*	0.089
k_{ro}^*	0.83	k_{ro}^*	0.83
S_{or}	0.267	S_{or}	0.163
S_{wirr}	0.3181	S_{wirr}	0.3181

In the previous equations, the words "high" and "low" in the superscripts indicate the value of the parameter at high and low trapping numbers, respectively. The values at high trapping number are usually assumed and the values at low trapping number can be considered as the values obtained through history-matching the effect of LSWI on the second cycle. It is worth mentioning that $k_{rl}^{o^{\text{high}}}$ was assumed to be 0.2 due to the low water endpoint relative permeability of initial seawater cycle (0.025). Table 8 and Figure 16 show two sets of relative permeability curves before and after exceeding the critical trapping number. The oil recovery and pressure drop match for trapping number effect on the second cycle are shown in Figures 17 and 18, respectively.

The k_r and S_{or} contributions curve for the third and fourth cycles is depicted in Figure 19. The cumulative oil recovery and the overall pressure drop curves using the third approach of the second method are shown in Figures 20 and 21, respectively. Sets of relative permeability curves used in history-matching using this approach are presented in Figure 22 and Table 9. The analysis showed that the coreflood of Chandrasekhar and Mohanty [16] was successfully matched using the third approach of the second method

Table 9: Summary of relative permeability parameters (second method-third approach).

Injection cycle	k_{rw}	k_{ro}	n_w	n_o
First cycle	0.025	0.203	1.30	3.50
Second cycle (LSWI Effect)	0.024	0.830	1.70	1.55
Second cycle (trapping number effect)	0.089	0.830	1.43	1.55
Third cycle	0.023	0.850	2.00	1.53
Fourth cycle	0.022	0.860	2.20	1.52

by tuning residual oil saturation and relative permeability curves, including endpoints and Corey's exponents.

5. Summary and Conclusions

Oil recovery and pressure drop data for the coreflood of Chandrasekhar and Mohanty [16] were matched successfully using UTCHEM. The main findings of this work are summarized as follows.

(i) Wettability alteration is still believed to be the contributor to the LSWI effect on oil recovery from carbonate rocks.

(ii) History-matching of the LSWI effect on oil recovery is sensitive to residual oil saturation and relative permeability curves.

(iii) Tuning both relative permeability endpoints and Corey's exponents is essential for good history-matching of both oil recovery and pressure drop data.

(iv) Neglecting capillary pressure effect on oil recovery and pressure drop history-matching in case of LSWI is a plausible assumption even if the coreflood is conducted at reservoir rate of 1 ft/day.

(v) Oil relative permeability parameters are more sensitive to LSWI compared to water relative permeability parameters.

(vi) The findings of this paper validate our previous findings [19] upon which the two corefloods of Yousef et al. [7] were history-matched.

Moreover, in light of the previous findings, a simple interpolation model can be implemented in UTCHEM and applied to history-match both works of Yousef et al. [7] and Chandrasekhar and Mohanty [16]. This is our next step to have more insight into the low salinity water injection (LSWI) mechanism before we propose our own mechanistic LSWI model.

Nomenclature

CPC: Parameter related to the maximum capillary pressure
EPC: Capillary pressure exponent
k_{rl}^*: Phase endpoint relative permeability
n_l: Phase Corey's exponent
S_l: Phase saturation
S_{lr}: Phase residual saturation
σ: Interfacial tension.

Conflict of Interests

The authors declare that there is no conflict of interests regarding the publication of this paper.

Acknowledgments

The authors wish to acknowledge useful discussions with K. K. Mohanty during the work. This work was funded by Abu Dhabi National Oil Company (ADNOC).

References

[1] E. J. Høgnesen, S. Strand, and T. Austad, "Waterflooding of preferential oil-wet carbonates: oil recovery related to reservoir temperature and brine composition," in *Proceedings of the 67th European Association of Geoscientists and Engineers (EAGE '05)*, pp. 815–823, Madrid, Spain, June 2005, SPE-94166.

[2] K. J. Webb, C. J. J. Black, and G. Tjetland, "A laboratory study investigating methods for improving oil recovery in carbonates," in *Proceedings of the International Petroleum Technology Conference*, pp. 785–791, Doha, Qatar, November 2005, SPE-10506.

[3] P. Zhang, M. T. Tweheyo, and T. Austad, "Wettability alteration and improved oil recovery by spontaneous imbibition of seawater into chalk: impact of the potential determining ions Ca^{2+}, Mg^{2+}, and SO_4^{2-}," *Colloids and Surfaces A: Physicochemical and Engineering Aspects*, vol. 301, no. 1–3, pp. 199–208, 2007.

[4] S. Strand, T. Austad, T. Puntervold, E. J. Høgnesen, M. Olsen, and S. M. F. Barstad, "'Smart Water' for oil recovery from fractured limestone: a preliminary study," *Energy and Fuels*, vol. 22, no. 5, pp. 3126–3133, 2008.

[5] I. Fjelde, "Low salinity water flooding experimental experience and challenges," in *Proceedings of the Force RP Work Shop: Low Salinity Water Flooding, the Importance of Salt Content in Injection Water*, Stavanger, Norway, 2008.

[6] S. Bagci, M. V. Kok, and U. Turksoy, "Effect of brine composition on oil recovery by waterflooding," *Petroleum Science and Technology*, vol. 19, no. 3-4, pp. 359–372, 2001.

[7] A. A. Yousef, S. Al-Saleh, A. Al-Kaabi, and M. Al-Jawfi, "Laboratory investigation of novel oil recovery method for carbonate reservoirs," in *Proceedings of the SPE Canadian Unconventional Resources and International Petroleum Conference*, pp. 1825–1859, Alberta, Canada, October 2010, SPE-137634.

[8] R. Gupta, P. Griffin, L. Hu et al., "Enhanced waterflood for middle east carbonates cores—impact of injection water composition," in *Proceedings of the SPE Middle East Oil and Gas Show and Conference*, Manama, Bahrain, 2011, SPE-142668.

[9] A. A. Yousef, S. Al-Saleh, and M. Al-Jawfi, "Improved/enhanced oil recovery from carbonate reservoirs by tuning injection water salinity and ionic content," in *Proceedings of the SPE Improved Oil Recovery Symposium*, Tulsa, Okla, USA, 2012, SPE-154076.

[10] A. S. Al-Harrasi, R. S. Al Maamari, and S. Masalmeh, "Laboratory investigation of low salinity waterflooding for carbonate reservoirs," in *Proceedings of the SPE Abu Dhabi International*

Petroleum Exhibition & Conference, Abu Dhabi, UAE, 2012, SPE-161468.

[11] J. Romanuka, J. P. Hofman, D. J. Ligthelm et al., "Low salinity EOR in carbonates," in *Proceedings of the SPE Improved Oil Recovery Symposium*, Tulsa, Okla, USA, 2012, SPE-153869.

[12] D. C. Standnes and T. Austad, "Wettability alteration in chalk 2. Mechanism for wettability alteration from oil-wet to water-wet using surfactants," *Journal of Petroleum Science and Engineering*, vol. 28, no. 3, pp. 123–143, 2000.

[13] P. Zhang, M. T. Tweheyo, and T. Austad, "Wettability alteration and improved oil recovery in chalk: the effect of calcium in the presence of sulfate," *Energy and Fuels*, vol. 20, no. 5, pp. 2056–2062, 2006.

[14] T. Puntervold, S. Strand, and T. Austad, "Water flooding of carbonate reservoirs: effects of a model base and natural crude oil bases on chalk wettability," *Energy & Fuels*, vol. 21, no. 3, pp. 1606–1616, 2007.

[15] A. A. Yousef, J. Liu, G. Blanchard et al., "Smart water flooding: industry's first field test in carbonate reservoirs," in *Proceedings of the SPE Annual Technical Conference and Exhibition*, San Antonio, Tex, USA, 2012, SPE-159526.

[16] S. Chandrasekhar and K. K. Mohanty, "Wettability alteration with brine composition in high temperature carbonate reservoirs," in *Proceedings of the SPE Annual Technical Conference and Exhibition*, New Orleans, La, USA, 2013, SPE-166280.

[17] G. A. Pope, W. Wu, G. Narayanaswamy, M. Delshad, M. M. Sharma, and P. Wang, "Modeling relative permeability effects in gas-condensate reservoirs with a new trapping model," *SPE Reservoir Evaluation & Engineering*, vol. 3, no. 2, pp. 171–178, 2000.

[18] M. Delshad, D. Bhuyan, G. A. Pope, and L. Lake, "Effect of capillary number on the residual saturation of a three-phase micellar solution," in *Proceedings of the SPE Enhanced Oil Recovery Symposium*, Tulsa, Okla, USA, 1986, SPE-14911.

[19] E. W. Al-Shalabi, K. Sepehrnoori, and M. Delshad, "Mechanisms behind low salinity water flooding in carbonate reservoirs," in *Proceedings of SPE Western Regional and AAPG Pacific Meeting*, Monterey, Calif, USA, 2013, SPE-165339.

3

Effect of Polymer Adsorption on Permeability Reduction in Enhanced Oil Recovery

Saurabh Mishra, Achinta Bera, and Ajay Mandal

Department of Petroleum Engineering, Indian School of Mines, Dhanbad 826004, India

Correspondence should be addressed to Ajay Mandal; mandal_ajay@hotmail.com

Academic Editor: Alireza Bahadori

In order to reduce the permeability to water or brine, there is a possibility of polymer injection into the reservoir. In the present work, special focus has been paid in polymer [partially hydrolyzed polyacrylamide (PHPA)] injection as a part of chemical method. Tests were conducted in the laboratory at the ambient temperature to examine the reduction in permeability to water or brine in the well-prepared sand packed after the polymer injection. The experiments were performed to study the effect of polymer adsorption on permeability reduction by analyzing residual resistance factor values with different concentrations of polymer solutions. The rheological behavior of the polymer has also been examined. The experimental results also indicate that the adsorption behavior of polymer is strongly affected by salinity, solution pH, and polymer concentration. To investigate the effect of polymer adsorption and mobility control on additional oil recovery, polymer flooding experiments were conducted with different polymer concentrations. It has been obtained that with the increase in polymer concentrations, oil recovery increases.

1. Introduction

With reservoir getting matured, the increased water production with producing oil is a major concern in the petroleum industry. Hydrocarbon production decreases, which affects recovery economics and disposal of the excessive high amount of underground water, which causes complex environmental problems [1]. Oil and gas reservoirs are often heterogeneous, having a different permeability in different-different layers. This causes channelling of excessive water production through high permeability layers, as a result of which large amount of movable oil and gas remains trapped in low permeability zones which results in poor recovery in primary and secondary stages of production. A significant part of the residual oil can be recovered by application of a polymer solution in the heterogeneous reservoirs [2]. The excess water production can be controlled without affecting oil production rate by using polymer to reduce the relative permeability to water more than the relative permeability to oil [3]. The injection of polymer solutions in production wells has proven to be an effective method to reduce or block excessive water production [4]. In a water-wet reservoir, oil flows inside pores while water through the annulus between the pore walls and the oil-water interface [5]. When the polymer solution is injected, it gets adsorbed on the rock surface to form a thin layer. As water flows through this zone, the adsorbed polymer swells and prevents the flow of water while allowing the oil to flow through. When the polymer is injected, it builds up a resistance to flow in the reservoir through permeability reduction. This increased flow resistance diverts the injected water into unswept areas [6]. Interestingly, the adsorbed polymer is sustained over a long time; this leads to a great reduction in permeability [7]. In 1974, some of the basic concerns pertaining polymer adsorption in porous medium and effect of polymer adsorption on flow properties have been investigated by Hirasaki and Pope [8]. They reported that polymer treatment reduces permeability by adsorption of polymer and reduction in pore radii. The permeability reduction caused by polymer adsorption was traditionally known as "residual resistance factor" which is tantamount to the endpoint relative permeability [9].

Laboratory experiments have been performed by many researchers to examine polymer retention and its effect on oil recovery. Zaitoun and Bertin showed in their work that oil and water permeability can be modified by wall effects due to adsorption of polymer [10]. Grattoni et al. reported atomic force microscopy study on polymer retention on microglass surface. They discussed the basic adsorption mechanism and flow of polyacrylamide in the formation of layer thickness [11]. Lai et al. have recently showed the permeability reduction performance of a hyperbranched polymer in high permeability porous media [12]. Al-Hashmi and Luckham studied the effect of salt concentration on the adsorption of high-molecular-weight nonionic and cationic polymers on glass surface and showed that with increase in salt concentration the adsorbed layer thickness also increases [13]. Al-Sharji et al. reported that polymer adsorption significantly influences the water permeability and enhances the oil recovery [14].

In polymer flooding technique, a high-molecular-weight water-soluble polymer is added as a thickening agent into injected water to increase viscosity of aqueous phase for mobility control which can improve sweep efficiency. One of the attractive properties of such polymers is their ability to reduce the relative permeability to water more than the relative permeability to oil in porous media. Liu et al. reported in their framework in the Daqing oil field that from polymer flooding water cut was significantly reduced and oil recovery was improved by 10% [15]. Polymer flooding projects yielded additional recoveries of 10–15% of the original oil in place. Since the efficiency of polymer flooding differs with the rock and fluid properties as it varies from field to field. Thus an efficient work is required for improvement of this technology in practical field [16].

In this work, Partially Hydrolyzed Polyacrylamide (PHPA) is used to examine the effect of polymer adsorption on the permeability modification and oil recovery in a sand pack system. Rheological property of the used polymer has also been investigated by measuring the viscosity of different concentrated polymer solutions with a variation of shear rate. Adsorption experiments have been carried out with a variation of polymer concentrations, salinity, and pH of the solution. Along with this, thickness of the adsorbed layer of polymer has also been calculated with the help of the permeability reduction data. Flooding experiments have been performed to investigate the additional oil recovery after conventional water flooding.

2. Experimental Sections

2.1. Materials Used. Partially hydrolyzed Polyacrylamide (PHPA) (Polymer Pusher 1000, SNF Floerger, France) with 98% purity was supplied by Central Drug House (CDH) Pvt. Ltd., India. Typical degree of hydrolysis of the used polymer is 30% to 35%. Sodium Chloride (NaCl) with 98% purity, procured from Qualigens Fine Chemicals, India, was used for preparation of brine. Reverse osmosis water from Millipore water system (Millipore SA, 67120 machine, France) was used for the preparation of solutions. Sand particles (60–70 mesh sized) were used as adsorbent in the adsorption study. The details compositional analysis of the used sand particles has been given in our previous work [17]. The crude oil used in the flooding experiments was collected from the Ahmedabad oil field (India). It was degassed and dehydrated before use. The physical properties of crude oil are shown as follows:

(i) viscosity: 5.12 Pa·s at 45°C at 20 s^{-1} shear rate,

(ii) total acid number: 0.038 mg KOH/g,

(iii) API gravity: 38.5,

(iv) color: blackish.

2.2. Experimental Procedures

2.2.1. Preparation and Rheology of Polymer Solution. Polymer solutions were prepared by dissolving the required amount of polymer in brine. Polymer solutions were prepared by the slow addition of the polymer powder to the brine in a vortex created by a magnetic stirrer. Gentle agitation was maintained over day for the complete polymer dissolution to obtain a homogeneous solution. Care was taken to rotate the magnetic stirrer at low rpm speed to avoid any mechanical degradation. In the present study, the rheological property of the used polymer was measured with the help of a rheometer (Physica MC1) at different concentrations. The flow curves were recorded between 10^{-1} and 10^{3} 1/s shear rate using a linear rate program and 2 min ramp time.

2.2.2. Permeability Reduction by Polymer Flooding. Sand packed was mounted horizontally in a sand packed holder and flow experiment was performed. First sand packed was flooded with brine and initial permeability to brine was calculated. Polymer solution was then injected, when sand packed was fully saturated with brine. The flow rate was measured during polymer flooding. When the flow rate attained a constant value, the polymer injection was stopped. The sand packed was then kept at rest for 50–56 hours for maximum interaction between sand particles and polymer molecules. Brine was then injected at constant pressure and flow rate curve was drawn. Above procedure was repeated for each concentration of polymer solution.

2.2.3. Polymer Adsorption and Thickness of Adsorbed Layers. A series of batch experiments were carried out to determine the adsorption of PHPA on the sand particles. 100 g of clean sand particles were added to 100 mL polymer solution in a 250 mL conical flux and allowed to conduct the experiment by constant shaking at 303 K for 24 h on a horizontal shaker mode. Core flooding is a dynamic process. So the dynamic condition was considered to maintain the polymer flooding process. During the polymer injection pressure was applied therefore the dynamic condition is maintained. Actually dynamic condition is an appropriate field relevant condition. It is important to mention that the ratio of adsorbent to adsorbate also varies for different studies [18, 19]. In the present study, as flooding experiment was conducted on sand packed, the ratio was taken in this pattern. Again a certain amount of adsorbent is required for maximum adsorption. So a large excess concentration of adsorbent was taken in this study. After adsorption, the polymer solutions were isolated by centrifugation. The equilibrium or residual concentrations of the polymer solutions were determined by UV-spectrophotometer. The equilibrium concentrations (C_e) were calculated by using Beer-Lambert equation:

$$A = \varepsilon * L * C, \quad (1)$$

where A is absorbent, ε is molar extinction coefficient (L/mole/cm), L is the path length (10 mm), and C is concentration (mole/L) of the solution.

The adsorption capacity of polymer on the adsorbent, Γ_i (mg/g), was calculated by a mass balance relation [20]:

$$\Gamma_i = (C_0 - C_e)\frac{V}{m}, \quad (2)$$

where C_0 and C_e are the initial and equilibrium concentrations of polymer solutions (mg/g), respectively, V is the volume of the polymer solution (L), and m is the weight of the sand particles (g) (adsorbent) used. The effect of salt concentration, pH, and polymer concentrations is investigated. These may greatly affect polymer adsorption hence permeability reduction and polymer flood efficiency.

The thickness of the adsorbed layer of polymer (χ_p) can be calculated from the permeability reduction data. The calculation is based on the assumption of Poiseuille's fluid flow through a capillary having its cross-section reduced by a uniform layer. The thickness of the adsorbed layer of polymer is calculated as follows [3]:

$$\chi_p = r_p\left(1 - \mathrm{RRF}^{-1/4}\right), \quad (3)$$

where RRF is the residual resistance factor and r_p is the average pore diameter, which can be calculated as [3]

$$r_p = 1.15\left(\frac{8k_w}{\varphi}\right)^{1/2}, \quad (4)$$

where k_w is the effective permeability to water and φ porosity of the sand pack.

2.2.4. Experimental Apparatus and Method for Polymer Flooding. The experimental apparatus is composed of a sand pack holder, cylinders for holding chemical slugs and crude oil, positive displacement pump, and measuring cylinders for collecting the samples. The detail of the apparatus and method of flooding is reported in our earlier paper [21]. The schematic diagram of flooding experiment is given in Figure 1. The model geometry size is $L = 35$ cm and $r = 3.5$ cm. The sand packed holder was tightly packed with uniform sands (60–100 mesh) and saturated with 1.0 wt% brine. The additional recoveries were calculated by material balance.

Darcy's law was used to calculate the effective permeability to oil (k_o) and effective permeability to water (k_w) at irreducible water saturation (S_{wi}) and residual oil saturation (S_{or}), respectively. For a horizontal linear system, flow rate is related to permeability as follows:

$$q = \frac{kA}{\mu}\frac{dp}{dx}, \quad (5)$$

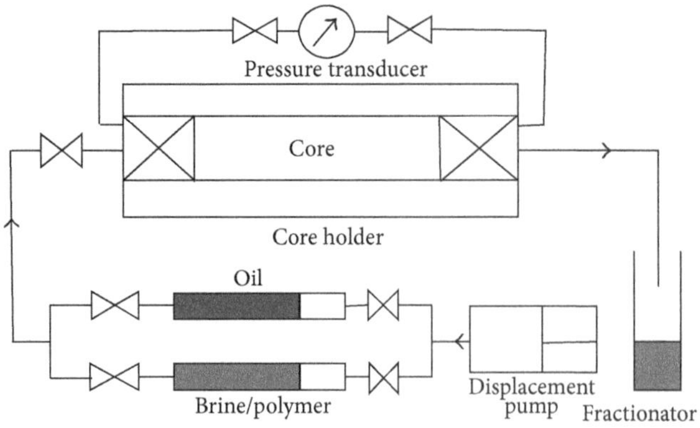

Figure 1: Schematic diagram of flooding experiment.

where q is the volumetric flow rate (cm^3/Sec.), A is the total cross-sectional area of the sand pack (cm^2), μ is the fluid viscosity (cp), dp/dx is the pressure gradient (atm/cm), and k is permeability in Darcy. Experiments were performed using 2500 ppm and 2000 ppm PHPA solutions.

3. Results and Discussion

3.1. Rheology of PHPA Solutions. Polymer rheology plays an important role in controlling the mobility ratio and hence the sweep efficiencies. The basic rheological behavior of aqueous solutions of partially hydrolyzed polyacrylamide has been investigated by varying polymer concentrations. Figure 2 shows the variation of apparent viscosity with shear rate. The viscosity of solutions increased with increasing polymer concentrations because of the increasing intermolecular entanglement. The viscosities of all samples decreased with the increasing shear rate, suggesting that the aqueous solution of PHPA exhibits non-Newtonian behavior. This is due to uncoiling and aligning of polymer chains when exposed to shear flow. As the shear rate approaches toward zero the solutions show maximum viscosity. The viscosity is quite significant at moderate shear rate. It has also been found that at higher shear rate the solution approaches towards Newtonian behavior.

3.2. Permeability Reduction. The flow behavior of polymer solutions of different concentrations and subsequent brine injection through the sand packs are shown in Figure 3. It may be found that as the cumulative injected volume of polymer solution increases the flow rate gradually decreases. This is because of adsorption of polymer on the sand particle. After a certain pore volume injected there is no further change in flow rate. This indicates that the surfaces of sand particles are now fully saturated with polymer and there is no further adsorption. The system is kept as it is for around 50 hours for maximum interaction between polymer and sand particles. The polymer coats on the rock surfaces with a hydrophilic film. As water passes over the polymer, the film swells, reducing the effective permeability. In presence of oil, the swelling does not occur, thereby improving potential oil recovery [22]. Then the sand packs

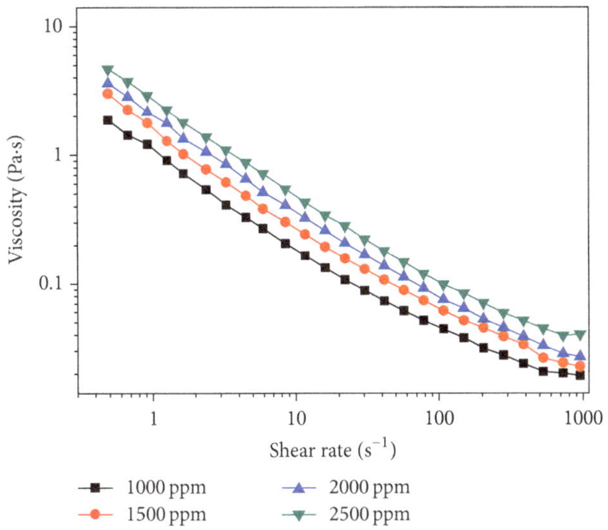

FIGURE 2: Viscosity of PHPA solutions with different concentrations at 303 K.

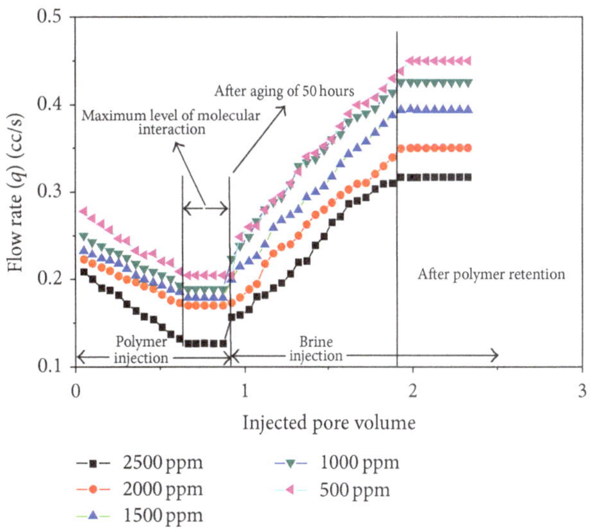

FIGURE 3: Flow rate curves during injection of brine after aging of polymer on sand grains.

are continuously injected with brine. Brine flow rate gradually increases due to desorption of polymer. But after a certain pore volume of brine injected it has been found that there is no further increase of flow rate. This suggests that even after a long time of chase water flooding after polymer flooding, a certain amount of polymer is retained and hence the benefits of permeability reduction can be achieved for a long time.

The absolute permeability and the residual permeabilities have been measured by Darcy's law and the results are shown in Figure 4. It has been found from the figure that absolute permeability of the sand pack is much higher than that of residual permeabilities. And the residual permeabilities increase with the increase in concentrations of polymer.

To examine restriction in flow due to polymer adsorption, resistance factor (RF) and residual resistance factor (RRF) ware calculated. Figure 5 shows the values of RF and RRF with a variation of pore volume injected at different polymer concentrations. Initially, when the polymer was injected, it prefers high permeability zones due to low resistance in the flow. Results show that the increment in the polymer concentrations brings to the high value of RF and RRF. Interaction between polymer and sand grains leads to profile modification. This profile modification depends on how polymer is adsorbed on the wall of sand grains in high permeability zones. Due to polymer adsorption pore throat radii reduces, this results in an unexpected decrement in mobility to polymer. The measure of the mobility reduction is known as "resistance factor." Resistance factor was calculated by the following equation:

$$\mathrm{RF} = \frac{\lambda_w}{\lambda_p}, \tag{6}$$

where λ_w is the mobility of water and λ_p is the mobility of polymer solution.

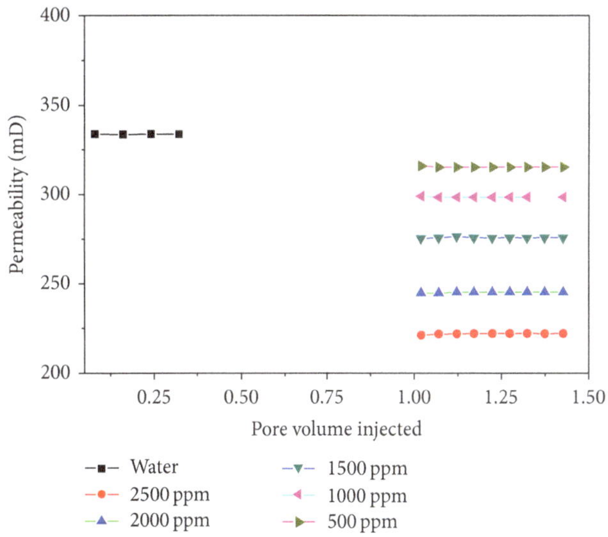

FIGURE 4: Reduction in permeability due to polymer retention.

Mobility is calculated by using Darcy's equation as follows:

$$\lambda = \frac{q * L}{A * \phi * \Delta p}. \tag{7}$$

Table 1 shows the calculated data of RF and RRF for PHPA with different concentrations.

3.3. Polymer Adsorption. Adsorption of polymer on rock surfaces is influenced by several factors. The influencing factors are polymer concentrations, salinity of the solution, the pH of the solution, temperature, and so forth. Some of the factors have been discussed here with experimental results.

Table 1: Calculated values of RF and RRF for different concentrated PHPA solutions.

Test samples	Mobility	Resistance factor (RF)	Residual resistance factor (RRF)
Water	1.027	—	—
C_1 (2500 ppm + 2 wt% NaCl)	0.231	4.461	1.504
C_2 (2000 ppm + 2 wt% NaCl)	0.284	3.621	1.361
C_3 (1500 ppm + 2 wt% NaCl)	0.327	3.143	1.211
C_4 (1000 ppm + 2 wt% NaCl)	0.353	2.905	1.119
C_5 (500 ppm + 2 wt% NaCl)	0.369	2.783	1.059

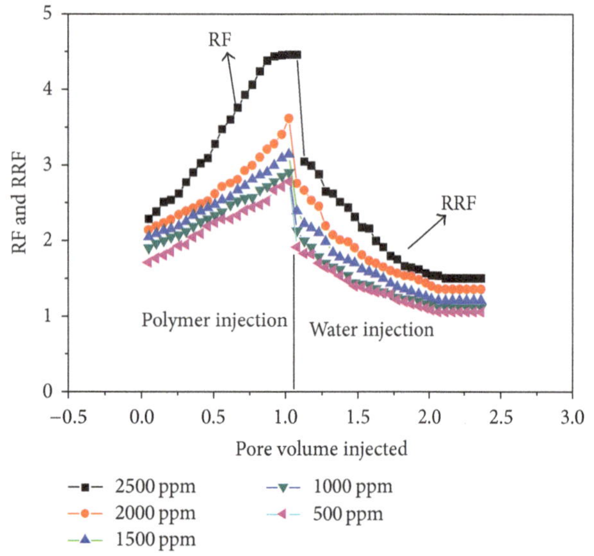

Figure 5: Plot of RF and RRF with different concentrations of polymer solution.

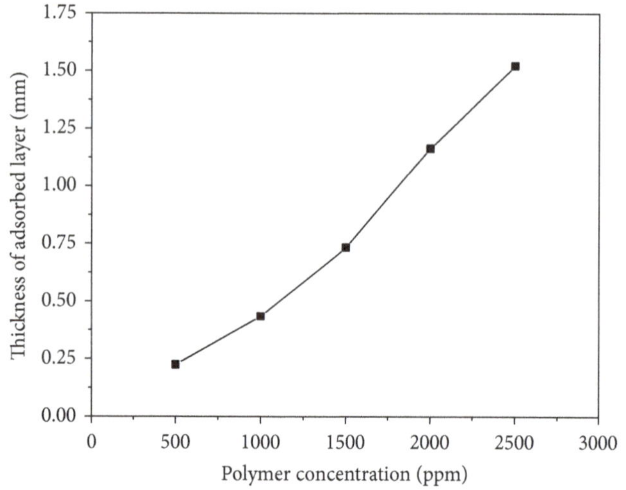

Figure 6: Relation between polymer concentrations and thickness of adsorbed layer.

3.3.1. Thickness of Adsorbed Layer. Figure 5 shows the relation between polymer concentrations and thickness of adsorbed layer. It is confirmed from Figure 6 that with the increase in polymer concentrations the thickness of adsorbed layer also increases. As concentration increases, the number of molecules increases, thereby the probability of interaction between polymer molecules and sand particles increases. This brought increment in adsorption capacity per unit area. Therefore, it is confirmed that the adsorption layer thickness increases with an increase in adsorption. The increase in the adsorption layer thickness as the polymer content in solution increases is due to a change in the conformation of the adsorbed molecules [23]. The change in the adsorption layer thickness in the initial segments of the isotherms suggests that at low concentrations of polymer, the molecules on the surface of the sand particles are distributed and bound with a small number of segments. The increase in concentrations of polymer causes a rearrangement of the structure of the adsorbed layer. The adsorbed molecules break previously formed bonds so that they straighten and unfold and the layer thickness increases.

3.3.2. Effect of Polymer Concentrations on Adsorption. Figure 7 shows the time dependent polymer adsorption on the sand surface at different concentrations. With increase in time polymer adsorption increases and after certain time adsorption remains constant with time. As time increases, the number of adsorption sites decreases due to formation of adsorbed polymer layer on the sand surface. When all the sites are covered with polymer, then further adsorption does not take place and time independent adsorption occurs. Figure 7 shows that with the increase in polymer concentrations the polymer adsorption capacity also increases. At higher concentrations of polymer solutions, number of polymer molecules increases, which in turn increase the probability of interaction between the sand surface and polymer molecule. It is also important to point out here that after certain concentration no further adsorption takes place due to saturation of the adsorption capacity of the active adsorption sites [9, 24, 25]. In the present case, only four different concentrations have been used to show the effect of concentrations on polymer adsorption onto the sand surface and it has been found that up to 2500 ppm concentration of PHPA polymer adsorption increases. Adsorption causes the permeability reduction. Ogunberu and Asghari have reported on their experimental work that increase in polymer concentrations results in an increase in permeability reduction caused by polymer adsorption and efficiency of polymer flooding [3].

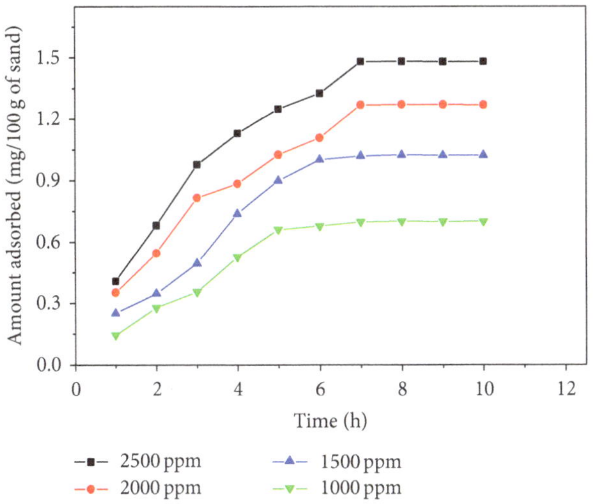

FIGURE 7: Effect of polymer concentrations on polymer adsorption.

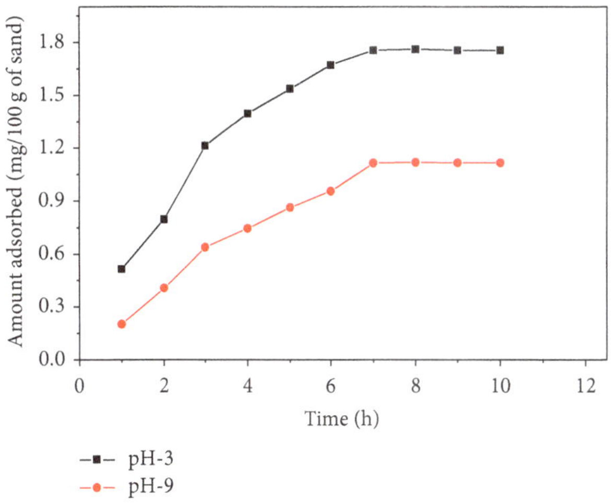

FIGURE 8: Effect of solution pH on polymer adsorption on sand surface.

3.3.3. Effect of Solution pH on Polymer Adsorption. The adsorption behavior of PHPA at different solution pH has been shown in Figure 8. The adsorption capacity increases when solution pH is decreased as shown in Figure 8. At low pH, sand surfaces become more positive due to the acidic nature of the polymer solutions and the interaction of sand surface with anionic polymer PHPA is high, hence, adsorption capacity is high. The negatively charged carboxyl group dissociated through electrostatic attraction to the positively charged ions (due to more H^+ ions). This causes more attraction between polymer molecules and rock surfaces. When the pH of the solution increases, that is, the solutions become alkaline in nature, the sand surfaces become more negative. The repulsion between PHPA with negatively charged carboxylic acid group and negatively charged sand surface takes place, which leads to lower adsorption of the polymer on sand surfaces. So, it is confirmed from the study of the effect of pH on adsorption that the adsorption of polymer on rock surfaces can be modified by fixing the solution pH which is important for polymer flooding.

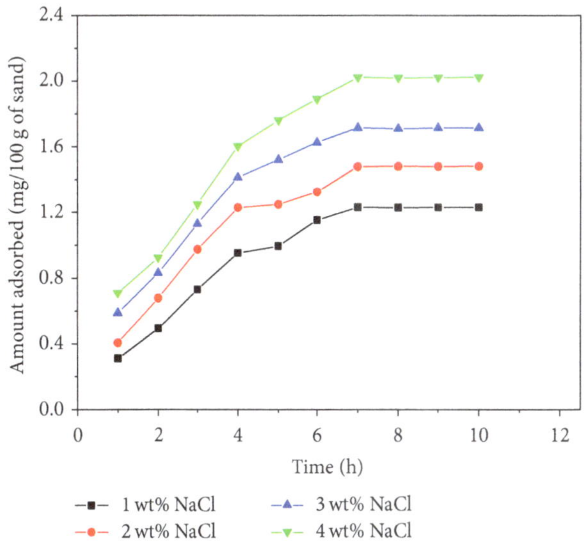

FIGURE 9: Effect of salt concentration on polymer adsorption.

3.3.4. Effect of Salt Concentration on Polymer Adsorption. Experimental results have been shown in Figure 9. It is clear from Figure 8 that polymer adsorption increases with increase in salinity. As the salt concentration increases, charge shielding takes place due to positively charged ions of the salt (Na^+), thus the hydrodynamic radius of polymer molecule reduces. Due to this intermolecular interaction, electrostatic repulsion in the polymer solution decreases. This causes an increase in adsorption capacity of polymer capacity of polymer solution. Moreover the salt concentration may affect the displacement efficiency and volumetric sweep efficiency of the polymer flood. High salt concentration may affect the polymer absorption by an increase in the viscosity of polymer due to electrostatic repulsion in the polymer solution and a reduction in the surface area of adsorbent, access to polymer molecules. Due to the above two parametrical effects, adsorption capacity increases as salt concentration increases.

3.4. Oil Recovery by Polymer Flooding. Injection of the polymer increases the recovery of oil by improving the mobility ratio, reducing the effective permeability to water and increasing the effective permeability to water. In the present study, two different concentrations of PHPA were used to measure the additional recovery of oil after conventional water flooding. Initially the sand pack was flooded with brine. When water cut reached around 95%, approximately 0.75 PV polymer solution was injected to investigate its effect on the efficiency of polymer flooding. Two sets of experiments were performed using 2000 ppm and 2500 ppm polymer solution. The recovery of oil while flooding the sand packed with brine and different concentrations of PHPA are shown in Figure 10. The main parameters and results of these flood tests are in Table 2, including the initial water saturations, the injected pore volumes and oil recoveries during water flooding and

Table 2: Porous media property and oil recovery by polymer flooding.

Experiment number	Porosity (%)	Permeability, k (mD)		Design of polymer slug for flooding	Recovery of oil by water flooding at 95% water cut (% OOIP)	Additional recovery (% OOIP)	% Saturation		
		k_w (S_w = 1)	k_o (S_{wi})				S_{wi}	S_{oi}	S_{or}
P1	32.5	333.5	58.33	0.75 PV polymer (2500 ppm) + chase water	43.9	21.00	12	88	23.5
P2	32.4	333.6	58.4	0.75 PV polymer (2000 ppm) + chase water	44.0	19.21	12	88	23.8

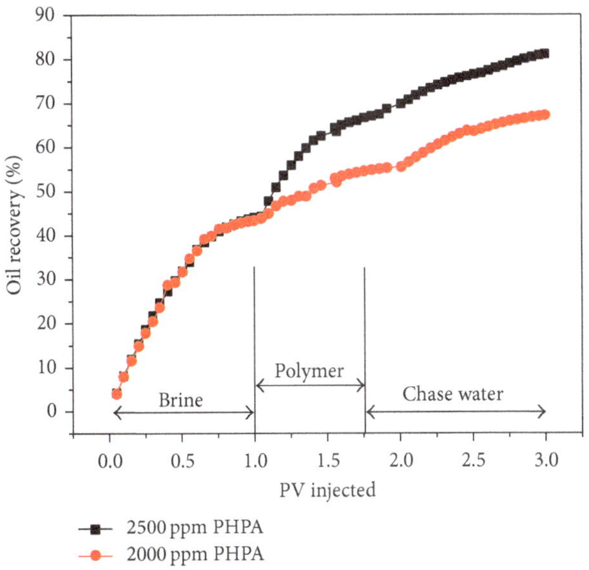

FIGURE 10: Oil recovery by polymer flooding using 2500 ppm and 2000 ppm PHPA solutions.

tertiary recoveries. All recoveries are calculated based on the original oil in place (OOIP).

The additional oil recoveries by polymer flooding were 19.21% and 21.00% OOIP for 2000 ppm and 2500 ppm PHPA solutions. These results were compared with another literature reported by Maitin, where the laboratory investigations, simulation studies, and a pilot test in the oil fields of a North German sediment basin have produced the incremental recovery of 8 to 22% OOIP [26].

4. Conclusions

Rheological properties of the PHPA solution have been determined and it has been found that with increase in polymer concentrations viscosity also increases. Adsorption of polymer on sand particles has been studied by polymer flooding through sand packed. Experimental results show that after aging, there was significant interaction between polymer molecules and sand grains, which plugged the flow path to inject brine at constant pressure; as a result, there was a significant decrement in flow rate. As brine injection increased, polymer retention decreased and reached a critical value which gives an absolute permeability reduction. The result shows that increase in polymer concentrations enhances the polymer adsorption capacity. At low pH, PHPA adsorption capacity on sand surface is high due to the acidic nature of the solution which makes the sand surface more positive and that is why the interaction of sand surface with anionic polymer PHPA is high, hence, adsorption capacity is high. When the pH of the solution increases the solution becomes alkaline and as a result repulsion between PHPA and negatively charged sand surface increases, therefore, adsorption of the polymer decreases. As the salt concentration increases, charge shielding takes place due to positively charged ions of the salt (Na^+), thus the hydrodynamic radius of polymer molecule reduces. Due to this intermolecular interaction, electrostatic repulsion in the polymer solution decreases. This causes an increase in adsorption capacity of polymer capacity of polymer solution. The result shows that with the increase in polymer concentrations, thickness of adsorbed layer also increases. To investigate the relationship between the incremental tertiary oil recovery and the effective viscosity of polymer solutions, two sets of sand packed flood tests were carried out by injecting polymer solutions of different viscosities. The additional oil recoveries by polymer flooding were 19.21% and 21.00% OOIP for 2000 ppm and 2500 ppm PHPA solutions, respectively.

Conflict of Interests

The authors declare that there is no conflict of interests regarding the publication of this paper.

Acknowledgments

The authors gratefully acknowledge the financial assistance provided by University Grant Commission [F. no. 37-203/2009 (SR)], New Delhi, to the Department of Petroleum Engineering, Indian School of Mines, Dhanbad, India. Thanks are also extended to all individuals associated with the project.

References

[1] P. L. J. Zitha, K. G. S. van Os, and K. F. J. Denys, "Adsorption of linear flexible polymers during laminar flow through porous media: effect of concentration," in *Proceedings of the SPE/DOE Improved Oil Recovery Symposium*, pp. 19–22, Tulsa, Okla, USA, April 1998, SPE paper no. 39675.

[2] O. L. Kouznetsov, E. M. Simkin, G. V. Chilingar, and S. A. Katz, "Improved oil recovery by application of vibro-energy to waterflooded sandstones," *Journal of Petroleum Science and Engineering*, vol. 19, no. 3-4, pp. 191–200, 1998.

[3] A. L. Ogunberu and K. Asghari, "Water permeability reduction under flow-induced polymer adsorption," in *Proceedings of the 5th Canadian International Conference*, Alberta, Canada, June 2004, paper no. 2004-236.

[4] P. D. Moffitt, "Long-term production results of polymer treatments in producing wells in western Kansas," *Journal of Petroleum Technology*, vol. 45, no. 4, pp. 356–362, 1993.

[5] A. Zaitoun and N. Kohler, "Two phase flow through porous Media: effect of an Adsorbed Polymer Layer," in *Proceedings of the SPE Annual Technical Conference and Exhibition*, Huston, Tex, USA, October 1998, SPE paper no. 18085.

[6] M. A. Mohammed, "Investigation of polymer adsorption on rock surface of high saline reservoirs," in *Proceedings of the Saudi Arabia Section Technical Symposium (SPE '08)*, Alkhobar, Saudi Arebia, May 2008, SPE paper no. 120807.

[7] H. T. Dovan and R. D. Hutchins, "New polymer technology for water control in gas wells," *SPE Production and Facilities*, vol. 9, no. 4, pp. 280–286, 1994.

[8] G. J. Hirasaki and G. A. Pope, "Analysis of factors influencing mobility and adsorption in the flow of polymer solution through porous media," *Society of Petroleum Engineering Journal*, vol. 14, no. 4, pp. 337–346, 1974.

[9] C. G. Zheng, B. L. Gall, H. W. Gao, A. E. Miller, and R. S. Bryant, "Effects of polymer adsorption and flow behavior on two-phase flow in porous media," *SPE Reservoir Evaluation & Engineering*, vol. 3, no. 3, pp. 216–223, 2000.

[10] A. Zaitoun and H. Bertin, "Two-phase flow property modification by polymer adsorption," in *Proceedings of the SPE Improved Oil Recovery Symposium*, Tulsa, Okla, USA, 1998, SPE Paper no. 39631.

[11] B. C. A. Grattoni, P. F. Luckham, X. D. Jing, L. Norman, and R. W. Zimmerman, "Polymers as relative permeability modifiers: adsorption and the dynamic formation of thick polyacrylamide layers," *Journal of Petroleum Science and Engineering*, vol. 45, no. 3-4, pp. 233–245, 2004.

[12] N. Lai, X. Qin, Z. Ye, C. Li, K. Chen, and Y. Zhang, "The study on permeability reduction performance of a hyperbranched polymer in high permeability porous medium," *Journal of Petroleum Science and Engineering*, vol. 112, pp. 198–205, 2013.

[13] A. R. Al-Hashmi and P. F. Luckham, "Characterization of the adsorption of high molecular weight non-ionic and cationic polyacrylamide on glass from aqueous solutions using modified atomic force microscopy," *Colloids and Surfaces A*, vol. 358, no. 1-3, pp. 142–148, 2010.

[14] H. H. Al-Sharji, C. A. Grattoni, R. A. Dawe, and R. W. Zimmerman, "Disproportionate permeability reduction due to polymer adsorptrion-entanglement," in *SPE European Formation Damage Conference*, The Hague, Netherlands, 2001, SPE Paper no. 68972.

[15] S. Liu, D. L. Zhang, W. Yan, M. Puerto, G. J. Hirasaki, and C. A. Miller, "Favorable attributes alkaline-surfactant-polymer flooding," *Society of Petroleum Engineering Journal*, vol. 13, no. 1, pp. 5–16, 2008.

[16] Y. Masuda, K.-C. Tang, M. Miyazawa, and S. Tanaka, "1D simulation of polymer flooding including the viscoelastic effect of polymer solution," *SPE Reservoir Engineering*, vol. 7, no. 2, pp. 247–252, 1992.

[17] A. Bera, T. Kumar, K. Ojha, and A. Mandal, "Adsorption of surfactants on sand surface in enhanced oil recovery: isotherms, kinetics and thermodynamic studies," *Applied Surface Science*, vol. 284, pp. 87–99, 2013.

[18] W. Lv, B. Bazin, D. Ma, Q. Liu, D. Han, and K. Wu, "Static and dynamic adsorption of anionic and amphoteric surfactants with and without the presence of alkali," *Journal of Petroleum Science and Engineering*, vol. 77, no. 2, pp. 209–218, 2011.

[19] M. Safian-Boldani, M. P. Shahri, M. Zargartalebi, and M. Arabloo, "New surfactant extracted from zizyphus spina christi for enhanced oil recovery: experimental determination of static adsorption isotherm," *Journal of the Japan Petroleum Institute*, vol. 56, no. 3, pp. 142–149, 2013.

[20] M. A. Ahmadi and S. R. Shadizadeh, "Experimental investigation of adsorption of a new nonionic surfactant on carbonate minerals," *Fuel*, vol. 104, no. 2, pp. 462–467, 2013.

[21] A. Samanta, K. Ojha, A. Sarkar, and A. Mandal, "Mobility control and enhanced oil recovery using partially hydrolyzed polyacrylamide (PHPA)," *International Journal of Oil, Gas Coal Technology*, vol. 6, no. 3, pp. 245–258, 2013.

[22] D. D. Sparlin, "Evaluation of polyacrylamides for reducing water production (includes associated papers 6561 and 6562)," *Journal of Petroleum Technology*, vol. 28, no. 5, pp. 906–914, 1976.

[23] Y. Cohen and A. B. Metzner, "Adsorption effects in the flow of polymer solutions through capillaries," *Macromolecules*, vol. 15, no. 5, pp. 1425–1429, 1982.

[24] N. Tekin, A. Dinçer, Ö. Demirbaş, and M. Alkan, "Adsorption of cationic polyacrylamide (C-PAM) on expanded perlite," *Applied Clay Science*, vol. 50, no. 1, pp. 125–129, 2010.

[25] N. Tekin, Ö. Demirbaş, and M. Alkan, "Adsorption of cationic polyacrylamide onto kaolinite," *Microporous and Mesoporous Materials*, vol. 85, no. 3, pp. 340–350, 2005.

[26] B. K. Maitin, "Performance analysis of several polyacrylamide floods in the North German oil fields," in *Proceedings of the SPE/DOE Enhance Oil Recovery Symposium*, Tulsa, Okla, USA, April 1992, SPE paper no. 24118.

4

Investigation of CO_2 Enhanced Oil Recovery Using Dimensionless Groups in Wettability Modified Chalk and Sandstone Rocks

Vahid Alipour Tabrizy[1,2]

[1] *Department of Reserve Replacement, Åsgard PTC, Statoil ASA, 7500 Stjørdal, Norway*
[2] *Department of Petroleum Engineering, Faculty of Science and Technology, University of Stavanger, 4036 Stavanger, Norway*

Correspondence should be addressed to Vahid Alipour Tabrizy; v.alipour@yahoo.com

Academic Editor: Guillaume Galliero

The paper addresses enhanced oil recovery in chalk and sandstone rocks by CO_2 injection, with different wettability, porosity, and permeability as well as injection rate and flooding conditions. Results indicate that an increase in Bond number has a positive effect on oil recovery whereas for capillary number, there is a limit in which recovery is improving. This limit is estimated when the pressure drop by viscous force is approximately equal to the threshold balance between capillary and gravity forces. A dimensionless group is proposed that combines the effect of capillarity, injection rate, permeability, and CO_2 diffusion on the oil recovery. Recovery from all experiments in this study and reported data in the literature shows a satisfactory relationship with the proposed group.

1. Introduction

CO_2 flooding is the most widely used method for medium and light oil recovery in sandstone and carbonate reservoirs during the last decades [1, 2]. In the past five decades, there have been extensive laboratory studies, numerical simulations, and field applications of CO_2 EOR processes [1, 3–6]. The mechanisms affecting the displacement of oil by CO_2 injection include oil swelling, IFT (interfacial tension), and viscosity reduction as well as increasing the injectivity index due to solubility of CO_2 in water and subsequent reaction of carbonic acid with the minerals [7–9]. For immiscible CO_2 flow in porous media, displacement efficiency could depend on the wetting properties of the fluids, the rate of CO_2 injection and oil production, the difference of oil and CO_2 density, the viscosity ratio of fluids, and the oil and CO_2 relative permabilities [7]. Miscible CO_2 EOR reduces viscosity and CO_2 can vaporize and extract intermediate hydrocarbons from the reservoir crude oil and/or condense to the oil for developing miscibility under certain pressures and temperatures [10]. Displacement efficiency for miscible CO_2 flooding is also influenced by wetting properties of the rock, injection and production rates, density difference between oil and gas, viscosity ratio of fluids, and oil/CO_2 relative permabilities [10].

Investigations in the laboratories and in the field [11–18] suggest the importance of the gas-oil gravity segregation in CO_2 EOR. Kulkarni and Rao [16] presented the effect of important dimensionless groups on the final oil recovery obtained from a number of miscible and immiscible CO_2 gravity drainage experiments. They stated that the main design parameters on a laboratory scale to evaluate the feasibility of gas injection depend on reservoir heterogeneity, rock type, fluid characteristics, injection gas, WAG ratio, and gravity considerations as well as miscibility development and compositions of oil and brine [16]. Wood et al. [19] introduced ten dimensionless groups to describe CO_2 flooding in a dipping water flooded reservoir and used them for experimental design purposes to develop screening criteria applicable to Gulf Coast reservoirs. Trivedi and Babadagli [20] proposed a new group incorporating the matrix-fracture diffusion transfer for scaling of the miscible displacement in fractured porous media.

The objective of this paper is to study CO_2 enhanced oil recovery in chalk and sandstone rocks. The wettability of calcite and silicate minerals has been changed using fatty acid (stearic acid), amine (N,N dimethyldodecylamine), and asphaltene dissolved in the base solvents. A number of CO_2 flood experiments are performed with various ranges of permeability, porosity, wetting state, and CO_2 injection pressure and rate. These parameters are changed extensively to obtain a wide range of dimensionless groups. The results have been analyzed using capillary, Bond, and diffusion numbers at different wetting states. A new dimensionless group is proposed that combines the effect of capillarity, injection rate, permeability, and CO_2 diffusion during flooding processes.

2. Theory

To quantify the relative magnitudes of the prevailing forces in CO_2 flooding processes, the following set of dimensionless groups is usually suggested: (I) the Bond number (N_B) which is the ratio of the pore scale hydrostatic pressure drop to the capillary pressure and (II) the capillary number (N_c) which is the ratio of the pore scale viscous pressure drop to the capillary pressure [22, 24]:

$$N_B = \frac{\Delta P_{\text{grav}}}{\Delta P_{\text{cap}}} = \frac{\Delta \rho g (k/\phi)}{\sigma}, \quad (1)$$

$$N_c = \frac{\Delta P_{\text{vis}}}{\Delta P_{\text{cap}}} = \frac{\mu_o v_{CO_2}}{\sigma \phi}, \quad (2)$$

where μ_o is the wetting phase viscosity, v_{CO_2} is the Darcy velocity of the injected CO_2 through porous medium, which is equal to CO_2 volume injection rate divided by cross-sectional area of the porous medium, σ is the interfacial tension between CO_2 and oil, k and ϕ are the permeability and porosity of the porous medium, respectively, $\Delta\rho$ is the density difference between two fluids, and g is the gravity acceleration in the direction of flow.

The interfacial tension between CO_2 and oil can be calculated from the following equation using Parachor method [25]:

$$\sigma^{1/4} = \left(\rho_M^L \Sigma P_i X_i - \rho_M^V \Sigma P_i Y_i\right). \quad (3)$$

In this equation, ρ_M is density (mol/cc), P is Parachor number, and X_i and Y_i are liquid and vapor mole fractions, respectively.

It should be noted that the above-defined capillary number, N_c, does not take into account the length and pore size for a porous medium [24], which affect the viscous force through the porous medium. In the literature, the complete capillary number, N_{ca}, for a circular capillary, is defined as [24]

$$N_{ca} = \frac{\Delta P_{\text{vis}}}{\Delta P_{\text{cap}}} = \frac{4\mu_{CO_2} v_{CO_2} l}{\phi \sigma r}. \quad (4)$$

In (4), l and r are the length and radius of the circular capillary, respectively. In consideration of the actual pore size and shape distributions in a reservoir rock, it is useful to relate the above-defined complete capillary number, N_{ca}, to some measurable reservoir characteristics, such as permeability and porosity of the oil reservoir, and Dullien [24] derived the following relation:

$$k = \frac{\phi r^2}{8}. \quad (5)$$

Here, k is the permeability, ϕ is the porosity, and r is the radius of the identical circular capillaries. Alternatively, the complete capillary number can be defined in terms of the permeability and porosity of the porous medium:

$$N_{ca} = \frac{\Delta P_{\text{vis}}}{\Delta P_{\text{cap}}} = \frac{\sqrt{2}\mu_{CO_2} v_{CO_2} l}{\sigma \sqrt{k\phi}}. \quad (6)$$

For drainage with high viscosity contrast between gas and oil, theoretical arguments based on percolation theory predict the following scaling law for the displacement under stabilized condition in three-dimensional (3D) pore structures:

$$l \approx \left(N_B^*\right)^{-1} \quad (7)$$

In this equation l is front width. The generalized Bond number, N_B^*, is also introduced by Løvoll et al. [26] as follows:

$$N_B^* = \frac{\Delta \rho g (k/\phi)}{\sigma} - \frac{\mu_o v_{CO_2}}{\sigma}. \quad (8)$$

They suggested that since the front width l in (7) diverges when $N_B^* \rightarrow 0$, the following criterion for interface stabilization is expected [26]:

$$N_B^* > 0. \quad (9)$$

After substitution and simplifications, the following criteria can be obtained:

$$\frac{\Delta \rho g k}{\mu_o} = v_c > v_{CO_2}. \quad (10)$$

This critical velocity also can be calculated during modeling of free gas gravity drainage as a rate of oil recovery before gas front reaches the capillary/gravity balance threshold. The appendix shows the analytical model describing free gravity drainage and the mathematical method to find the critical oil recovery velocity. The calculation of critical oil recovery velocity with another approach has been described also by Marle [27].

The diffusion number is defined as ratio of diffusive and convective dispersion (bulk flow or viscous force):

$$N_d = \frac{D\phi}{v_{CO_2} l}. \quad (11)$$

In (11), D is diffusivity coefficient in Fick's law between CO_2 and oil and ϕ is the porosity; l is the length of capillary tube and v_{CO_2} is the Darcy velocity of the injected CO_2 through porous medium, which is equal to CO_2 volume injection rate divided by cross-sectional area of the porous medium.

The number can scale the role of diffusive and convective forces in miscible processes.

Diffusivity coefficient (D) of CO_2 in oil can be estimated from the following equation [28]:

$$D_{CO_2,oil} = 10^{-9} {\mu_{oil}}^{-0.4562} {M_{CO_2}}^{-0.6898} \times {V_{CO_2}}^{-1.706} P^{-1.831} T^{4.524}. \quad (12)$$

In this equation, $D_{CO_2,oil}$ is diffusivity coefficient of CO_2 gas in oil (m^2/sec), μ_{oil} is viscosity of oil (cp), M_{CO_2} is the molecular weight of CO_2, V_{CO_2} is the molar volume of CO_2 (cm^3/mol), P (psia) is the pressure of CO_2, and oil system in equilibrium and T (Kelvin) is the temperature of CO_2 and oil system in equilibrium.

3. Experimental Section

3.1. Materials

3.1.1. Solid Phase. The core flood experiments were done in outcrop Stevns klint chalks near Copenhagen and outcrop Benthiemer sandstones. The dimension of cores varied between 5–7 cm in length and 3.8 cm in diameter. The out crop chalks have approximate porosity of 44–48% and absolute permeability of 3–7 mD. Sandstone cores have the same porosity near 22–25%, while they are classified based on the permeability in the two groups with permeability range of 600–800 mD and with the permeability range of 1500–1700 mD. The capillary threshold of chalk cores for oil and gas system is measured near 2 bar by porous plate technique. The capillary threshold of Benthiemer sandstones was reported as 0.05 atm [29]. Table 1 is the cores detailed descriptions.

3.1.2. Oil Phase. The investigation is done for model oil containing asphaltene dissolved in toluene (95% purity), 0.01 M stearic acid (SA) dissolved in n-decane (95% purity) for chalk cores, and 0.01 M N,N-dimethyldodecylamine (NN-DMDA) dissolved in n-decane (95% purity) for sandstone cores. Synthetic seawater (SSW) has been used as irreducible water saturation. Tables 2 and 3 show the polar components used in this work and also the ionic composition of synthetic seawater.

3.1.3. Asphaltene Preparation Procedure. The model oil system is prepared from asphaltene precipitated from crude oil in excess of n-heptane (1 : 40). The mixture was shaken for at least twice a day and left for 48 hours to equilibrate. The mixture solution was then centrifuged and filtered through a 0.22 micrometer filter (Millipore) and dried for 1 day using a vacuum oven at room temperature. The dried asphaltene was then dissolved in toluene.

3.2. Experimental Setup. The detailed description of the experimental set up has been addressed elsewhere [30]. The major components of the experimental setup consist of a core holder, pressure regulators, two gas flow meters, pressure manometers, Gilson pump, CO_2 piston cell, graduated gas/oil separator, and a lab view version 7.1 connected to a digital data acquisition system. Saturated core samples with synthetic seawater were inserted into a horizontally placed core holder that consists of a steel cylindrical body and rubber/nylon sleeve. A net overburden pressure of 20 bar is applied on the sleeve. The chalk core is flooded with oil sample at flow rate of 0.1 mL/min until oil breakthrough is occurred and then it is increased to 5 mL/min to overcome the capillary threshold using Gilson pump (806 manometric modules) to establish the irreducible water. The sandstone core is flooded with oil sample at flow rate of 3 mL/min until oil breakthrough is occurred and then it is increased to 5 mL/min to overcome the capillary threshold. Table 1 shows the irreducible water saturation, which is varying between 10 and 13% for chalks and 20 and 25% for sandstones. The cores were aged for at least two weeks to change the wettability of cores toward oil wet, when the saturating oil contains polar components. Then, CO_2 injection was carried out in two modes, namely, injection at constant rate and constant pressure. At constant rate, CO_2 is injected from piston connected to Gilson pump (806 manometric modules) with desired rate. At constant pressure modes, CO_2 is injected in immiscible and miscible processes. In the immiscible process, the pressure of CO_2 injection is lower than minimum miscibility pressure. Whereas, in the miscible mode, the oil saturated core was flooded with CO_2 at pressure equal to minimum miscibility pressure. CO_2 was injected into the core at constant pressures of 40 ± 0.2 bar and temperature of 25°C for immiscible process and 90, 70, and 140 $\pm$ 0.2 bar and at 50, 70, and 80°C for miscible processes. The constant pressure of CO_2 is provided by Gilson pump when the upper and lower pressure limits of pump are adjusted to the desired injection pressure plus/minus 0.5 bar. CO_2 is injected to the core passing mass flow meter (1) that records the inflow properties of CO_2 (mass flow rate, density, and total mass). A back pressure regulator is connected downstream the core to control the pressure during CO_2 flooding. The outflow properties (mass flow rate, density, and total mass) of the evolved gas were, also, recorded using flow meter (2) connected to the separator. The CO_2 injection continued for at least 3 pore volumes until there was no oil production. When the injection was terminated, the core was then removed from the core holder and dried using vacuum oven at a temperature of 120°C until a constant weight was obtained.

4. Results and Discussion

The experimental results with the relation to the selected dimensionless groups are discussed in this section.

4.1. Bond Number. The variations in Bond number were primarily performed by varying the permeability of the porous medium, where reduction in permeability leads to a reduction in Bond number. As it is seen in (2), permeability does not affect the capillary number, thus allowing investigation of the effect of Bond number on the oil recovery at a constant capillary number. Recovery factor after two pore volumes of injected CO_2 as a function of the Bound number is shown in

Table 1: Core descriptions of immiscible and miscible CO_2 flooding experiments.

EXP. number	S_{wi}	Phi	K (cm^2)	L (cm)	q (cc/min)	Type	Core
1a	0.22	0.23	6.5E − 09	6	0.2	Immiscible	Aged sandstone
2a	0.2	0.24	7.5E − 09	5.5	0.5	Immiscible	Aged sandstone
3a	0.25	0.25	7.3E − 09	5.8	2	Immiscible	Aged sandstone
4a	0.2	0.25	8.3E − 09	6.2	6	Immiscible	Aged sandstone
5a	0.25	0.23	6.7E − 09	6	8	Immiscible	Aged sandstone
5a*	0.21	0.22	6E − 09	5.5	8	Immiscible	Aged sandstone
5a**	0.22	0.24	7.6E − 09	6	8	Immiscible	Aged sandstone
1b	0.2	0.23	1.5E − 08	6	0.5	Immiscible	Aged sandstone
2b	0.25	0.24	1.6E − 08	5.5	2	Immiscible	Aged sandstone
3b	0.22	0.24	1.7E − 08	5.5	8	Immiscible	Aged sandstone
4b	0.25	0.25	1.7E − 08	5.8	10	Immiscible	Aged sandstone
1c	0.14	0.44	5E − 11	6	0.2	Immiscible	Aged chalk
2c	0.15	0.45	3.5E − 11	6.5	0.5	Immiscible	Aged chalk
3c	0.12	0.48	6E − 11	6.7	2	Immiscible	Aged chalk
4c	0.13	0.47	5.5E − 11	5.8	2.5	Immiscible	Aged chalk
5c	0.1	0.48	6.5E − 11	5.9	3.5	Immiscible	Aged chalk
5c*	0.13	0.45	6.2E − 11	6.5	3.5	Immiscible	Aged chalk
5c**	0.15	0.47	6E − 11	6	3.5	Immiscible	Aged chalk
1d	0.17	0.44	5.5E − 11	6	0.01	Immiscible	Aged chalk
2d	0.15	0.45	4E − 11	6.5	0.05	Immiscible	Aged chalk
3d	0.11	0.48	2E − 11	6.7	0.1	Immiscible	Aged chalk
1e#	0.16	0.45	6.5E − 11	6	0.05	Miscible	Aged chalk
2e#	0.14	0.46	4E − 11	6.5	0.05	Miscible	Aged chalk
3e#	0.12	0.48	2.5E − 11	6.8	0.05	Miscible	Aged chalk
1e	0.27	0.25	6.5E − 09	6	0.05	Miscible	Unaged sandstone
2e	0.3	0.23	7.7E − 09	6.5	0.05	Miscible	Unaged sandstone
3e	0.125	0.22	8.2E − 09	6.8	0.05	Miscible	Unaged sandstone
1g	0.22	0.23	1.8E − 08	6	0.5	Immiscible	Unaged sandstone
2g	0.23	0.26	1.9E − 08	5.5	2	Immiscible	Unaged sandstone
3g	0.24	0.3	2.1E − 08	5.5	8	Immiscible	Unaged sandstone
4g	0.24	0.25	1.8E − 08	5.8	10	Immiscible	Unaged sandstone
1h	0.14	0.44	5.5E − 11	6	0.05	Miscible	Unaged chalk
2h	0.15	0.45	5.5E − 11	6.5	0.05	Miscible	Unaged chalk
3h	0.17	0.47	6E − 11	6.8	0.05	Miscible	Unaged chalk

*Repetition of experiment (first time).
**Repetition of experiments (second times).
#Previously reported in Hamouda, A.A., Alipour Tabrizy, V., 2013, the effect of light gas on miscible CO_2 flooding to enhance oil recovery from sandstone and chalk reservoir, and journal of Petroleum Science and Engineering, volume 108, pages 259–266.

Table 2: Polar components.

Components	Supplier and purity	Structural formula
Stearic acid (SA)	Aldrich ≥99%	$CH_3(CH_2)_{16}COOH$
N,N-dimethyldodecylamine (NN-DMDA)	Fulka ≥99%	$CH_3(CH_2)_{11}N(CH_3)_2$

Figure 1. Results indicate that recovery at a constant capillary number increases with increasing the Bond number. This behavior is in line with the expectation about the effect of capillarity on oil drainage by gas injection. To examine the relationship between oil recovery and Bond number, the recovery between all experiments is plotted versus Bond numbers in Figure 2. No satisfactory relation was found between the oil recovery and the balance between gravity and capillary forces.

The scaling of oil recovery using Bond number has three major limitations: (I) the effect of viscous force is not included. For example, by changing the gas injection rate, the oil recovery may be changed drastically, while the Bond number is remained unchanged; (II) the effect of viscosity ratio between displaced and displacing fluids is not included. For example, when the gas/oil viscosity ratio changes leading to a change in stability of the front and recovery, the Bond

TABLE 3: Synthetic seawater composition.

Ions	Concentration (mol/L)
NaCl	0.4
KCl	0.01
$CaCl_2(2H_20)$	0.013
$MgCl_2(6H_20)$	0.045
$NaHCO_3$	0.002
$Na_2(SO_4)$	0.024
TDS	0.494

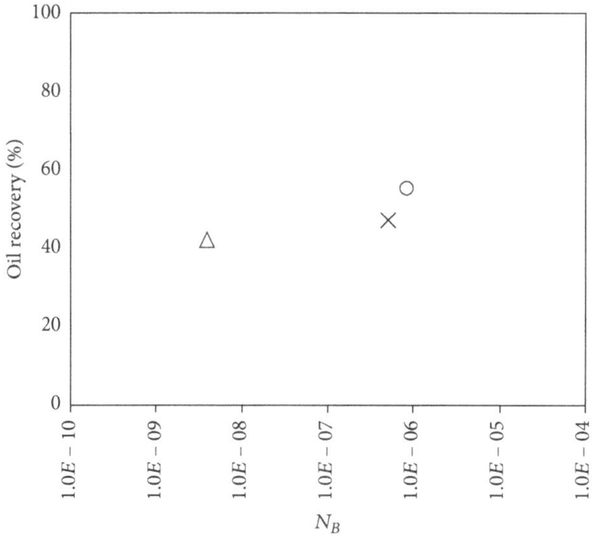

FIGURE 1: CO_2 enhanced oil recovery at constant capillary number ($N_c \cong 3.50E - 07$) as a function of Bond number.

number remains unchanged. The effect of CO_2 diffusion in oil and miscibility parameters that is, pressure, temperature, and oil composition, are not considered also in Bond number. The effects of viscous forces, viscosity ratio, and CO_2 diffusion are incorporated in the following sections.

4.2. Capillary Number. Large viscous forces in high-rate gas flooding, characterized by large capillary numbers, may result in unstable flooding fronts and the oil bypassing/trapping mechanism. From this point of view, an increase in the viscous force will result in reduction in the oil recovery because the microscopic displacement efficiency decreases. However, the low-rate gas flooding may lead to considerable residual oil since the pressure drop across the core may be lower than the capillary threshold. In the work reported by Rostami et al. [31], it has been observed that in unconsolidated bead pack porous media and high permeable water-wet outcrop sandstones, the oil recovery reduces as capillary number increases. This may not be applicable in the real reservoir cores, especially for tight oil-wet carbonate rocks with

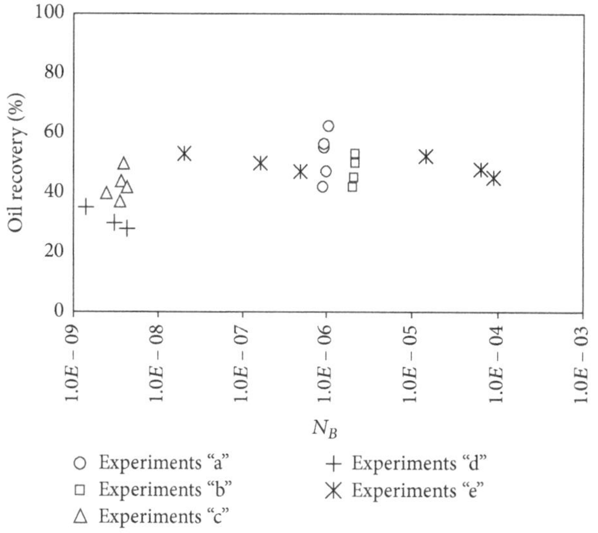

FIGURE 2: CO_2 enhanced oil recovery from all experiments as a function of Bond number.

considerable capillary threshold. Experimental results are shown in Figure 4(a) to Figure 5(d), where oil recovery after two-pore volume CO_2 injection is plotted versus capillary number. Results in each figure are characterized by a Bond number that remains approximately constant. The interfacial tension between CO_2 and oil is estimated near to 20 mN/m for immiscible process and 0.1 mN/m for miscible process according to (3). Results indicate that oil recovery improves as capillary number increases up to the critical limit where afterward decreases. In other words, the recovery increases when the pressure drop due to viscous force increases near to the capillary threshold and then decreases due to flow instability. From Figure 3, it is clear that for experiments "a" and "c," the oil recovery is increasing up to a certain value and then it is decreasing by increasing the injection rate and viscous force. The maximum oil recovery may be linked to the pressure drop in which $\Delta P \cong P_{c_{th}}$. To observe the reliability of results, experiment "a5" and experiment "c5" are repeated and they are shown in Figure 4. The variations of oil recovery for similar conditions are near to 5%.

It can be concluded from Figures 3 and 4 that the oil recovery is not directly proportional to the capillary number; hence, the following modified capillary number is suggested:

$$N_c^* = N_c \left(1 - \text{abs}\left(1 - \frac{\Delta P}{Pc_{th}}\right)\right). \quad (13)$$

In this equation, abs is showing the absolute value of number. The other symbols are defined previously. Figure 5 shows the relation between modified capillary number and CO_2 enhanced oil recovery. Applying this number shows linear relationship of oil recovery to the number in semilog scale; hence, it might be more appropriate for scaling of oil recovery in gas flooding projects. The associated regression parameter for each figure is presented also in the graph.

The effect of injection rate on immiscible oil recovery can be also investigated by definition of displacement efficiency

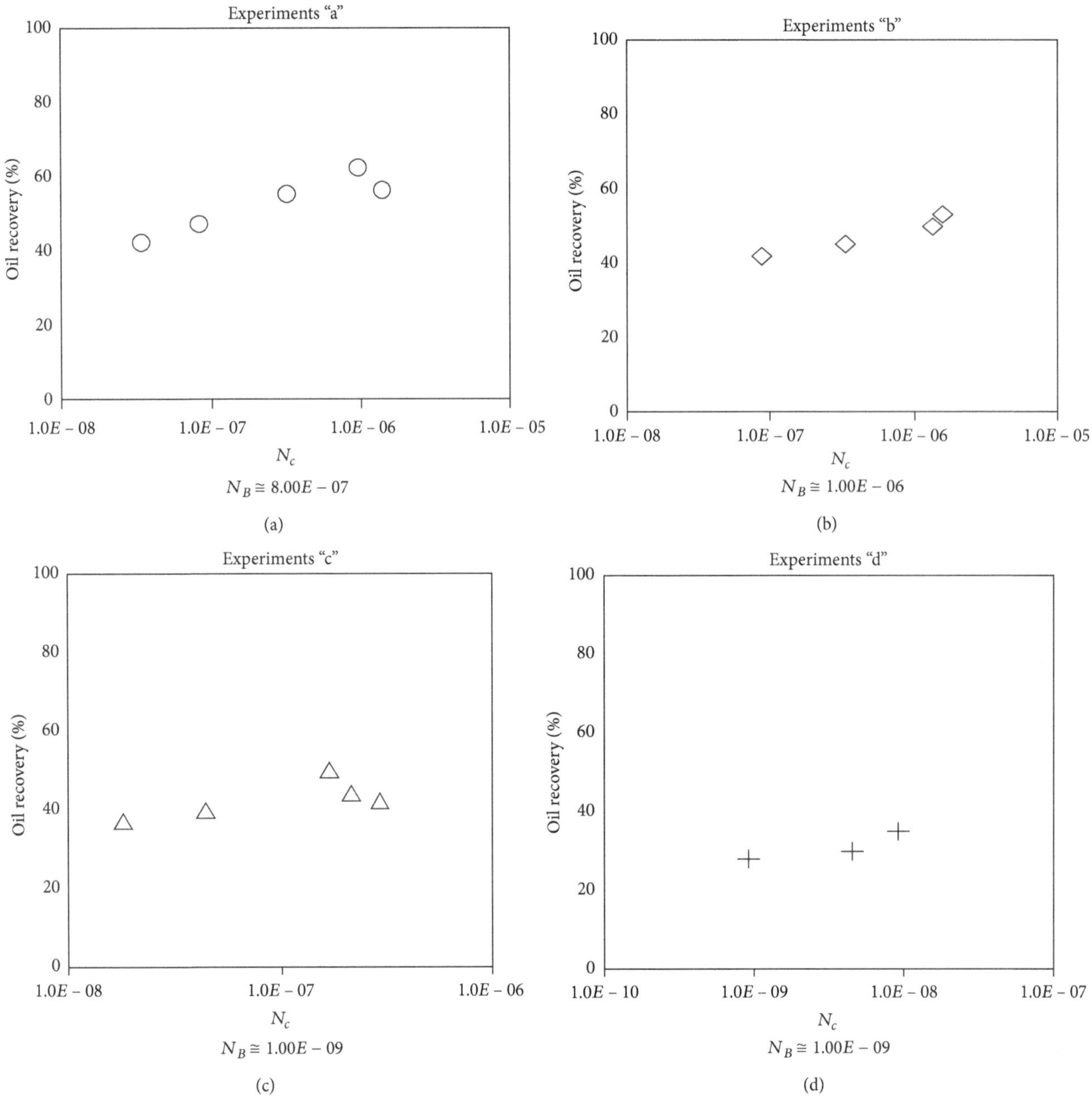

FIGURE 3: CO_2 enhanced oil recovery at constant Bond number ($N_c \cong 3.50E - 07$) as a function of capillary number.

number using the critical velocity derived in the appendix as follows:

$$N_{DE} = \frac{v}{v_c}\left(1 - \text{abs}\left(1 - \frac{\Delta P}{Pc_{\text{th}}}\right)\right). \quad (14)$$

In Figure 6, the oil recovery as function of displacement efficiency number is shown. This number takes into account the effect of injection rate, capillary threshold, and density difference between injected and displaced fluids and may be alternatively used instead of modified capillary number.

To explore an appropriate dimensionless number that describes the dynamics of the oil recovery, recoveries from all experiments are plotted versus their modified capillary numbers (Figure 7). Again, no satisfactory relation is observed. This confirms that the capillary number alone is insufficient to predict the oil recovery behavior during the CO_2 flooding.

4.3. Diffusion Number. Diffusion number expresses the ratio of diffusivity to convective dispersion (bulk flow or viscous force). The number is important in miscible processes, when the diffusion of solvent in oil is considerable and the oil production is slow; hence, the viscous force and convective dispersion are small and the diffusion number should be taken into account according to (6). Experiments in porous media “e” and “f” are done in miscible mode in chalk and

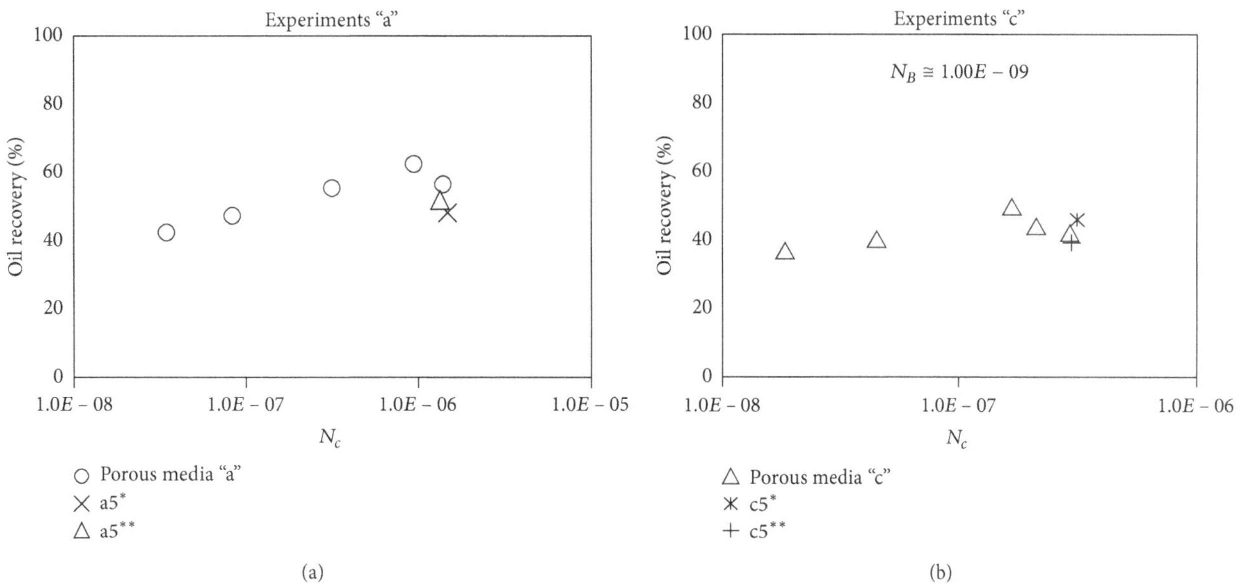

FIGURE 4: Reproducibility of CO_2 enhanced oil recovery experiments.

sandstone cores, respectively. The minimum miscible pressure (MMP) for model oil composition at different temperatures is calculated according to SRK-Peneloux EOS, PVTsim version 17. The oil recovery for miscible CO_2 flooding experiments for porous media "e" and "f" is shown in Figure 8. The CO_2 flooding has been done at 50°C, 70°C, and 90°C and pressures 90 bar, 120 bar, and 140 bar for experiments "e1," "f1," "e2," "f2," "e3," and "f3," respectively. Form the graph, it can be seen that as temperature and pressure are increasing, the diffusion number (N_d) increases and the recovery by CO_2 flooding is increasing. It is important to mention that the oil recovery rate before CO_2 breakthrough is kept near 0.05 cc/min for all experiments; accordingly the viscous force and capillary number of miscible experiments are small ($N_c \cong 10^{-7}$) and close to each other.

The dependency of all miscible and immiscible experiments to diffusion number is shown in Figure 9. As it can be seen, no relation could be drawn down for results, since the diffusion number is not playing a critical major role in all experiments.

4.4. Combined Dimensionless Group. Examination of oil recoveries as a function of dimensionless numbers (capillary number, Bond number, modified capillary number, and diffusion number) suggests that none of the above scaling groups individually can define the oil recovery relationship to affecting parameters. In the following section, a combined dimensionless group that could show the relationship of oil displacement over the whole range of variables is investigated. Figure 10 shows the results of all experiments, when ratio of the modified capillary number over Bond number is used. Results indicate that the oil recovery by CO_2 flooding is directly proportional to the ratio of modified capillary number over Bond number; however, no specific relation may explain this relationship.

In order to take into account the effect of diffusion number and appropriate ratio of the capillary number to Bond number, a combined dimensionless group, $N_{\text{Co.}}$, is proposed as follows:

$$N_{\text{Co.}} = \frac{N_d \times N_c^*}{N_B^a} = \frac{N_d \times N_c \times \left(1 - \text{abs}\left(1 - \Delta P/P_{C_{\text{th}}}\right)\right)}{N_B^a}, \quad (15)$$

where subscript Co. stands for the term "combined," and the parameter "a" is the scaling factor. It is found that in this research, the value of $a = 0.01$ provides a good relation. The small value of exponent suggests that the effect of the Bond number is much more than the importance of the capillary number. The results for the above dimensionless group are shown in Figure 11, where the recoveries of all experiments are plotted versus this number. This dimensionless group combines the effect of the three main forces that affect CO_2 enhanced oil recovery. The results suggest that within the range of experiments performed here, there is a logarithmic relationship between the oil recovery and proposed group.

In order to validate the applicability of the new group proposed for gas injection processes, the experimental results reported by Hudigins et al., (1990), Darvish [23], and Nobakht et al. [22] are used. Table 4 summarizes the properties of literature experimental data used in this study.

The first two experimental data are related to miscible N_2 injection in the slim tube and core, respectively; hence, the diffusion number and other relevant parameters are calculated for nitrogen. The third and fourth experiments were done for miscible CO_2 flooding in sandstone cores and the fifth experiment was reported for CO_2 flooding in saturated chalk core with live reservoir oil. Figure 12 compares the experimental data reported in this research and literature data reported previously.

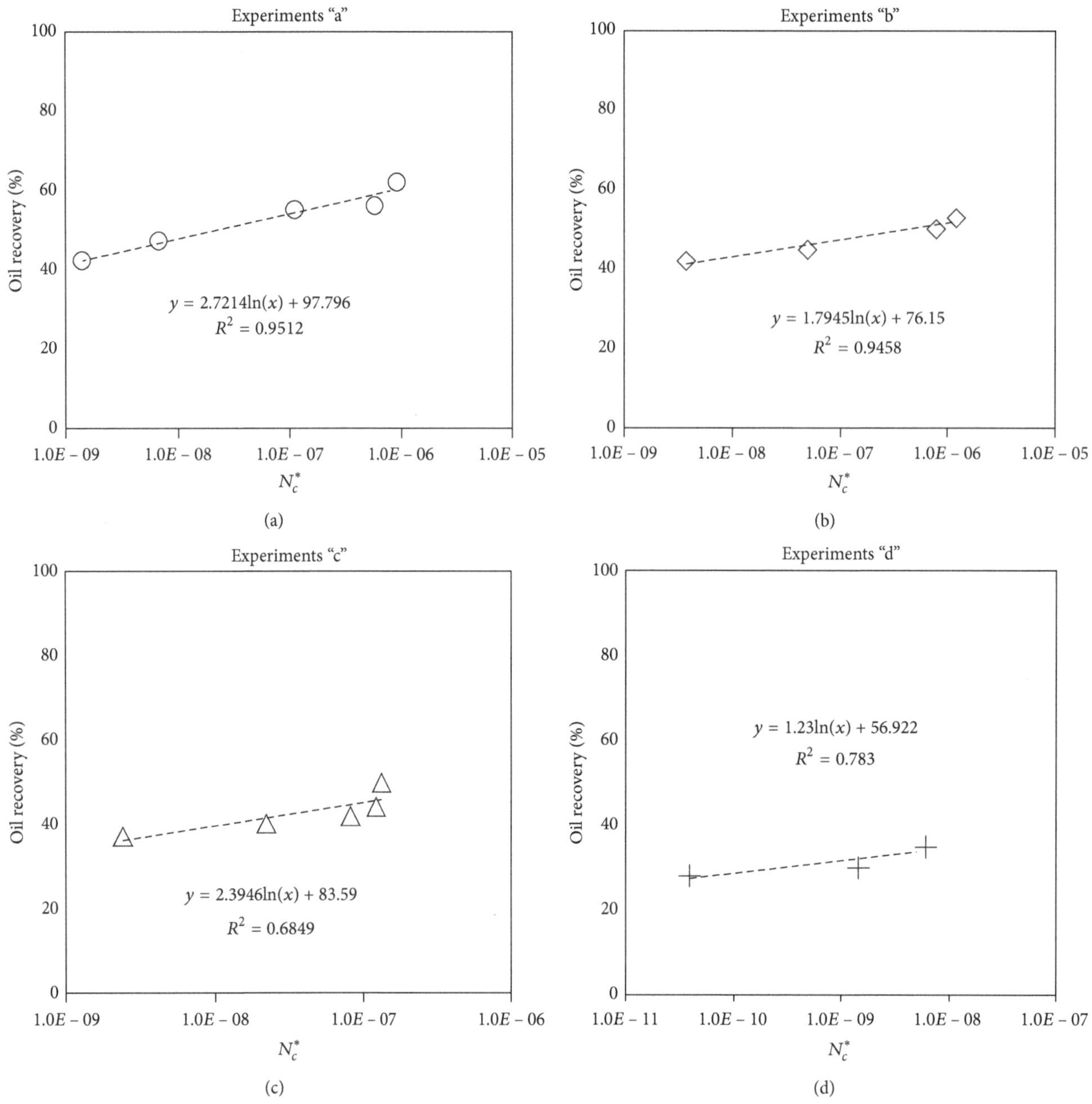

FIGURE 5: CO_2 enhanced oil recovery as a function of modified capillary number.

TABLE 4: Literature experimental data used in this study.

Number	Type	S_{wi}	Phi	K (cm^2)	L (cm)	q (cc/min)	σ gas oil (mN/m)	Reference
11	Slim tube (N_2 injection)	0	0.379	1.2E − 07	2720	4.76	0.1	Hudgins et al. [21]
21	Core (N_2 injection)	0.4	0.198	4.6E − 09	60.96	0.8	0.1	Hudgins et al. [21]
31	Core (CO_2 injection)	0.07	0.372	1.8E − 05	28.8	0.5	4.22	Nobakht et al. [22]
41	Core (CO_2 injection)	0.025	0.369	1.8E − 05	28.8	0.5	1.13	Nobakht et al. [22]
51	Core (CO_2 injection)	0	0.444	4E − 11	60	0.02	0.05	Darvish (2007) [23]

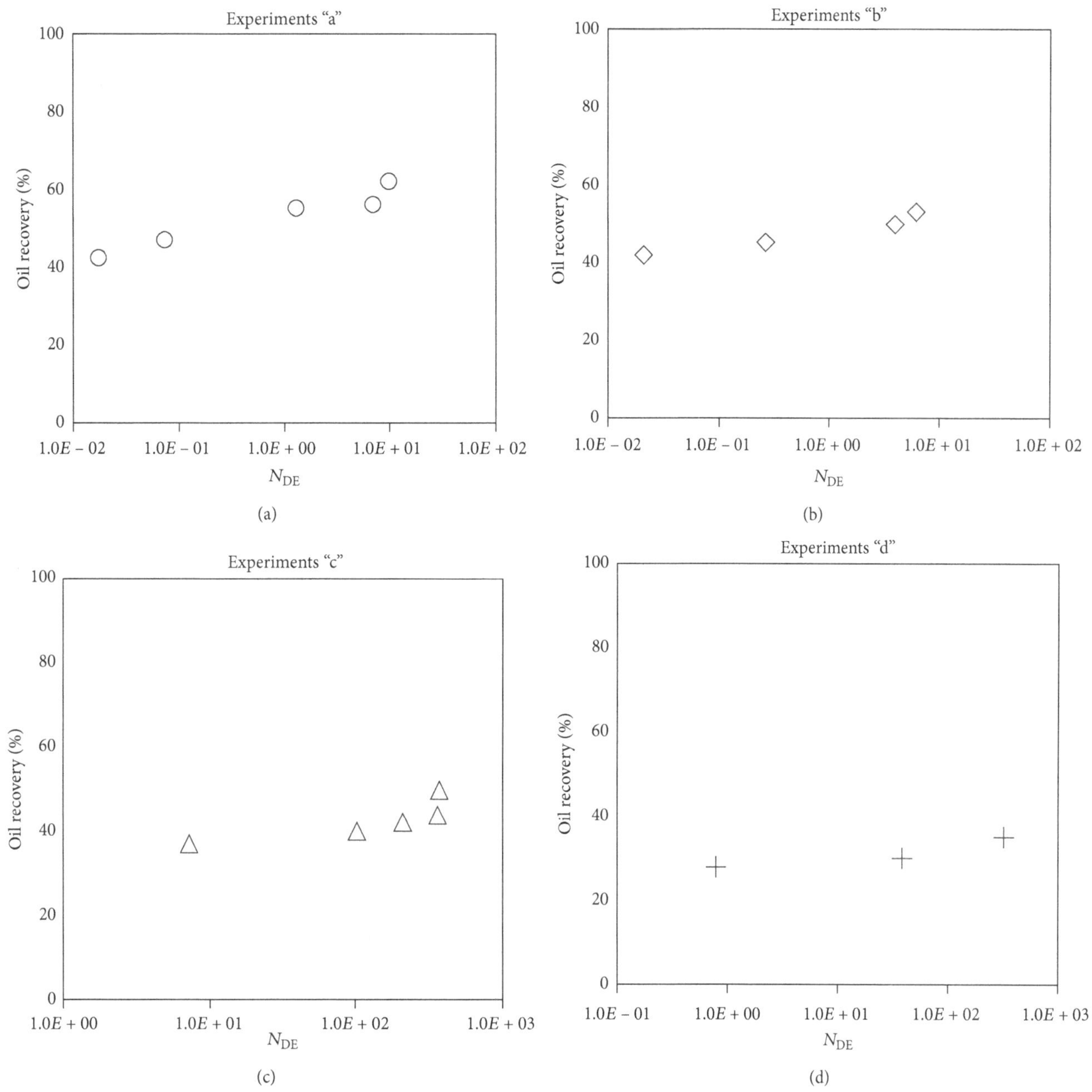

FIGURE 6: CO_2 enhanced oil recovery as a function of displacement efficiency number.

4.5. Effect of Wettability. To understand the role of wettability in miscible and immiscible CO_2 flooding, some experiments are done for unaged cores and the results of oil recovery at CO_2 breakthrough and 2 PV of injected CO_2 are compared in Figures 13 and 14. Figure 13 shows that the oil recovery after 2 PV of injected CO_2 for unaged samples is higher than the aged samples. In fact, in aged sample (oil-wet porous media), oil is the wetting phase and occupies the corner of pores; water is the nonwetting phase which occupies the center of pores. During the flowing of oil in an oil-wet media, oil flows through wetting films, gas flows principally as continuous nonwetting filaments, and water is in the form of isolated ganglia; hence, the existing residual water plays a blocking role during the displacement and the recovery of aged cores after 2 PV of injected CO_2 is significantly lower than unaged core. In contrast, for unaged core (water-wet porous media), water is not mobilized but wets the pore walls and the continuity of oil is maintained along the porous media because the spreading of oil within water and gas phases is positive. The spreading coefficient of gas, oil water system is calculated based on the equation proposed by Blunt et al. [32] as follows:

$$S = \sigma_{gw} - \left(\sigma_{go} + \sigma_{wo}\right). \quad (16)$$

Assuming $\sigma_{gw} = 70\,\text{mN/m}$ [32], $\sigma_{ow} = 19\,\text{mN/m}$ (measured by pendent drop technique) and $\sigma_{go} = 20\,\text{mN/m}$ (calculated

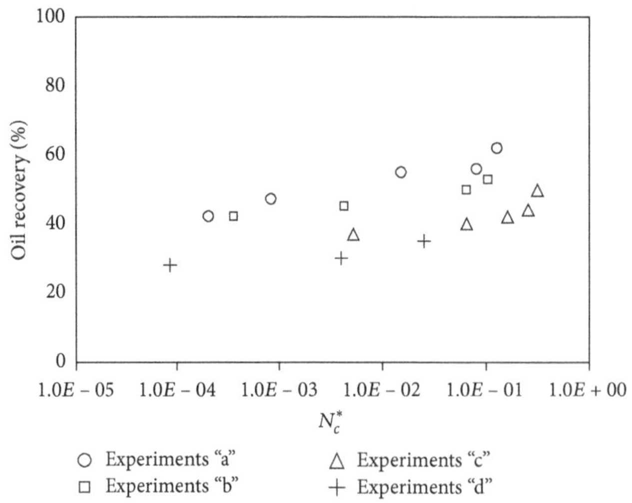

FIGURE 7: CO_2 enhanced oil recovery from all experiments as a function of modified capillary number.

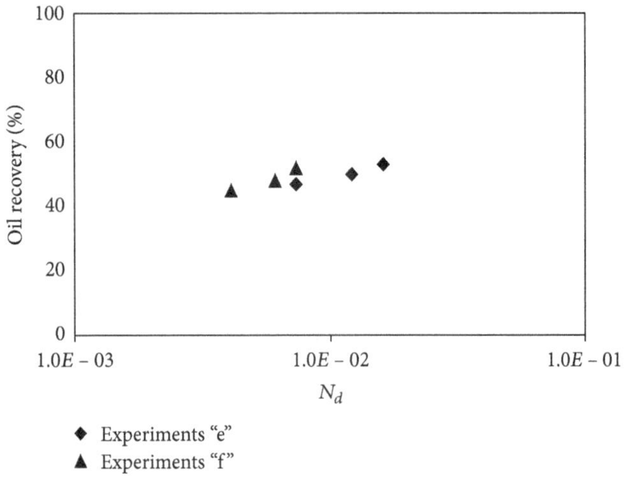

FIGURE 8: CO_2 enhanced oil recovery as a function of diffusion number in miscible experiments.

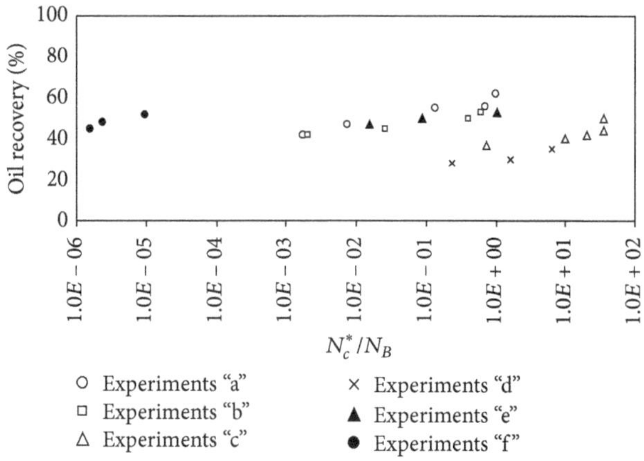

FIGURE 10: CO_2 enhanced oil recovery in all experiments as a function of modified capillary number to Bond number.

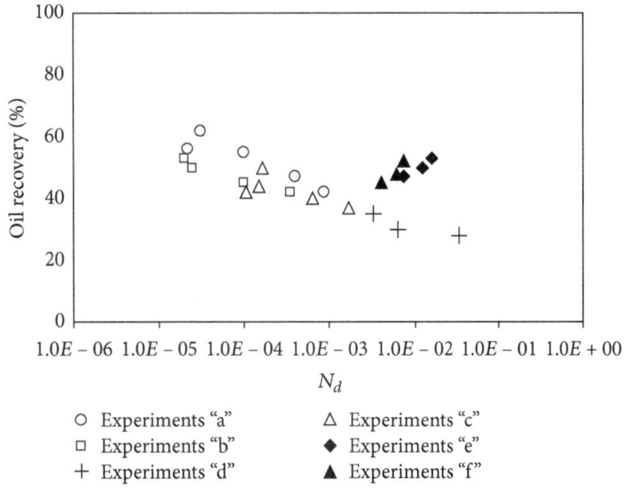

FIGURE 9: CO_2 enhanced oil recovery as a function of diffusion number in miscible and immiscible experiments.

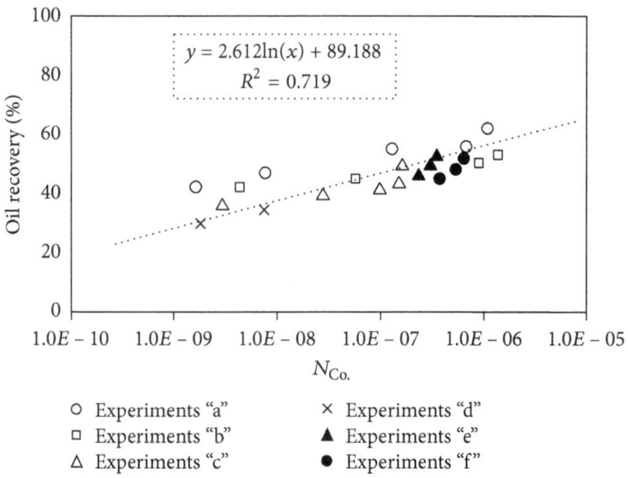

FIGURE 11: CO_2 enhanced oil recovery in all experiments as a function of combined dimensionless number.

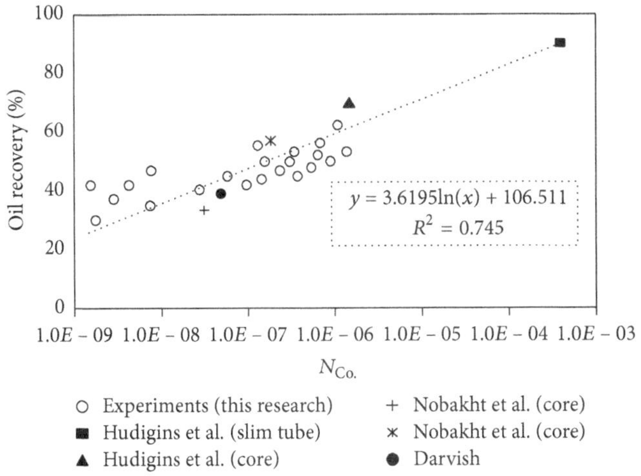

FIGURE 12: CO_2 enhanced oil recovery versus combined dimensionless number in all experiments and the experimental data reported in the literature.

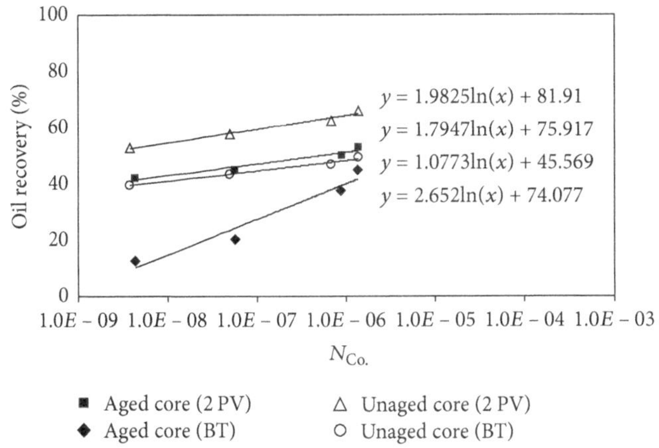

FIGURE 13: Immiscible CO_2 enhanced oil recovery versus combined dimensionless number in aged and unaged experiments.

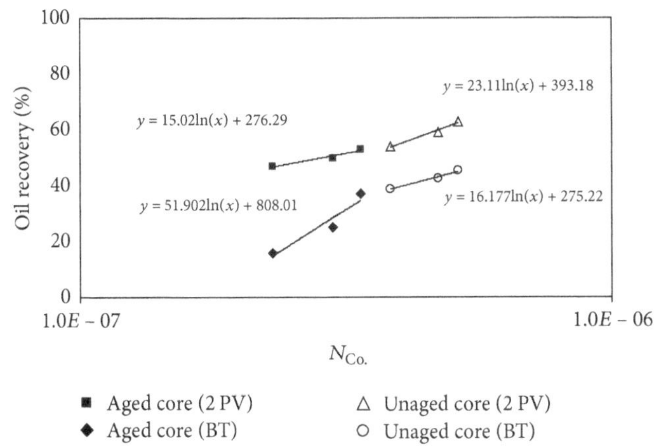

FIGURE 14: Miscible CO_2 enhanced oil recovery versus combined dimensionless number in aged and unaged experiments.

by (3)) the value of spreading coefficient is +31 mN/m. According to the classification reported by Blunt et al. [32] when $S > 0$, the oil can flow as continuous phase among initial water and injected gas and the film flow contributes significantly to the recovery but with slow rate. From Figure 13, it is also clear that the oil recovery is decreasing as $N_{Co.}$ is decreasing both for aged and unaged cores at 2 PV and breakthrough of injected CO_2. However, the slop of this relation for aged core (oil wet) at breakthrough time is larger than unaged cores, which indicates that the recovery of aged cores is strongly dependent on the flooding parameters. This has been observed previously by Shahidzadeh-Bonn et al. [33] for forced and free gravity drainage of gas in glass beak pack media, where the wettability of glass beads was changed by chemical treatment with TCMS (Three-Chloro-Methylsilane).

Figure 14 shows that the oil recoveries after 2 PV of injected CO_2 for unaged samples are higher than the aged samples. Similar trend to immiscible CO_2 flooding can be seen for miscible CO_2 flooding.

In unaged core (water-wet porous media), water is not mobilized and wets the pore walls and oil is maintained in the center of pores; hence, the CO_2 diffuses to oil and the recoveries at CO_2 breakthrough and 2 PV injected CO_2 are higher than aged core. In aged core (oil wet), the water droplet occupies the center of pore and the injected CO_2 will dissolve in water and then the mass transfer between oil and CO_2 may occur; hence, the water will act as a barrier. Figure 15 shows the schematic of CO_2 enhanced oil recovery in hydrophobic (aged) and hydrophilic (unaged) porous media. In this figure, oil is shown by brown; water is shown by blue and CO_2 is represented by white color.

The diffusivity of CO_2 in seawater can be found based on the following equations [34]:

$$\text{Log}\left(D_{CO_2,w}\right) = -4.1764 + \frac{712.52}{T} - \left(\frac{2.5907 \times 10^{-5}}{T}\right), \tag{17}$$

$$\text{Log}\left(\frac{D_{CO_2,sw}}{D_{CO_2,w}}\right) = 0.87 \log\left(\frac{\mu_{sw}}{\mu_w}\right), \tag{18}$$

where T is temperature in Kelvin, μ_{sw} is viscosity of seawater (cp), μ_w is viscosity of distilled water, and $D_{CO_2,sw}$ is diffusivity of CO_2 in seawater (cm^2/sec). In Figure 16 the diffusivity of CO_2 in seawater based on (17) and diffusivity of CO_2 in model oil are calculated based on the (7) and are compared at different temperatures and corresponding minimum miscibility of CO_2 within oil. The figure clearly indicates that the diffusion of CO_2 in seawater is higher than oil; hence, most of CO_2 will diffuse in water rather than oil.

5. Conclusion

Analysis of oil recovery with Bond number, at constant capillary number, shows a direct relation, but when the capillary number is changing, dependency of oil recovery with the

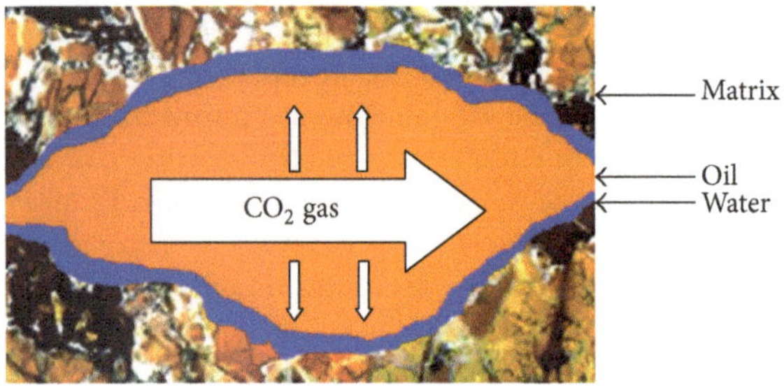

(a) Unaged porous medium (hydrophilic)

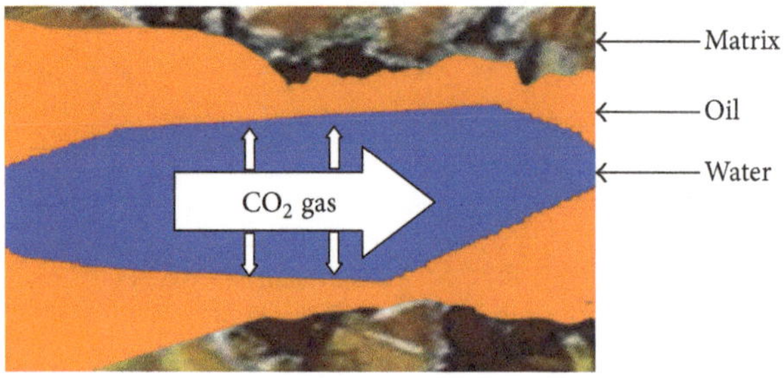

(b) Aged porous medium (hydrophobic)

FIGURE 15: Schematic of rock pore in miscible CO_2 enhanced oil recovery for hydrophobic (aged) and hydrophilic (unaged) porous media [30].

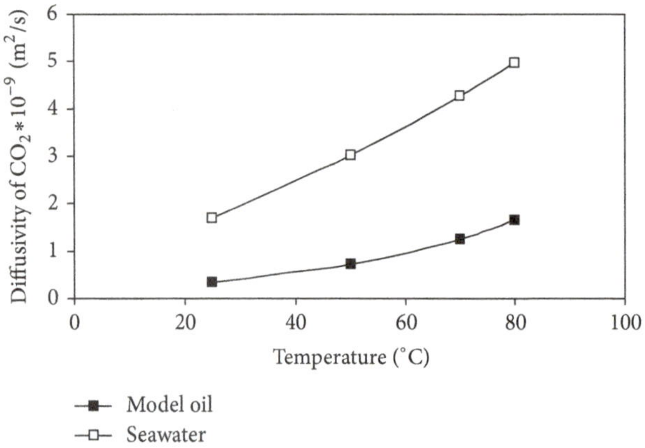

FIGURE 16: Comparison of CO_2 diffusion in seawater and model oil at different temperatures and corresponding minimum miscibility pressures [30].

Bond number is varying also. In addition, at constant Bond number, oil recovery increases as capillary number increases up to a certain limit and then decreases drastically as a result of higher viscous forces and fluid flow instability. Results presented in this work suggest that the Bond, capillary, and diffusion numbers could not scale the oil recovery, individually. A dimensionless group was defined to relate the competition between three forces affecting oil recovery over a wide range of petrophysical and experimental conditions. A logarithmic relationship was observed between the proposed group and oil recovery. Applicability of the proposed relation beyond the range of experiments reported in this work needs to be further examined. It is observed also that the recovery by CO_2 injection for unaged rocks (water wet) is higher than the aged (oil wet) rocks due to higher diffusivity of CO_2 in water (occupying the center of pores in oil-wet cores) compared to that in oil.

Appendix

By using the continuity equation and Darcy's law, and the concepts of relative permeability and capillary pressure, one-dimensional displacement of multiphase flow can be described mathematically. The one-dimensional Darcy's law for two-phase flow in vertical direction is

$$u_o = \frac{KK_{ro}}{\mu_o}\left(\frac{\partial P_o}{\partial z} - \gamma_o\right), \tag{A.1}$$

$$u_g = \frac{KK_{rg}}{\mu_g}\left(\frac{\partial P_g}{\partial z} - \gamma_g\right), \tag{A.2}$$

where the z axis is vertically downward, u_o and u_g are the oil and gas velocity, and g is the acceleration due to gravity. The gas-oil capillary pressure (P_c) is defined by

$$P_c = P_g - P_o. \tag{A.3}$$

By substituting the flow equation of oil in the continuity equation and combining it with gas-oil capillary pressure, the equation of 1D displacement of oil by gas in vertical direction is gained:

$$\frac{\partial}{\partial z}\left(\frac{KK_{ro}}{\mu_o}\left(\frac{\partial\left(P_g - P_c\right)}{\partial z} - \gamma_o\right)\right) = -\frac{\partial}{\partial t}\left(\phi S_o\right). \tag{A.4}$$

Gas has a very low viscosity with respect to oil; therefore, the assumption of infinite gas mobility is viewed as a reasonable assumption. Consider the Darcy law for gas phase in (A.2), when u_g is a real number and mobility of gas (i.e., (K_{rg}/μ_g)) approaches infinity, the value of driving force, that is, $(dP_g/dz - \rho_g g)$, is relatively small. This implies that hydrostatic gas pressure gradient is a good estimation in free gravity drainage modeling. By substituting the gas pressure in (A.4) and assuming constant porosity the following equation is obtained:

$$\frac{\partial}{\partial z}\left(\frac{KK_{ro}}{\mu_o}\left(\frac{\partial P_c}{\partial z} + \left(\rho_o - \rho_g\right)g\right)\right) = -\phi\frac{\partial}{\partial t}\left(S_o\right). \tag{A.5}$$

Since $P_c = P_c\left(S_o\right)$, the above equation can be written as

$$\frac{\partial}{\partial z}\left(\frac{KK_{ro}}{\mu_o}\left(\frac{\partial P_c}{\partial S_o}\frac{\partial S_o}{\partial z} + \left(\rho_o - \rho_g\right)g\right)\right) = -\phi\frac{\partial}{\partial t}\left(S_o\right). \tag{A.6}$$

Equation (A.6) is written in diffusion-convection form for simplifying:

$$\frac{\partial}{\partial z}\left[D\left(S_o\right)\frac{\partial S_o}{\partial z}\right] + U\left(S_o\right)\frac{\partial S_o}{\partial z} = \frac{\partial S_o}{\partial t}. \tag{A.7}$$

In this equation $U(S_o)$ is the coefficient of first derivative term and is defined in (A.9). This is a convective term and is affected by oil-gas density difference and derivative of oil relative permeability. $D(S_o)$ is a coefficient of second derivative term and is defined in (A.8). This diffusive term is a function of oil relative permeability and first derivative of capillary pressure with respect to the oil saturation:

$$D(S_o) = -\frac{KK_{ro}}{\mu_o\phi}\frac{dP_c}{dS_o}, \quad (A.8)$$

$$U(S_o) = -\frac{K}{\mu_o\phi}(\rho_o - \rho_g)\, g\frac{dK_{ro}}{dS_o}. \quad (A.9)$$

The boundary and initial conditions for (A.7) are as follows.

Initial Condition: $S(z,0) = 1.0 - S_{wi}$.

Inlet Boundary Condition:

$$u_o = \frac{KK_{ro}}{\mu_o\phi}\left(\left(\frac{dP_c}{dS_o}\right)\left(\frac{\partial S_o}{\partial z}\right) + (\rho_{\text{oil}} - \rho_{\text{gas}})\, g\right) = 0 \qquad z = 0, \quad t > 0. \quad (A.10)$$

The inlet boundary condition occasionally can be rewritten as $S_o = S_{\text{org}}$ @ $z = 0, t > 0$. This assumption only is valid when viscosity of displaced phase (i.e., oil phase) is relatively small.

Outlet Boundary Condition: $S(z,t) = 1$ @$l < z < l$. L is the length of rock column and H is the height of capillary/gravity equilibrium. Since the governing partial differential equation of free gravity drainage is nonlinear it is not possible to find an exact solution analytically. Instead, if the capillary pressure is neglected, (A.7) is converted to (A.11) which is linear and can be solved by separation of variable method:

$$-\frac{K}{\mu_o\phi}(\rho_o - \rho_g)\, g\left(\frac{dK_{ro}}{dS_o}\right)\left(\frac{\partial S_o}{\partial z}\right) = \frac{\partial S_o}{\partial t}. \quad (A.11)$$

Equation (A.12) shows the Corey-type function for oil relative permeability; using this correlation, Equation (A.11) is written as (A.13):

$$K_{ro} = K_{ro\max}\left(\frac{S_o - S_{\text{org}}}{1 - S_{\text{org}} - S_{wi}}\right)^{no}. \quad (A.12)$$

In this correlation $K_{ro\max}$ is the end point oil relative permeability, no is Corey exponent and affects the degree of curvature in relative permeability curve, and S_{org} is the residual gas-oil saturation:

$$-\left[\frac{KK_{ro\max}no(\rho_o - \rho_g)\, g}{\mu_o\phi(1 - S_{\text{org}} - S_{wi})}\right]\left(\frac{S_o - S_{\text{org}}}{1 - S_{\text{org}} - S_{wi}}\right)^{no-1} \times \left(\frac{\partial S_o}{\partial z}\right) = \frac{\partial S_o}{\partial t}. \quad (A.13)$$

Assuming the above equation and following dimension less parameter this equation is simplified as (A.15):

$$S_{OD} = \frac{S_o - S_{\text{org}}}{1 - S_{\text{org}} - S_{wir}}, \quad (A.14)$$

$$-\left(\frac{KK_{ro\max}no(\rho_o - \rho_g)\, g}{\mu_o\phi(1 - S_{\text{org}} - S_{wi})}\right)(S_{OD})^{no-1}\left(\frac{\partial S_{OD}}{\partial z}\right) = \frac{\partial S_{OD}}{\partial t}. \quad (A.15)$$

Assuming $s_{OD} = F(z) * G(t)$ and using (A.15) the following differential equation for $F(z)$ and $G(t)$ can be deduced:

$$\left(\frac{KK_{ro\max}no(\rho_o - \rho_g)\, g}{\mu_o\phi(1 - S_{\text{org}} - S_{wi})}\right)F(z)^{no-2}\left(\frac{\partial F(z)}{\partial z}\right) = -\frac{\partial G(t)}{\partial t}G(t)^{-no}. \quad (A.16)$$

Equation (A.16) can be solved by calculation of $F(z)$ and $G(t)$ using separation of variable method. The saturation function can be found by multiplication of these two functions and can be written as

$$S_{OD} = \left[\frac{\mu_o\phi(1 - S_{\text{org}} - S_{wi})\, z}{KK_{ro\max}no(\rho_o - \rho_g)\, gt}\right]^{1/(no-1)}. \quad (A.17)$$

Occasionally this equation can be simplified as (A.19) using the following definition of dimensionless time:

$$t_{Dg} = \frac{\Delta\rho g K t}{\mu\phi(1 - S_{wir} - S_{\text{org}})\, l}, \quad (A.18)$$

$$S_{OD} = \left[\frac{z}{no\, t_{Dg}K_{ro\max}l}\right]^{1/(no-1)}. \quad (A.19)$$

It should be noted that the above saturation function is valid for z values which are less than the gas penetration depth from top of the block at specific time. For larger z values the oil column is not affected by gas invasion and oil saturation is equal to $1 - S_{wi}$. Consequently, the solution of (A.16) should be written as

$$S_{OD} = \left[\frac{z}{no\, t_{Dg}K_{ro\max}l}\right]^{1/(no-1)} \quad @0 \le z \le l(t), \quad (A.20)$$

$$S_{OD} = 1 \quad @l(t) \le z \le l. \quad (A.21)$$

In this solution $l(t)$ represents the penetration depth of gas from top of the block and can be found when the dimensionless saturation (s_{OD}) in (A.21) is equal to 1:

$$h(t) = no\, K_{ro\max}t_{DG}l. \quad (A.22)$$

At early stage, gas invades in the column and the invasion depth can be calculated by (A.22). The gas moves down the oil in the column until it approaches capillary/gravity equilibrium. For this transition time, l ($t = t_{\text{transition}}$) becomes equal to l. From (A.22) one may conclude that the gas front moves linearly with time. For later stage, the gas cannot displace the oil beyond the capillary/gravity equilibrium height due to capillary end effect. The value of threshold height can be calculated by (A.23). In this equation $r_{\min}$ represents the smallest pore throat of porous media:

$$Z_{\text{threshhold}} = \frac{Pc_{\text{Threshold}}}{\Delta\rho g} = \frac{2\sigma_{og}\cos\theta_{og}}{r_{\min}\Delta\rho g}. \quad \text{(A.23)}$$

The oil recovery factor can be found using saturation function as it is defined by following equation:

$$R = 1 - \frac{\overline{S_o}}{1 - S_{wi}} = 1 - \frac{(1/l)\int_o^l S_o dz}{1 - S_{wi}}. \quad \text{(A.24)}$$

In early stage when the gas does not reach the gravity/capillary equilibrium the average residual oil saturation should be calculated as follows:

$$\overline{S_o} = \left(\frac{1}{l}\right)\int_o^l S_o dz = \left(\frac{1}{l}\right)\left[\int_0^{l(t)} S_o dz + \int_{l(t)}^{l} (1 - S_{wi})\, dz\right]. \quad \text{(A.25)}$$

Using (A.25), (A.26), and (A.21) the value of average residual oil saturation and recovery factor are

$$\overline{S_o} = 1 - S_{wi} - \left[\left(1 - S_{wi} - S_{\text{org}}\right)\left(K_{ro\max} t_{Dg}\right)\right], \quad \text{(A.26)}$$

$$R = \left(\frac{1 - S_{wi} - S_{\text{org}}}{1 - S_{wi}}\right)\left(t_{Dg} K_{ro\max}\right). \quad \text{(A.27)}$$

The value of cumulative oil production is also

$$Q_p = (Al\phi)\left(1 - S_{wi} - S_{\text{org}}\right)\left(t_{Dg} K_{ro\max}\right). \quad \text{(A.28)}$$

This equation shows that the cumulative oil production increased linearly as a function of time until the gas front reaches capillary/gravity equilibrium height. The velocity of oil in this stage is constant and is equal to

$$v_c = \frac{dQ_p}{Adt} = \frac{K K_{ro\max}\Delta\rho g}{\mu_o}. \quad \text{(A.29)}$$

Nomenclature

A: Cross-sectional area
S_o: Oil saturation
S_g: Gas saturation
S: Saturation
S_i: Initial water saturation
S_{wi}: Irreducible water saturation
S_{OD}: Dimensionless saturation
φ : Porosity
μ: Viscosity
ρ: Density
q: Injection rate, volumetric flux
q_c: Vertical flow rate
P: Capillary pressure/parachor confident
P_c: Pressure
$P_{C_{th}}$: Threshold pressure
t_{Dg}: Dimensionless time
N_c: Capillary number
N_g: Gravity number
N_B: Bond number
N_d: Diffuision number
$N_{\text{Co.}}$: Combined number
N_{DE}: Displacement efficiency number
$K_{ro\max}$: End point oil permeability
$K_{rg\max}$: End point gas permeability
K_r: Relative permeability
K_{ro}: Relative oil permeability
K_{rg}: Relative gas permeability
ng: Corey exponent in gas relative permeability function
no: Corey exponent in oil relative permeability function
σ: Interfacial tension
S: Spreading coefficient
R: Recovery factor
M: Mobility ratio
PV: Pore volume
l: Length or height.

Subscripts

g: Gas
w: Water
o: Oil
abs: Absolute
wg: Water-gas system
wo: Water-oil system
og: Oil-gas system.

Conflict of Interests

The author declares that there is no conflict of interests regarding the publishing of this paper. The paper will not reflect the viewpoints of the affiliated institutions by any means.

Acknowledgments

The author wish to thank the University of Stavanger for the financial support, K. P. Dziadosz and Professor Hamouda for their comments/advices in the experimental lab works, and Svein Myrhen for Lab view support and Inger Johanne for her positive contribution in getting the chemicals used in this work. This research paper is lovingly dedicated to author's parents who have been constant source of inspiration. At time of publication of this paper, author's father (Ibrahim Alipour Tabrizy) sadly passed away. The author never forgets his moral love and supports which without it, this research would not have been made possible.

References

[1] G. Moritis, "EOR survey," *Oil and Gas Journal*, vol. 104, pp. 37–57, 2006.

[2] V. Alvarado and E. Manrique, "Enhanced oil recovery: an update review," *Energies*, vol. 3, no. 9, pp. 1529–1575, 2010.

[3] E. A. Chukwudeme and A. A. Hamouda, "Enhanced Oil Recovery (EOR) by miscible CO_2 and water flooding of asphaltenic and non-asphaltenic oils," *Energies*, vol. 2, no. 3, pp. 714–737, 2009.

[4] A. A. Hamouda, E. A. Chukwudeme, and D. Mirza, "Investigating the effect of CO_2 flooding on asphaltenic oil recovery and reservoir wettability," *Energy & Fuels*, vol. 23, no. 2, pp. 1118–1127, 2009.

[5] A. Werner, F. Behar, J. C. De Hemptinne, and E. Behar, "Thermodynamic properties of petroleum fluids during expulsion and migration from source rocks," *Organic Geochemistry*, vol. 24, no. 10-11, pp. 1079–1095, 1996.

[6] R. B. Grigg and D. S. Schecter, "State of the industry in CO_2 floods," in *Proceedings of the SPE Annual Technical Conference and Exhibition*, SPE 38849, San Antonio, Tex, USA, October 1997.

[7] P. M. Jarrell, C. E. Fox, M. H. Stein, and S. L. Webb, *Practical Aspects of CO_2 Flooding*, vol. 22 of *SPE Monograph*, Society of Petroleum Engineers, 2002.

[8] L. W. Holm and V. A. Josendal, "Mechanisms of oil displacement by carbon dioxide," *Journal of Petroleum Technology*, vol. 26, pp. 1427–1438, 1974.

[9] F. M. Orr Jr., M. K. Silva, C. L. Lien, and M. T. Pelletier, "Laboratory experiments to evaluate field prospects for CO_2 flooding," *Journal of Petroleum Technology*, vol. 34, no. 4, pp. 888–898, 1982.

[10] G. A. Rojas, T. Zhu, S. B. Dyer, S. Thomas, and S. M. F. Ali, "Scaled model studies of CO_2 floods," *SPE Reservoir Engineering*, vol. 6, no. 2, pp. 169–178, 1991.

[11] J. Hagoort, "Oil recovery by gravity drainage," *Society of Petroleum Engineers Journal*, vol. 20, no. 3, pp. 139–150, 1980.

[12] I. Chatzis, A. Kantzas, and F. A. L. Dullien, "On the investigation of gravity-assisted inert gas injection using micromodels, long berea sandstone cores, and computer-assisted tomography," in *Proceedings of the SPE Annual Technical Conference and Exhibition (ATCE '88)*, SPE 18284, Houston, Tex, USA, October 1988.

[13] W. J. da Sie and D. S. Guo, "Assessment of a vertical hydrocarbon miscible flood in the Westpem Nisku D reef," *SPE Reservoir Engineering*, vol. 5, no. 2, pp. 8–17354, 1990.

[14] D. N. Rao, S. C. Ayirala, M. M. Kulkarni, and A. P. Sharma, "Development of Gas Assisted Gravity (GAGD) process for improved light oil recovery," in *Proceedings of the SPE/DOE Symposium on Improved Oil Recovery*, SPE 89357, Tulsa, Okla, USA, April 2004.

[15] M. Verlaan and P. Boerrigter, "Miscible gas-oil gravity drainage," in *Proceedings of the 1st International Oil Conference and Exhibition*, SPE 10399, Cancun, Mexico, September 2006.

[16] M. M. Kulkarni and D. N. Rao, "Characterization of operative mechanisms in gravity drainage field projects through dimensional analysis," in *Proceedings of the SPE Annual Technical Conference and Exhibition (ATCE '06)*, pp. 4512–4530, San Antonio, Tex, USA, September 2006.

[17] P. S. Jadhawar and H. K. Sarma, "Scaling and sensitivity analysis of gas-oil gravity drainage EOR," in *Proceedings of the SPE Asia Pacific Oil and Gas Conference and Exhibition*, SPE 115065, pp. 508–521, Perth, Australia, October 2008.

[18] V. K. Bangla, F. Yau, and G. R. Hendricks, "Reservoir performance of a gravity stable vertical CO_2 miscible flood: wolf camp reservoir," in *Proceedings of the 66th Annual Technical Conference and Exhibition*, SPE 22898, Wellman Unit, 1991.

[19] D. J. Wood, L. W. Lake, R. T. Johns, and V. Nunez, "A screening model for CO_2 flooding and storage in Gulf Coast reservoirs based on dimensionless groups," *SPE Reservoir Evaluation and Engineering*, vol. 11, no. 3, pp. 513–520, 2008.

[20] J. J. Trivedi and T. Babadagli, "Efficiency of diffusion controlled miscible displacement in fractured porous media," *Transport in Porous Media*, vol. 71, no. 3, pp. 379–394, 2008.

[21] D. A. Hudgins, F. M. Liave, and F. T. H. Chung, "Nitrogen miscible displacement of light crude oil: a laboratory study," *SPE Reservoir Engineering Journal*, vol. 5, no. 1, SPE 17372, pp. 100–106, 1990.

[22] M. Nobakht, S. Moghadam, and Y. Gu, "Effects of viscous and capillary forces on CO_2 enhanced oil recovery under reservior conditions," *Energy & Fuels*, vol. 21, no. 6, pp. 3469–3476, 2007.

[23] G. H. Darvish, *Physical effects controlling mass transfer in matrix fracture system during CO_2 injection into chalk fractured reservoirs [Ph.D. thesis]*, Norwegian University of Science and Technology, Trondheim, Norway, 2007.

[24] F. A. L. Dullien, *Porous Media—Fluid Transport and Pore Structure*, Academic Press, San Diego, Calif, USA, 2nd edition, 1992.

[25] A. Danesh, *PVT and Phase Behaviour of Petroleum Reservoir Fluids*, Elsevier Science, Amsterdam, The Netherlands, 1998.

[26] G. Løvoll, Y. Méheust, K. J. Måløy, E. Aker, and J. Schmittbuhl, "Competition of gravity, capillary and viscous forces during drainage in a two-dimensional porous medium, a pore scale study," *Energy*, vol. 30, no. 6, pp. 861–872, 2005.

[27] C. Marle, *Multiphase Flow in Porous Media*, Institut Français du Petrole (IFP) Publications, TECHNIP, Paris, France, 1981.

[28] T. A. Renner, "Measurement and correlation of diffusion coefficients for CO_2 and rich-gas applications," *SPE Reservoir Engineering*, vol. 3, no. 2, pp. 517–523, 1988.

[29] O. Helland and O. I. Frette, "Computation of fluid configurations and capillary pressures in mixed-wet 2d pore spaces from rock images," in *Proceedings of the International Conference on Water Resources (CMWR '10)*, J. Carrera, Ed., Barcelona, Spain, 2010.

[30] V. A. Tabrizy, *Investigated miscible CO_2 flooding for enhancing oil recovery in wettability altered chalk and sandstone rock [Ph.D. thesis]*, University of Stavanger, Stavanger, Norway, 2012.

[31] B. Rostami, R. Kharrat, M. Pooladi-Darvish, and C. Ghotbi, "Identification of fluid dynamics in forced gravity drainage using dimensionless groups," *Transport in Porous Media*, vol. 83, no. 3, pp. 725–740, 2010.

[32] M. Blunt, D. Zhou, and D. Fenwick, "Three-phase flow and gravity drainage in porous media," *Transport in Porous Media*, vol. 20, no. 1-2, pp. 77–103, 1995.

[33] N. Shahidzadeh-Bonn, A. Tournié, S. Bichon et al., "Effect of wetting on the dynamics of drainage in porous media," *Transport in Porous Media*, vol. 56, no. 2, pp. 209–224, 2004.

[34] A. Al-Rawajfeh, *Modeling and simulation of CO_2 release in multiple-effect distillers for seawater desalination [Ph.D. thesis]*, Martin-Luther-University Halle-Wittenberg, Halle, Germany, 2008.

5

CFD Method for Predicting Annular Pressure Losses and Cuttings Concentration in Eccentric Horizontal Wells

Titus N. Ofei,[1] Sonny Irawan,[1] and William Pao[2]

[1] *Petroleum Engineering Department, Universiti Teknologi PETRONAS, Bandar Seri Iskandar, 31750 Tronoh, Malaysia*
[2] *Mechanical Engineering Department, Universiti Teknologi PETRONAS, Bandar Seri Iskandar, 31750 Tronoh, Malaysia*

Correspondence should be addressed to Titus N. Ofei; titusofei@hotmail.com

Academic Editor: Markus Kraft

In oil and gas drilling operations, predictions of pressure losses and cuttings concentration in the annulus are very complex due to the combination of interacting drilling parameters. Past studies have proposed many empirical correlations to estimate pressure losses and cuttings concentration. However, these developed correlations are limited to their experimental data range and setup, and hence, they cannot be applicable to all cases. CFD methods have the advantages of handling complex multiphase flow problems, as well as, an unlimited number of physical and operational conditions. The present study employs the inhomogeneous (Eulerian-Eulerian) model to simulate a two-phase solid-fluid flow and predict pressure losses and cuttings concentration in eccentric horizontal annuli as a function of varying drilling parameters: fluid velocity, diameter ratio (ratio of inner pipe diameter to outer pipe diameter), inner pipe rotation speed, and fluid type. Experimental data for pressure losses and cuttings concentration from previous literature compared very well with simulation data, confirming the validity of the current model. The study shows how reliable CFD methods can replicate the actual, yet complex oil and gas drilling operations.

1. Introduction

Predictions of pressure losses and cuttings concentration in annular wells are strongly affected by varying drilling parameters such as fluid velocity, fluid properties (density, viscosity), cuttings size and density, hole-pipe eccentricity, drill pipe rotation, and annular diameter ratios. There are few attempts made by some investigators to estimate pressure losses and cuttings concentration in annular geometries with and without drill pipe rotation by employing either experimental or numerical approaches.

Among the first authors to conduct extensive experimental study on cuttings transport at varying angles of inclinations is Tomren et al. [1]. The authors studied the effects of fluid velocity, fluid rheological properties, pipe-hole eccentricity, drill pipe rotation, and flow regimes on cuttings concentration at steady state condition. They concluded that fluid velocity, hole inclination, and mud rheological properties were the major factors affecting mud carrying capacity. Becker and Azar [2] also investigated experimentally the effects of mud weight and annular diameter ratio on the performance of hole cleaning in inclined wellbores. The authors observed that variations in the drill pipe have minimum effect on particle concentration for the same fluid velocity. According to Adari et al. [3], the practical use of drilling factors in controlling cuttings transport is much dependent on their controllability in the field. It is believed that, cuttings transported in the annulus are not always affected by a single parameter, but a combination of parameters to ensure efficient hole cleaning [4]. Other studies have also confirmed that increase in fluid velocity results in a decrease in cuttings accumulation in the wellbore [5–7]. Ozbayoglu and Sorgun [8] also conducted cuttings transport experiment and developed empirical correlations for estimating pressure losses with the presence of cuttings and drill pipe rotation in horizontal and inclined wellbores. They observed that the influence of drill pipe rotation on pressure loss is more significant if fluid is more non-Newtonian. The annular test section has diameter ratio of 0.62. Similar cuttings transport experiment study was carried out by Sorgun et al. [9] in horizontal and inclined annular geometry of diameter ratio of 0.62. The authors observed that, the existence of cuttings in the system caused an increase in pressure loss due to a decrease in flow area inside the annular gap. Further observation was that, drill pipe rotation decreases the pressure loss significantly if the drill pipe is making orbital motion in eccentric annulus. Another experimental study was conducted [10] to analyse the effects of some "very difficult to identify" data on the estimation of total pressure loss and cuttings concentration in horizontal and inclined annulus. Results from this study indicate that drill pipe rotation does not have significant influence on pressure loss for constant rate of penetration (ROP) and fluid velocity. The annular test section has diameter ratio of 0.64.

One of the pioneering works by Markatos et al. [11] modelled single phase Newtonian flow in nonuniform narrow annular gaps using finite difference technique. The velocity flow fields as well as static pressure were predicted in a two-dimensional flow.

Han et al. [12] is among the first to conduct experimental and CFD studies on solid-fluid mixture flow in vertical and highly deviated slim hole annulus. They concluded that, annular pressure losses increase with mixture fluid velocity, annular angle of inclination, and drill pipe rotation speed. The annular test section has diameter ratio of 0.70. Mokhtari et al. [13] employed CFD method to model the effects of eccentricity and flow behaviour index on annular pressure loss and velocity profile for varying diameter ratios from 0.30 to 0.90. The authors, however, did not include cuttings in the annular mainstream. Recently, Ofei et al. [14] also employed CFD technique to analyse the influence of diameter ratio, fluid velocity, fluid type, fluid rheology, and drill pipe rotation speed on pressure loss in eccentric horizontal wellbore with the presence of cuttings.

The present study also utilises a CFD technique to examine the effects of fluid velocity, annular diameter ratio (ranging from 0.64 to 0.90), drill pipe rotation, and fluid type on the prediction of pressure loss and cuttings concentration for solid-fluid flow in eccentric horizontal wellbore. Contours of cuttings volume fraction, 3D cuttings velocity streamlines and radial cuttings velocity profiles are also presented to give further insight on cuttings transport. The new findings from this study would provide better understanding and guide in the selection of optimum drilling parameters in narrow annuli drilling such as casing drilling and slim holes.

2. Materials and Methods

Multiphase component of CFD software ANSYS-CFX 14.0 is adopted in this study. In ANSYS-CFX, a multiphase flow containing dispersed particles may be modelled using either Lagrangian Particle Tracking model or Eulerian-Eulerian model. The inhomogeneous (Eulerian-Eulerian) model, sometimes called the two-fluid model, regards both continuous and dispersed phases as continuous media. In this study, the Eulerian-Eulerian model is preferred to the Lagrangian Particle Tracking model due to its ability to handle high solid volume fractions. Furthermore, it accounts for solid particle-particle interaction and includes turbulence automatically [15]. A drawback of this model is, however, that they need complex closure relations. The following continuity and momentum equations representing the two-phase flow model are described for the sake of brevity.

2.1. Continuity Equations. The fluid phase continuity equation assuming isothermal flow condition can be expressed as [16, 17]

$$\frac{\partial}{\partial t}(\hbar_l) + \nabla(\hbar_l U_l) = 0. \quad (1)$$

Similarly, for a solid phase,

$$\frac{\partial}{\partial t}(\hbar_s) + \nabla(\hbar_s U_s) = 0, \quad (2)$$

where the solid and fluid phase volume fraction sum up as follows:

$$\hbar_s + \hbar_l = 1. \quad (3)$$

At steady state condition, $\partial/\partial t = 0$.

2.2. Momentum Equations. The forces acting on each phase and interphase momentum transfer term that models the interaction between each phase are given below [16, 17].

For fluid phase,

$$\rho_l \hbar_l \left[\frac{\partial U_l}{\partial t} + U_l \cdot \nabla U_l\right] = -\hbar_l \nabla p + \hbar_l \nabla \cdot \overline{\overline{\tau}}_l + \hbar_l \rho_l g - M. \quad (4)$$

Similarly, for solid phase,

$$\rho_s \hbar_s \left[\frac{\partial U_s}{\partial t} + U_s \cdot \nabla U_s\right] = -\hbar_s \nabla p + \hbar_s \nabla \cdot \overline{\overline{\tau}}_l + \nabla \cdot \overline{\overline{\tau}}_s - \nabla P_s + \hbar_s \rho_s g + M. \quad (5)$$

At steady state condition, $\partial/\partial t = 0$.

2.3. Other Closure Models

2.3.1. Interphase Drag Force Model. For spherical particles, the drag force per unit volume is given as

$$M_d = \frac{3C_D}{4d_s} \hbar_s \rho_l \left|U_s - U_l\right| (U_s - U_l). \quad (6)$$

For densely distributed solid particles, where the solid volume fraction $\hbar_s < 0.2$, the Wen and Yu [18] drag coefficient model may be utilised. This model is modified and implemented in ANSYS-CFX to ensure the correct limiting behaviour in the inertial regime as

$$C_D = \hbar_l^{-1.65} \max\left[\frac{24}{N'_{\mathrm{Re}_p}}\left(1 + 0.15 N'^{\,0.687}_{\mathrm{Re}_p}\right), 0.44\right], \quad (7)$$

where $N'_{\mathrm{Re}_p} = \hbar_l N_{\mathrm{Re}_p}$ and $N_{\mathrm{Re}_p} = \rho_l |U_l - U_s| d_s / \mu_l$.

For large solid volume fraction, $\mathscr{k}_s > 0.2$, the Gidaspow drag model may be used with the interphase drag force per unit volume defined as [19]

$$M_D = \frac{150(1-\mathscr{k}_l)^2 \mu_l}{\mathscr{k}_l d_s^2} + \frac{7}{4}\frac{(1-\mathscr{k}_l)\rho_l |U_l - U_s|}{d_s}. \quad (8)$$

In this study, both Wen and Yu and Gidaspow drag models were employed depending on the computed solid volume fraction. An approximate method for computing the solid volume fraction is presented in (18).

2.3.2. Lift Force Model. For spherical solid particles, ANSYS-CFX employs the Saffman and Mei lift force model as

$$M_L = \frac{3}{2\pi}\frac{\sqrt{\nu_l}}{d_s\sqrt{|\nabla \times U_l|}} C'_L \mathscr{k}_s \rho_l (U_s - U_l) \times (\nabla \times U_l + 2\Omega). \quad (9)$$

Saffman [20, 21] correlated the lift force for low Reynolds number past a spherical solid particle where $C'_L = 6.46$, and $0 \le N_{\mathrm{Re}_p} \le N_{\mathrm{Re}_\omega} \le 1$. For higher range of solid particle Reynolds number, Saffman's correlation was generalised by Mei and Klausner [22] as follows:

$$C'_L = \begin{cases} 6.46 \cdot f\left(N_{\mathrm{Re}_p}, N_{\mathrm{Re}_\omega}\right) & \text{for: } N_{\mathrm{Re}_p} < 40 \\ 6.46 \cdot 0.0524 \cdot \left(\beta N_{\mathrm{Re}_p}\right)^{1/2} & \text{for: } 40 < N_{\mathrm{Re}_p} < 100, \end{cases} \quad (10)$$

where $\beta = 0.5(N_{\mathrm{Re}_\omega}/N_{\mathrm{Re}_p})$,

$$f\left(N_{\mathrm{Re}_p}, N_{\mathrm{Re}_\omega}\right) = \left(1 - 0.3314\beta^{0.5}\right) \cdot e^{-0.1 N_{\mathrm{Re}_p}} + 0.3314\beta^{0.5}, \quad (11)$$

and $N_{\mathrm{Re}_\omega} = \rho_l \omega_l d_s^2/\mu_l$, $\omega_l = |\nabla \times U_l|$.

2.3.3. Turbulence k-ε Model in Multiphase Flow. The k-ε turbulence model offers a good compromise in terms of accuracy and robustness for general purpose simulations. It is a semiempirical model based on transport equation for the estimation of turbulent length scale and velocity scale from the turbulent kinetic energy (k) and dissipation rate (ε) [23]. In multiphase flow, the transport equations for k and ε are phase dependent and assume a similar form to the single-phase transport equations, respectively, as

$$\frac{\partial}{\partial t}(C_\alpha \rho_\alpha k_\alpha) + \nabla \cdot \left(C_\alpha \left(\rho_\alpha U_\alpha k_\alpha - \left(\mu + \frac{\mu_{t\alpha}}{\sigma_k}\right)\nabla k_\alpha\right)\right) = C_\alpha (P_\alpha - \rho_\alpha \varepsilon_\alpha) + T_{\alpha\beta}^{(k)}, \quad (12)$$

$$\frac{\partial}{\partial t}(C_\alpha \rho_\alpha \varepsilon_\alpha) + \nabla \cdot \left(C_\alpha \rho_\alpha U_\alpha \varepsilon_\alpha - \left(\mu + \frac{\mu_{t\alpha}}{\sigma_\varepsilon}\right)\nabla \varepsilon_\alpha\right) = C_\alpha \frac{\varepsilon_\alpha}{k_\alpha}(C_{\varepsilon 1} P_\alpha - C_{\varepsilon 2}\rho_\alpha \varepsilon_\alpha) + T_{\alpha\beta}^{(\varepsilon)}. \quad (13)$$

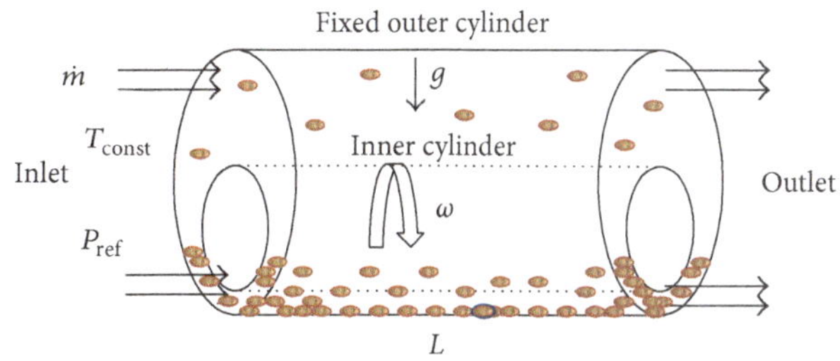

FIGURE 1: Physical model for solid-fluid flow.

Diffusion of momentum in phase α is governed by an effective viscosity as

$$\mu_{\mathrm{eff}} = \mu + \mu_{t\alpha}. \quad (14)$$

The k-ε model assumes that the turbulence viscosity is linked to the turbulence kinetic energy and dissipation by the following relation:

$$\mu_{t\alpha} = C_\mu \rho_\alpha \frac{k_\alpha^2}{\varepsilon_\alpha}. \quad (15)$$

The governing sets of partial differential equations were discretized using finite volume technique. The discretized equations together with initial and boundary conditions are solved iteratively for each control volume of pressure drop and cuttings concentration using ANSYS CFX 14.0 solver.

2.4. Physical Model and Carrier Fluid. Two-phase solid-fluid flow in eccentric horizontal annulus with stationary outer pipe and rotating inner pipe is presented. The inner pipe represents the actual drill pipe while the outer pipe represents the hole. Four annular 3D geometries are modelled with diameter ratios $\kappa = 0.64, 0.70, 0.80$ and 0.90 using ANSYS Workbench.

In order to eliminate end effects and ensure fully developed flow, the length of the annular pipe must be longer than the hydrodynamic entrance length. For a single phase Newtonian fluid flowing in a pipe, the hydrodynamic length is presented by Shook and Roco [24] as

$$L_h = 0.062 N_{\mathrm{Re}}(D). \quad (16)$$

However, for a two-phase flow in annular gap with a non-Newtonian fluid, such expression as in (16) does not yet exist in literature. As a rule of thumb, the authors have adopted (16) by replacing the pipe diameter D with a hydraulic diameter $D_h = D_2 - D_1$. It should be noted that a much longer annular length would only result in a computationally expensive CFD simulation. Figure 1 shows the physical model for solid-fluid flow. The fluid is considered incompressible, steady state, and isothermal. The rheology of the fluid is described by both Newtonian (water) and Power-Law model. The apparent viscosity for Power-Law model is given as

$$\mu_a = K\dot{\gamma}^{n-1}, \quad (17)$$

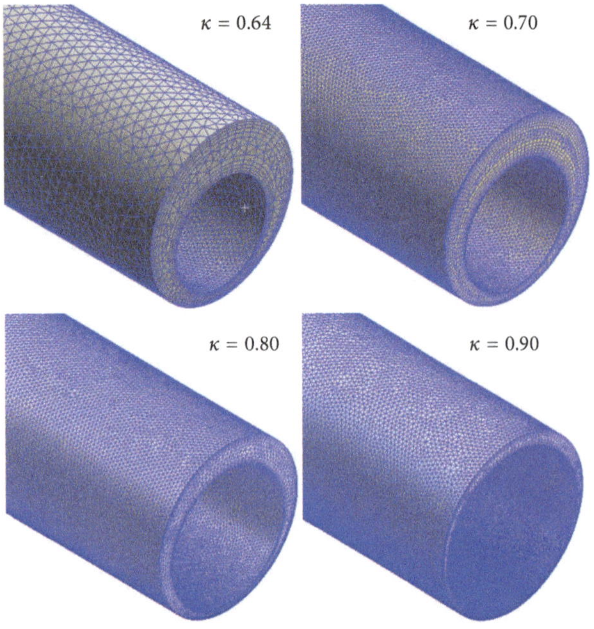

Figure 2: 3D section of meshed annular geometry.

where K is consistency index, $\dot{\gamma}$ is shear rate, and n is flow behaviour index. For $n = 1$, (17) reduces to Newtonian model, $n < 1$, fluid is shear thinning, and $n > 1$, fluid is shear thickening. In this study, $0.3 < n \leq 1$.

2.5. Boundary Conditions and Meshing. A mixture mass flow rate boundary condition was specified at the inlet while zero gauge pressure specified at the outlet. No-slip boundary conditions were imposed on both inner and outer pipe walls for both fluid and particles. The 3D annular geometries were meshed into unstructured tetrahedral grids of approximately $0.66–2.15 \times 10^6$ elements. Inflation layers were created near the walls covering about 20% of the inner and outer radii for resolving the mesh in the near-wall region as well as accurately capturing the flow effects in that region. Figure 2 shows the 3D section of the meshed annular geometries.

2.6. Grid Independence Study. To optimise the mesh sizes until results were insignificantly dependent on mesh size, grid independence study was conducted for all diameter ratios. The carrier fluid used is water flowing at a velocity of 2.743 m/s and the inner pipe rotation speed is 80 rpm. The cuttings feed concentration which gives an idea of the amount of particles in motion that are introduced to the annular space. This is computed as a function of area of bit, fluid velocity, and rate of penetration (ROP) as [6]

$$C_{\text{cf}} = \frac{(\text{ROP})\, A_{\text{bit}}}{R_T Q}, \tag{18}$$

where R_T is defined as the ratio of the particle transport velocity to the average annular fluid velocity. For the purpose of this study, R_T is taken as 0.5 based on experimental findings [9]. Figures 3(a) to 3(d) show the variation of pressure losses as a function of element sizes. In Figures 3(a) and 3(b), element size of 0.003 m and below would result in insignificant changes in pressure losses; however, more computational time is required for elements sizes below 0.003 m. In Figures 3(c) and 3(d), element size of 0.003 m and above also shows no significant changes in pressure losses. An optimum element size of 0.003 m is chosen for all diameter ratios resulting in approximately $0.66–2.15 \times 10^6$ number of elements with increasing diameter ratio from 0.64 to 0.90. The CPU time recorded in this study ranges between 7.2×10^3 s to 5.4×10^4 s. The simulations were run on a computer with the following specifications: Windows 7 64-bit operating system, with 4 GB RAM, and Pentium Dual-Core processor at 2.3 GHz.

2.7. Simulation Model Validation. The simulation model setups were validated against experiment data available from previous studies. Pressure loss and cuttings concentration data for cuttings-water flow in a horizontal wellbore were adopted from Osgouei [25]. Also, pressure loss data using non-Newtonian fluid of 0.4% CMC solution for cuttings transport experiment were adopted from Han et al. [12]. Table 1 summarises the rheological properties and operating parameters for the experimental studies.

From Figure 4(a), the calculated pressure loss slightly overpredicted the experimental data by a mean percentage error of 0.84%. Similarly, the calculated cuttings concentration data slightly overpredicted the experimental data by a mean percentage error of 12% as shown in Figure 4(b). The total cuttings concentration, C_{cT}, is defined as

$$C_{\text{cT}} = \frac{\text{Net volume occupied by particles}}{\text{Total volume of annlulus}} \times 100. \tag{19}$$

Moreover, Figure 4(c) shows the calculated pressure loss deviating slightly from the experimental data by a mean percentage error of 2.5%. The analyses show a good agreement between calculated and experimental data confirming the validity of the current model setup.

2.8. Simulation Study. Table 2 summarises the simulation setup including fluid rheological properties and drilling parameters. The present study adopts the Eulerian-Eulerian model to simulate a two-phase solid-fluid flow in eccentric horizontal annuli. ANSYS-CFX solver, which is based on a finite volume method [26], is used to solve the continuity and momentum equations with the appropriate initial and boundary conditions. The solution is assumed to be converged when the root mean square (RMS) of the normalised residual error reached 10^{-4} for all simulations. Both Newtonian (water) and non-Newtonian (Power-Law model) fluids are used as carrier fluids. Variations in annular pressure losses and cuttings concentration as a function of fluid velocity, diameter ratio, inner pipe rotation speed, and fluid type are analysed and results are presented. In addition, contours of cuttings volume fraction and cuttings velocities, streamlines of cuttings velocities, as well as profiles of cuttings velocities are also presented.

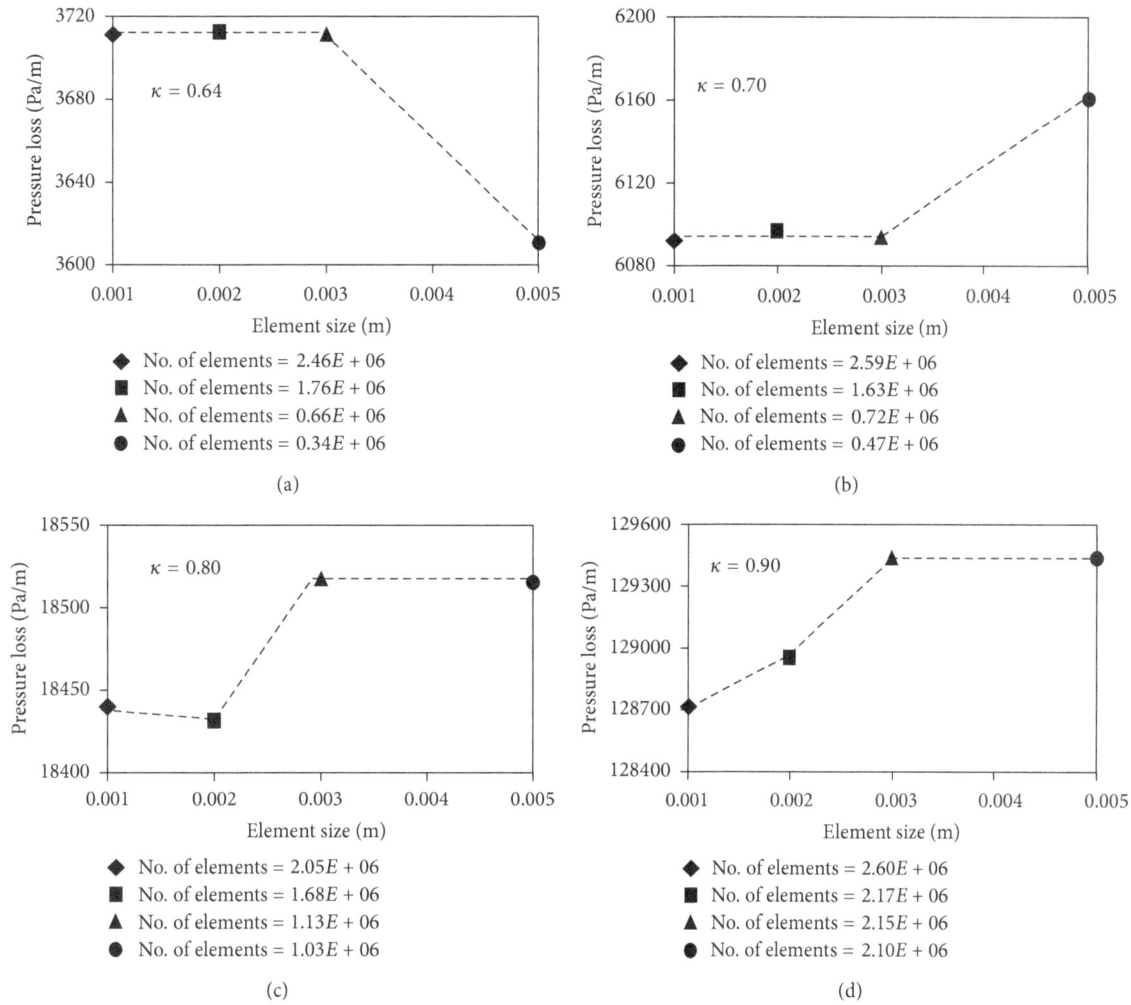

FIGURE 3: Grid independence study: (a) $\kappa = 0.64$, (b) $\kappa = 0.70$, (c) $\kappa = 0.80$, (d) $\kappa = 0.90$ at 80 rpm.

TABLE 1: Summary of experimental data from previous studies.

Experimental data source	Fluid density (kg/m^3)	n (—)	K (Pa s^n)	Pipe rotation (rpm)	κ (D_1/D_2)	e (—)	Fluid velocity (m/s)	Avg. cuttings size (m)	Cuttings density (kg/m^3)	ROP (m/s)
Osgouei [25]	998.5	1.0	0.001	80	0.64	0.623	1.524–2.7432	0.00201	2761.4	0.00508
Han et al. [12]	998.5	0.75	0.048	0	0.70	0	0.327–0.654	0.001	2550	0.00526

3. Results and Discussion

3.1. Effect of Fluid Velocity. Previous studies [1, 4] have revealed that fluid velocity is a dominant factor during cuttings transport. This phenomenon is also observed in this study. Figure 5 presents the variations in pressure loss and cuttings concentration as a function of increasing annular fluid velocity at constant diameter ratio and 80 rpm. Using both water and mud as carrier fluids, increasing fluid velocity significantly increases pressure losses, while a decrease in cuttings concentration occurs for each constant diameter ratio. This effect is however more pronounced for $\kappa = 0.90$. Figures 5(a)–5(d) depict these observations. For instance, when using mud as carrier fluid in Figure 5(c) and for $\kappa = 0.90$, annular pressure loss was dramatically increased by 97% when the flowing fluid velocity increased from 1.524 m/s to 2.749 m/s. Similarly, as shown in Figure 5(d), the cuttings concentration decreased by 37% in the annulus as fluid velocity increased from 1.524 m/s to 2.749 m/s for $\kappa = 0.90$. Another observation is that, in Figure 5(b), where the carrier fluid is water, there is almost no variation in cuttings concentration as fluid velocity increases for $\kappa = 0.90$. This indicates that in extreme narrow annuli, lower fluid velocities are capable of transporting enough cuttings from the annulus,

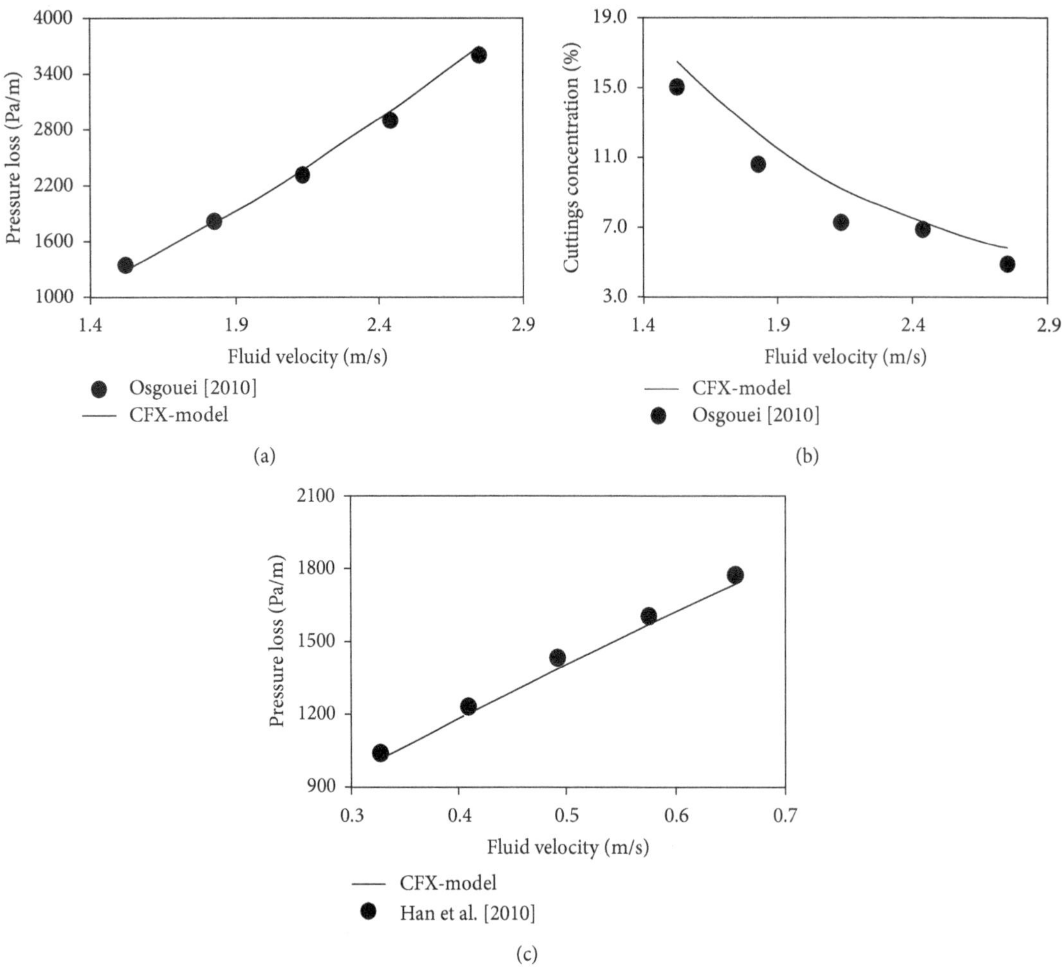

FIGURE 4: Experimental and simulation data comparison: (a) pressure loss data for cuttings-water flow, (b) cuttings concentration data for cuttings-water flow, (c) pressure loss data for cuttings—0.4% CMC flow.

TABLE 2: Simulation data for cuttings fluid flow.

Rheological and drilling parameter	Case 1 Water	Case 2 Mud
Fluid density (kg/m^3)	998.5	1006.3
Cuttings density (kg/m^3)	2761.4	
Avg. cuttings size (m)	0.00201	
Flow behaviour index, n	1	0.51
Viscosity consistency, K, (Pa s^n)	0.001	0.289
Fluid velocity (m/s)	1.524–2.7432	
Inner pipe rotation speed (rpm)	0, 80, 120	
Diameter ratio ($\kappa = D_1/D_2$)	0.64, 0.70, 0.80, 0.90	
Eccentricity (e)	0.623	
ROP (m/s)	0.00508	

as higher fluid velocities will not improve on the amount of cuttings transport, but will also increase the annular pressure dramatically, which may adversely affect the formation pressure. A slight variation in cuttings concentration could however be observed in Figure 5(d) as fluid velocity increases from 1.524 m/s to 2.749 m/s for $\kappa = 0.90$ when the carrier fluid is mud.

3.2. Effect of Diameter Ratio. Figure 6 presents the influence of diameter ratio on pressure loss and cuttings concentration at constant fluid velocity and 80 rpm. Analyses are shown for both water and mud as carrier fluids. For all cases, as diameter ratio increases from $\kappa = 0.64$ to 0.90, an increase in pressure loss also occurs, whereas a decrease in cuttings concentration is observed for each constant fluid velocity. This influence is however more pronounced for $\kappa = 0.90$. As the annular gap becomes narrower, there are more interactions between cuttings-fluid and pipe walls which results in an increase in friction, and hence, pressure losses. It is worth noting that while the pressure loss difference between $\kappa = 0.64$ and $\kappa = 0.90$ could result in extreme increase by over 3600%, a decrease of about 86% could be realised for cuttings concentration as water flows with a velocity of 1.524 m/s (see Figures 6(a) and 6(b)). Moreover, in Figure 6(b), where the carrier fluid is water, there is almost no disparity in cuttings concentration when $\kappa = 0.90$ for each constant fluid velocity. Although better cuttings transport could be observed in very

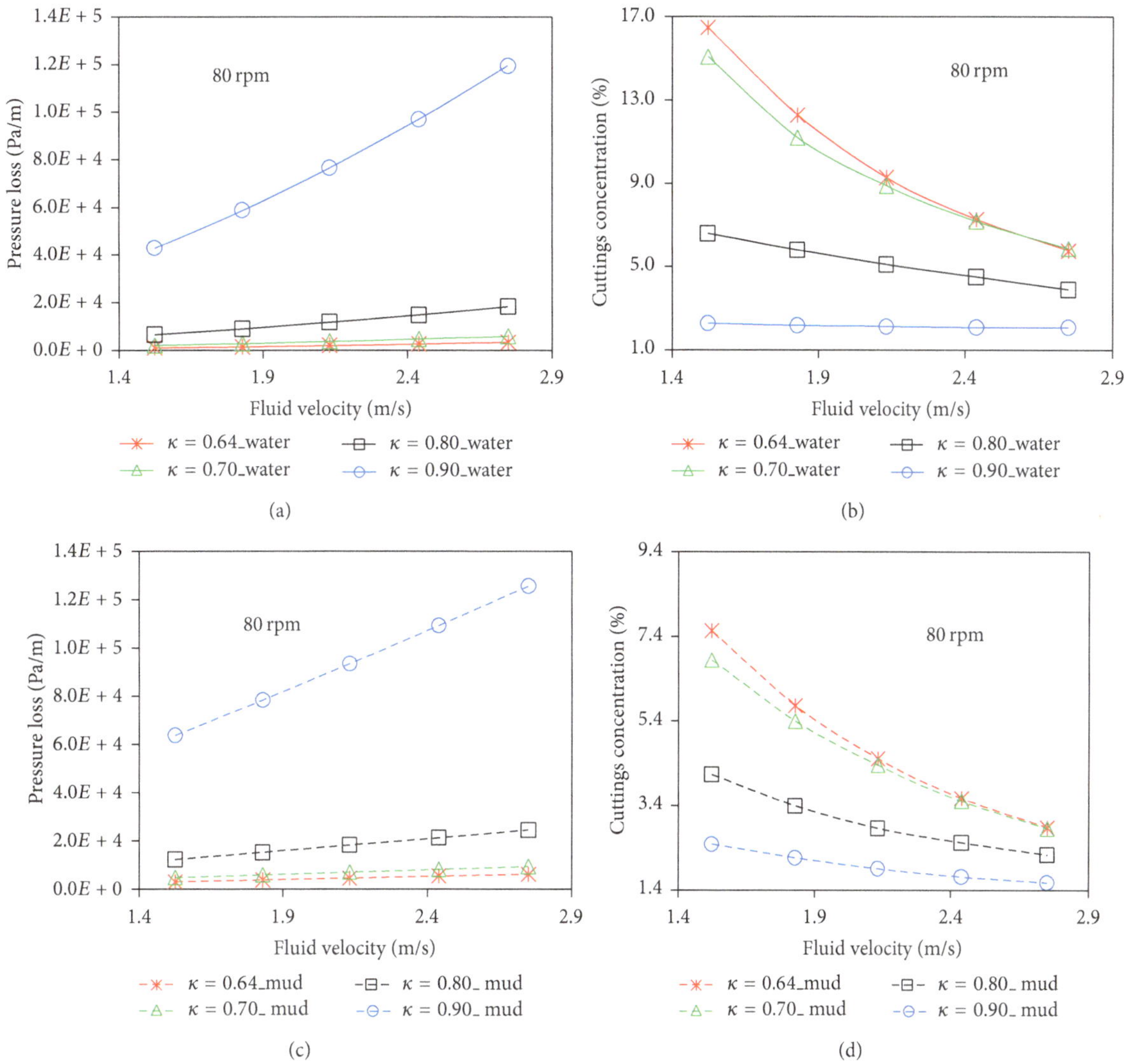

FIGURE 5: Effect of fluid velocity at constant diameter ratio on (a) pressure loss with water as carrier fluid, (b) cuttings concentration with water as carrier fluid, (c) pressure loss with mud as carrier fluid, (d) cuttings concentration with mud as carrier fluid.

narrow annuli, optimum drilling parameters must be selected to prevent excessive damage to the formation.

3.3. Effect of Drill Pipe Rotation. The effect of increasing drill pipe rotation on pressure loss and cuttings concentration is shown in Figure 7 when using both water and mud as carrier fluids. In Figures 7(a) and 7(c), an increase in drill pipe rotation speed from 80 rpm to 120 rpm did not result in any significant increment in pressure losses with both water and mud as carrier fluids. The effect on cuttings concentration is quite predominant especially in annular gaps with diameter ratio below $\kappa = 0.70$. For water as carrier fluid, as shown in Figure 7(b), the influence of increasing drill pipe rotation speed from 80 rpm to 120 rpm had a negative impact where the cuttings concentration increased when the diameter ratio is below $\kappa = 0.70$. To explain this behaviour, the low viscous water would generate high turbulence as a function of both axial and rotational flows; which in addition to gravity could cause rapid settling of cuttings in the annulus. Above $\kappa = 0.70$, the influence is virtually the same on cuttings concentration for each constant fluid velocity. On the contrary, when the carrier fluid is mud, as shown in Figure 7(d), increasing drill pipe rotation speed from 80 rpm to 120 rpm shows a decrease in cuttings concentration for a diameter ratio range of $0.64 \leq \kappa < 0.80$. Above $\kappa = 0.80$, the influence is relatively negative on cuttings concentration. In all cases, the rotation effect is dominant at lower fluid velocities.

3.4. Effect of Fluid Type. The effect of Newtonian (water) and non-Newtonian Power-Law fluid (mud) on pressure loss and cuttings concentration are analysed in Figures 8(a) and 8(b), respectively, at 120 rpm. With mud as carrier fluid, high pressure losses were recorded compared to water especially at low fluid velocity and $\kappa = 0.90$ (see Figure 8(a)). Similarly, the mud transported much cuttings compared to water especially

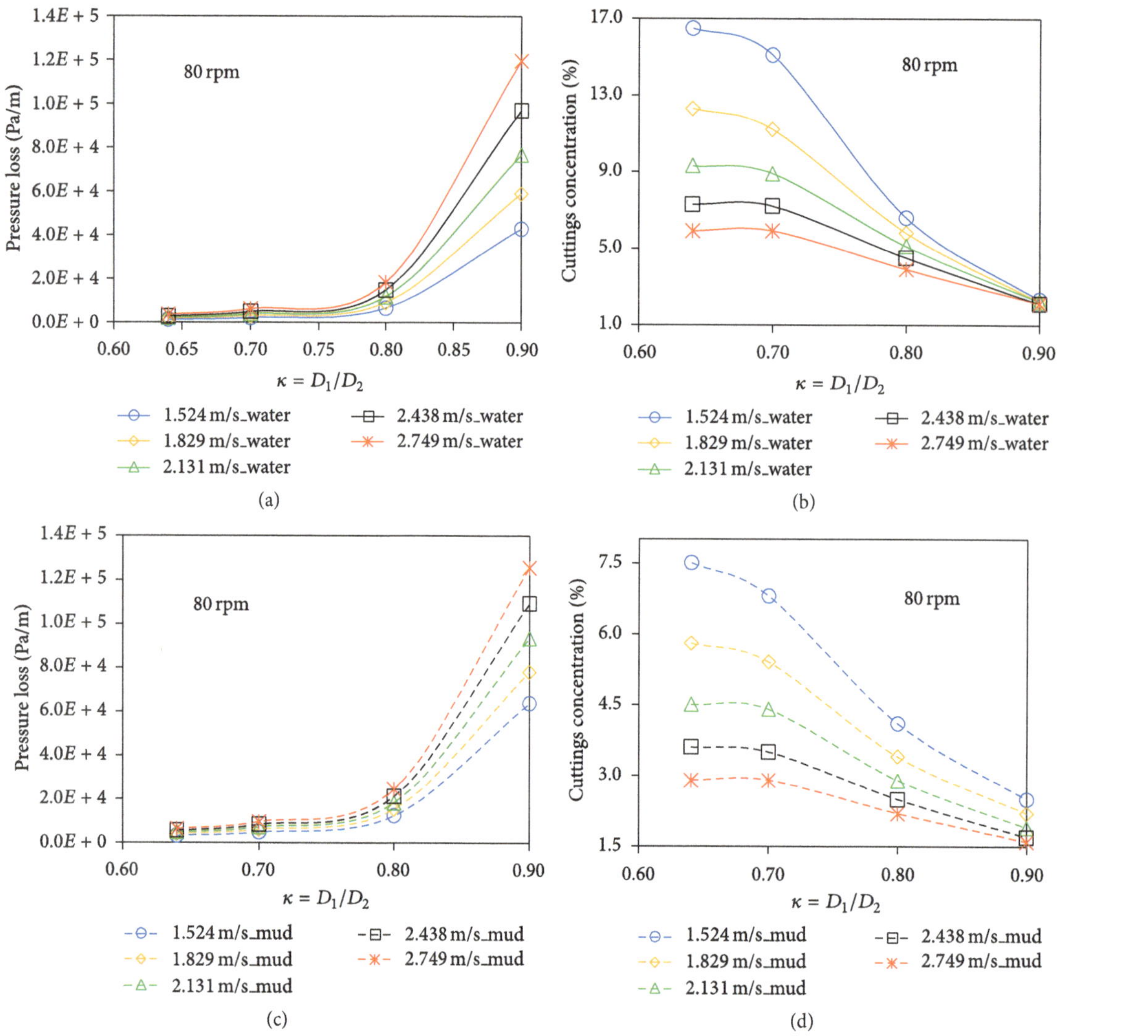

FIGURE 6: Effect of diameter ratio at constant fluid velocity on (a) pressure loss with water as carrier fluid, (b) cuttings concentration with water as carrier fluid, (c) pressure loss with mud as carrier fluid, (d) cuttings concentration with mud as carrier fluid.

at a constant diameter ratio of $\kappa = 0.64$ and low fluid velocities (see Figure 8(b)). For example, 19.2% and 6.0% concentration of cuttings remained in the annulus after flowing with water and mud, respectively, for $\kappa = 0.64$ and fluid velocity of 1.524 m/s. The performance of both fluids on cuttings concentration is quite similar at high diameter ratios.

3.5. Cuttings Volume Fraction, Velocity, and Profiles with Water as Carrier Fluid. Figures 9–11 show the contours of cuttings volume fraction, 3D streamlines of cuttings velocities, and radial measurements of cuttings velocity profiles, respectively, flowing with water at 1.524 m/s. As shown in Figure 9, the cuttings concentration accumulates in the narrowest gap of the eccentric annuli forming a bed due to gravity and the low viscosity of the carrier fluid. However, the rotation of the drill pipe from 0 rpm to 120 rpm reduces the cuttings bed by sweeping it into the widest gap where the fluid velocity is high to transport them to the surface. This observation is evident for all diameter ratios and shows the significance of drill pipe rotation in minimising differential pipe sticking, cuttings bed erosion, as well as excessive pressure losses. Figure 10 also depicts 3D streamlines of cuttings velocity. From the colour legend, the velocity of cuttings is high at some distance from the annular inlet and decreases to a minimum velocity towards the exit of the annular geometries. The decrease in cuttings velocity is an indication of cuttings settling to form a bed due to the low viscous nature of the carrier fluid and gravity. Drill pipe rotation induces a rotational flow on the cuttings bed into the annular mainstream and carries them to the surface. This rotation effect reduces the annular bed area for all diameter ratios. The radial measurements of cuttings velocity profiles at 1.524 m/s and 120 rpm are also presented in Figure 11. The radial distance is normalised. In the widest gap of the annular area, as shown in Figure 11(a), cuttings velocity increases with increasing diameter ratio where the peak velocities calculated are 1.896 m/s, 1.970 m/s, 2.043 m/s, and 1.999 m/s for κ =

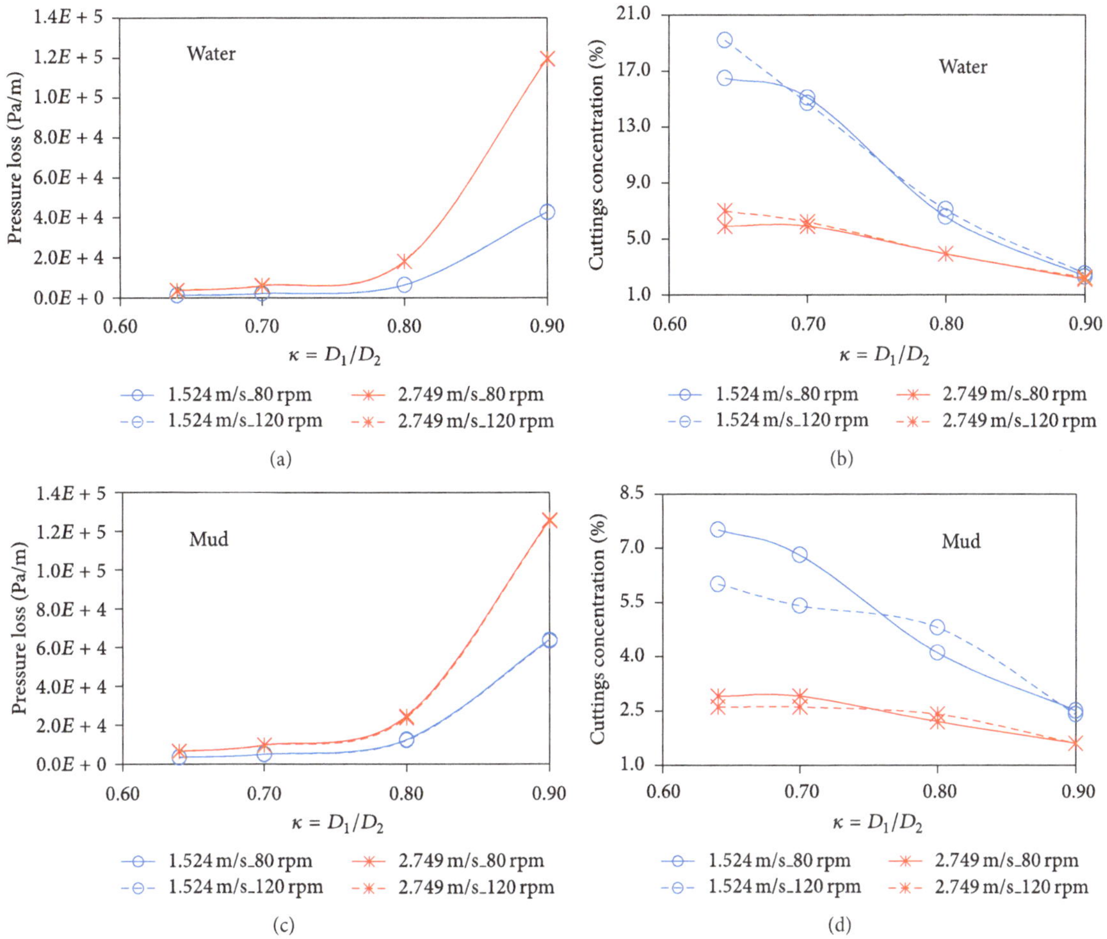

Figure 7: Effect of drill pipe rotation speed on (a) pressure loss with water as carrier fluid, (b) cuttings concentration with water as carrier fluid, (c) pressure loss with mud as carrier fluid, (d) cuttings concentration with mud as carrier fluid.

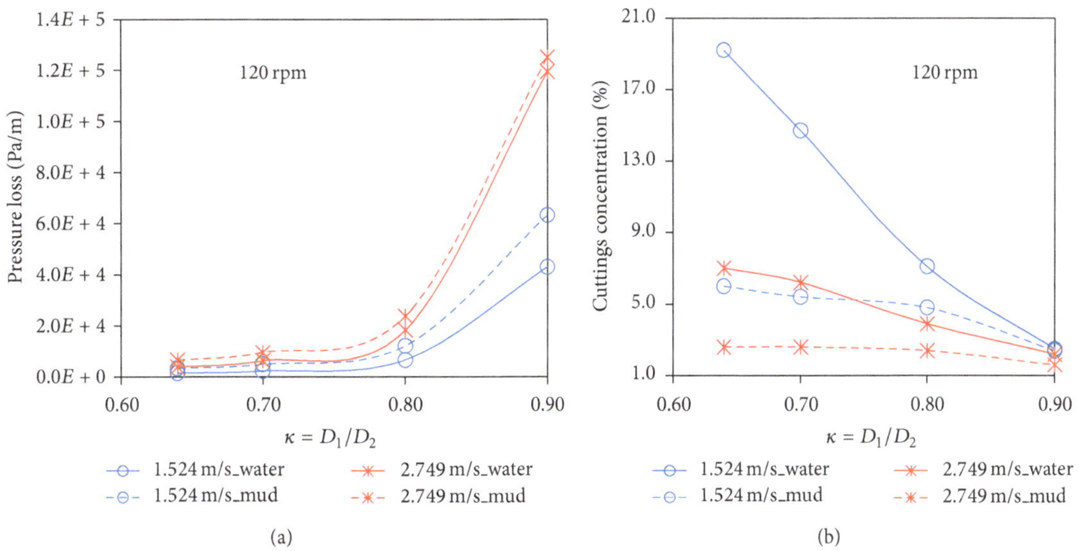

Figure 8: Effect of fluid type on (a) pressure loss and (b) cuttings concentration.

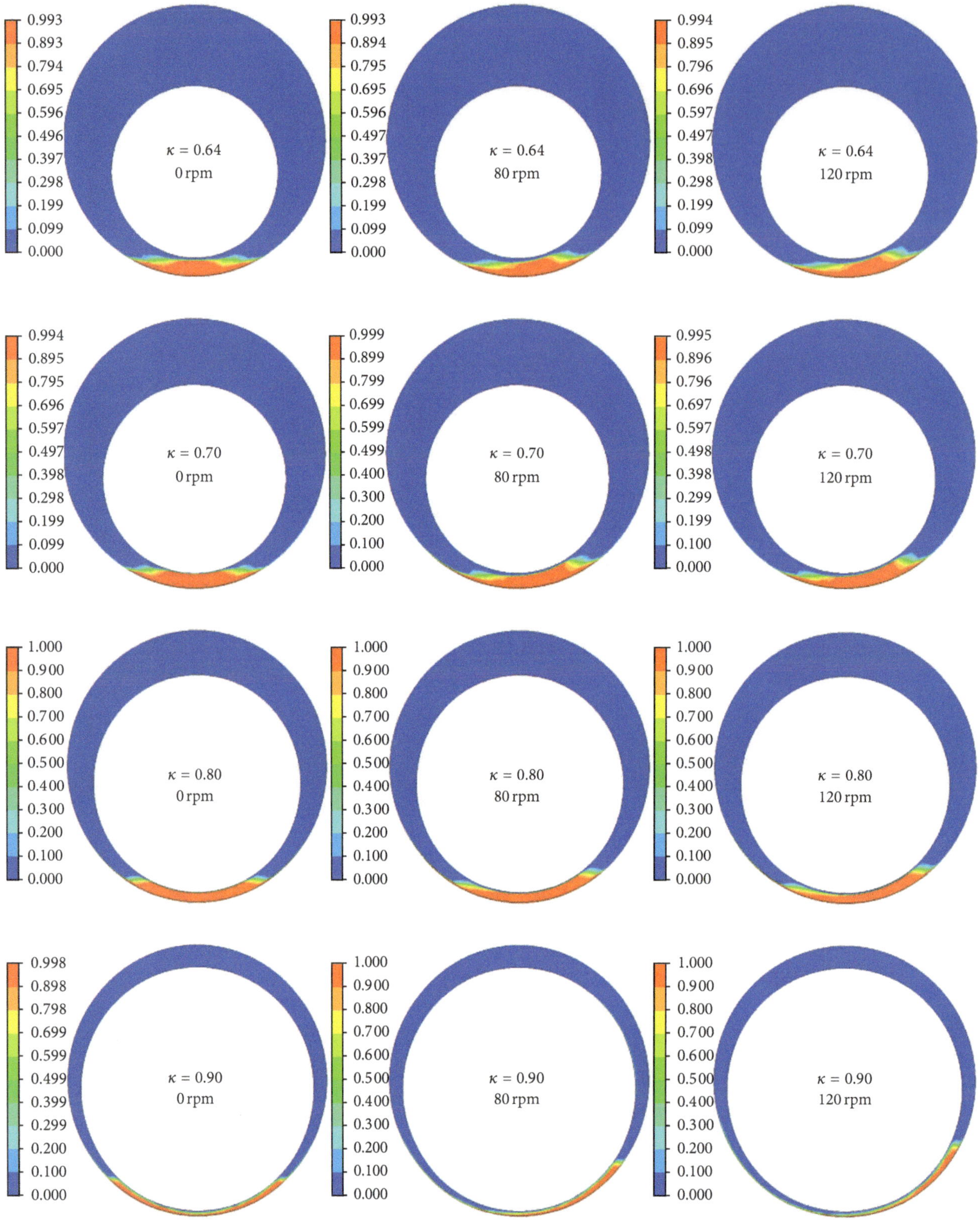

FIGURE 9: Contours of cuttings volume fraction for varying diameter ratios and inner pipe rotation with water as carrier fluid at 1.524 m/s.

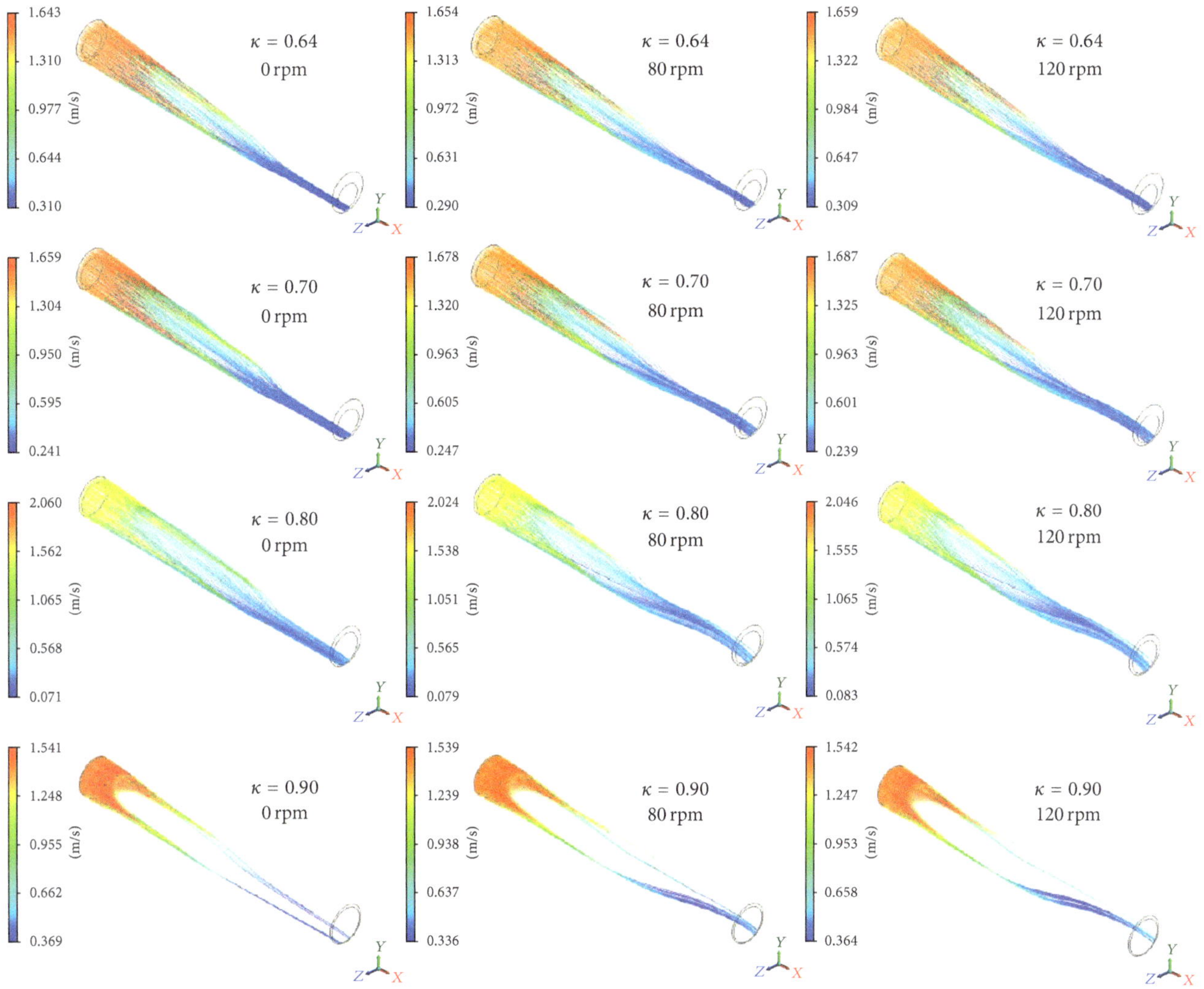

FIGURE 10: 3D streamlines of cuttings velocity for varying diameter ratios and inner pipe rotation with water as carrier fluid at 1.524 m/s.

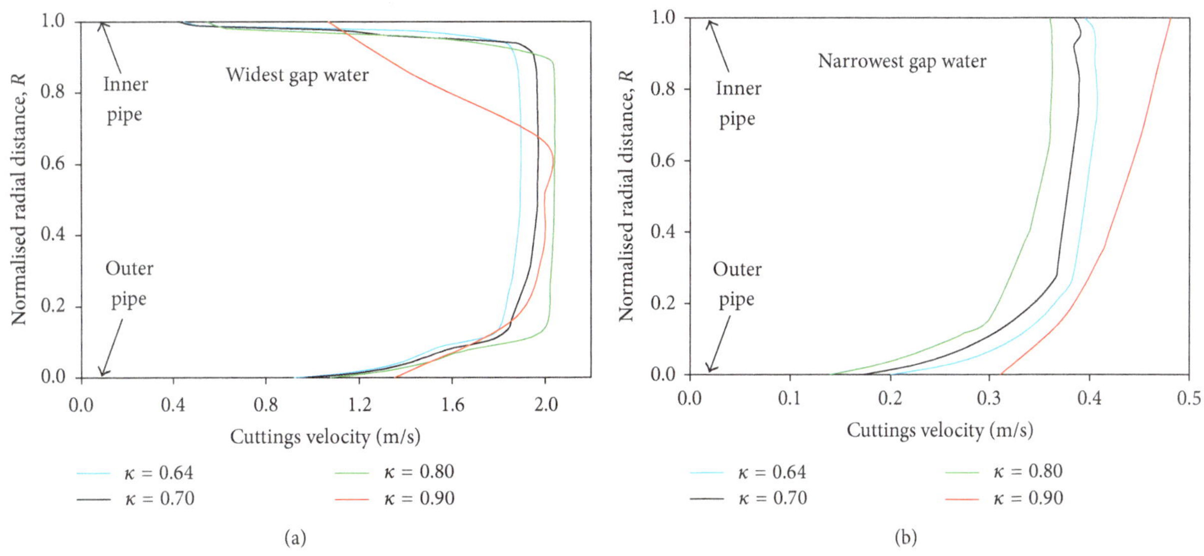

FIGURE 11: Cuttings velocity profiles with water as carrier fluid for varying diameter ratios at 1.524 m/s and 120 rpm: (a) gap above inner pipe and (b) gap below inner pipe.

0.64, 0.70, 0.80 and 0.90, respectively. On the contrary, in Figure 11(b), the cuttings velocity in the narrowest annular gap show irregular profiles as diameter ratio increases. The effect of drill pipe rotation is seen to have greater impact on the cuttings velocity especially near the vicinity of the drill pipe where there is high shear. For example, at $\kappa = 0.90$, the peak cuttings velocity recorded was 0.481 m/s, and it occurred at the vicinity of the drill pipe.

3.6. Cuttings Volume Fraction, Velocity, and Profiles with Mud as Carrier Fluid. With mud as carrier fluid and flowing at 1.524 m/s and a drill pipe rotating at 120 rpm, Figure 12 shows a very small cuttings volume fraction within the annular gap. Due to the high viscous nature of the mud, many cuttings are able to be suspended in the mud and then transported to the surface. This reduces the cuttings tendency to slip to the bottom of the wellbore to form a bed. The cuttings velocity presented in 3D streamlines (see Figure 13) shows how the cuttings travel in almost the entire annular space for all diameter ratios. This indicates better carrying capacity of the mud in transporting the cuttings to the surface. The radial measurements of the cuttings velocity profiles as shown in Figure 14 further illustrate the mud's carrying capacity in both the widest and narrowest annular gaps. The peak cuttings velocity also increases with increasing diameter ratio and is recorded in the widest gap as 1.698 m/s, 1.758 m/s, 1.838 m/s, and 1.840 m/s for $\kappa = 0.64, 0.70, 0.80$, and 0.90, respectively, as shown in Figure 14(a). In the narrowest gap, as shown in Figure 14(b), the cuttings velocity profiles show irregular behaviours and are also very similar in magnitude for all diameter ratios. The peak cuttings velocities calculated are 1.000 m/s, 1.304 m/s, 1.025 m/s, and 1.071 m/s for $\kappa = 0.64, 0.70, 0.80$, and 0.90, respectively (see Figure 14(b)).

4. Conclusions

The present study employs a CFD method to analyse the effects of fluid velocity, annular diameter ratio (ranging from 0.64 to 0.90), drill pipe rotation, and fluid type on the prediction of pressure losses and cuttings concentration for solid-fluid flow in eccentric horizontal annular geometries. The following can be inferred from this study.

(1) Using water as carrier fluid, simulation data for pressure loss and cuttings concentration are in good agreement with experimental data with mean percentage errors of 0.84% and 12%, respectively. Similarly, with mud as carrier fluid, only 2.5% mean error exists between simulation and experimental pressure data, confirming the validity of the current model setup.

(2) Increasing annular fluid velocity significantly increases pressure losses, while a decrease in cuttings concentration occurs for each constant diameter ratio. This effect is however more pronounced for $\kappa = 0.90$ when using both water and mud as carrier fluids. Annular pressure loss is dramatically increased by 97% while cuttings concentration is decreased by 37% when the flowing mud velocity increased from 1.524 m/s to 2.749 m/s for $\kappa = 0.90$.

(3) When other drilling parameters are kept constant, increasing diameter ratio increases pressure loss, whereas a decrease in cuttings concentration is observed for each constant fluid velocity. This influence is however pronounced for $\kappa = 0.90$. Over 3600% increase in pressure loss could be realised while a decrease of about 86% in cuttings concentration is observed between diameter ratios of $\kappa = 0.64$ and $\kappa = 0.90$ for water flowing at a velocity of 1.524 m/s.

(4) Increasing drill pipe rotation speed from 80 rpm to 120 rpm did not result in any significant increment in pressure losses with both water and mud. The rotation effect on cuttings concentration is quite predominant especially in annular gaps with diameter ratio below $\kappa = 0.70$ and at low fluid velocities. Contours of cuttings volume fraction show how rotation effect sweeps cuttings bed into the annular mainstream and transports them to the surface.

(5) Although mud recorded higher pressure losses compared to water, it has better carrying capacity as opposed to water especially at smaller diameter ratios. The performance of both fluids on cuttings concentration is quite similar at high diameter ratios.

Nomenclature

A_{bit}: Area of bit (m^2)
C_D: Drag coefficient (—)
C_{cf}: Cuttings feed concentration (—)
C_{cT}: Total cuttings concentration (—)
C_α: Volume fraction of phase α
$C_{\varepsilon 1}$: (k-ε) turbulence model constant (1.44)
$C_{\varepsilon 2}$: (k-ε) turbulence model constant (1.92)
C_μ: (k-ε) turbulence model constant (0.09)
d_s: Solid particle mean diameter (m)
D_1: Outer diameter of inner pipe (m)
D_2: Inner diameter of outer pipe (m)
D_h: Hydraulic diameter, $D_2 - D_1$ (m)
e: Eccentricity $(2\delta/(D_2 - D_1))$
g: Gravity vector (m/s$_2$)
$\hbar_l$: Fluid phase volume fraction (—)
$\hbar_s$: Solid phase volume fraction (—)
K: Consistency index (Pa·s^n)
k_α: Turbulence kinetic energy (m$_2$/s$_2$)
L: Annular geometry length (m)
L_h: Hydrodynamic length (m)
$\dot{m}$: Mass flow rate (kg/s)
M: Interphase momentum transfer
M_d: Drag force per unit volume (N/m$_3$)
M_L: Lift force per unit volume (N/m$_3$)
n: Flow behaviour index (—)
N_{Re}: Fluid Reynolds number (—)
N_{Re_p}: Solid particles Reynolds number (—)
N_{Re_ω}: Vorticity Reynolds number (—)

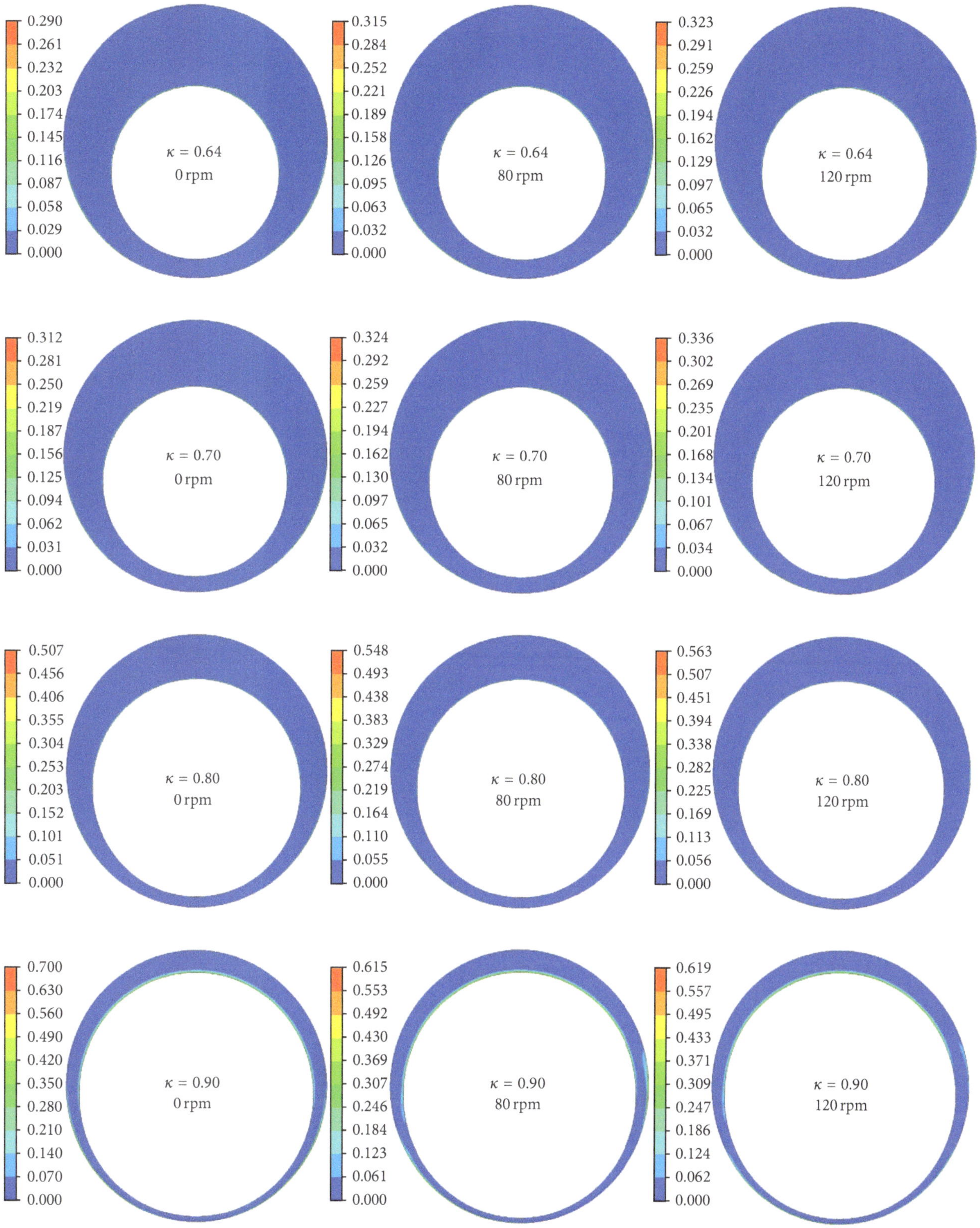

FIGURE 12: Contours of cuttings volume fraction for varying diameter ratios and inner pipe rotation with mud as carrier fluid at 1.524 m/s.

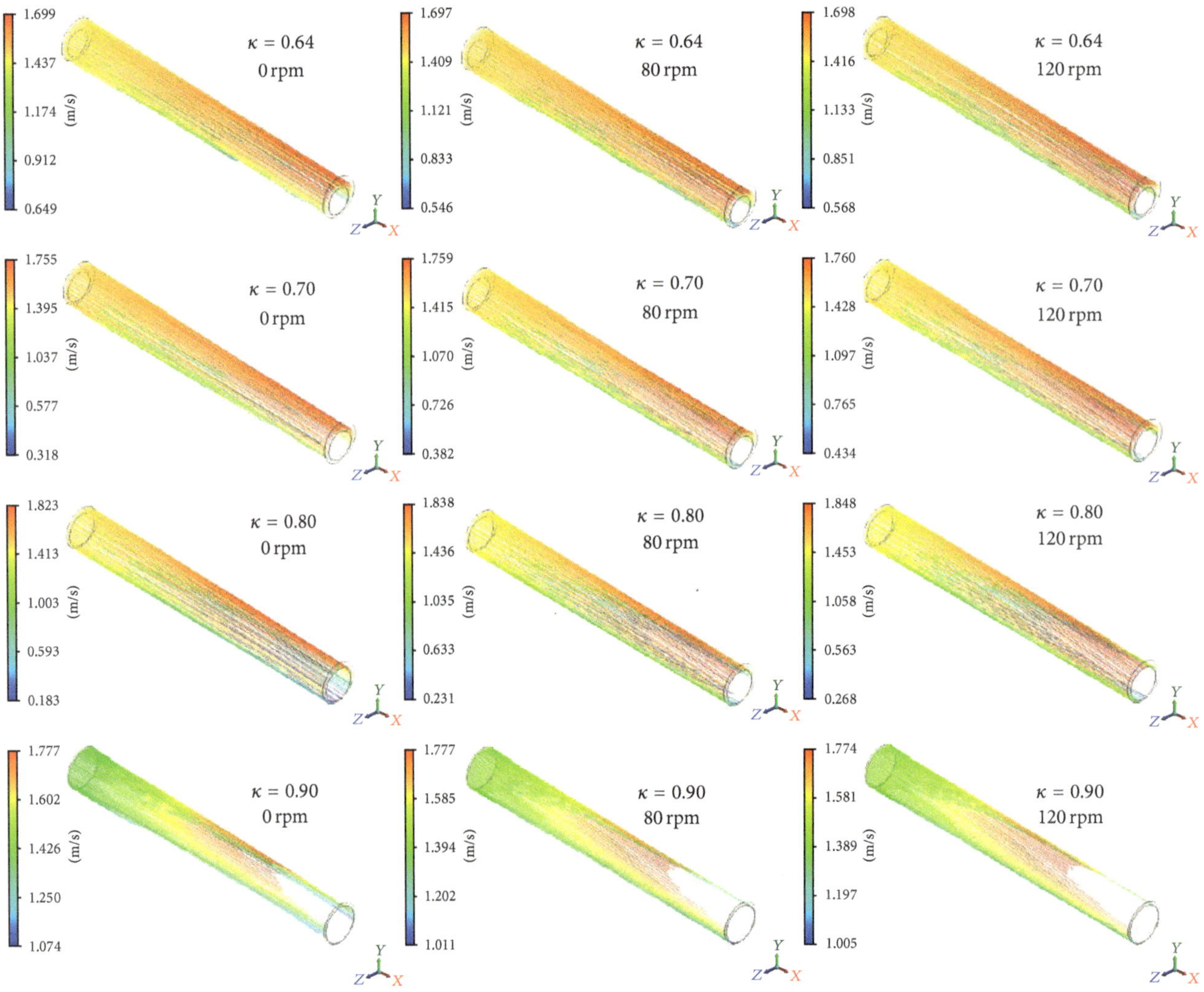

FIGURE 13: 3D streamlines of cuttings velocity for varying diameter ratios and inner pipe rotation with mud as carrier fluid at 1.524 m/s.

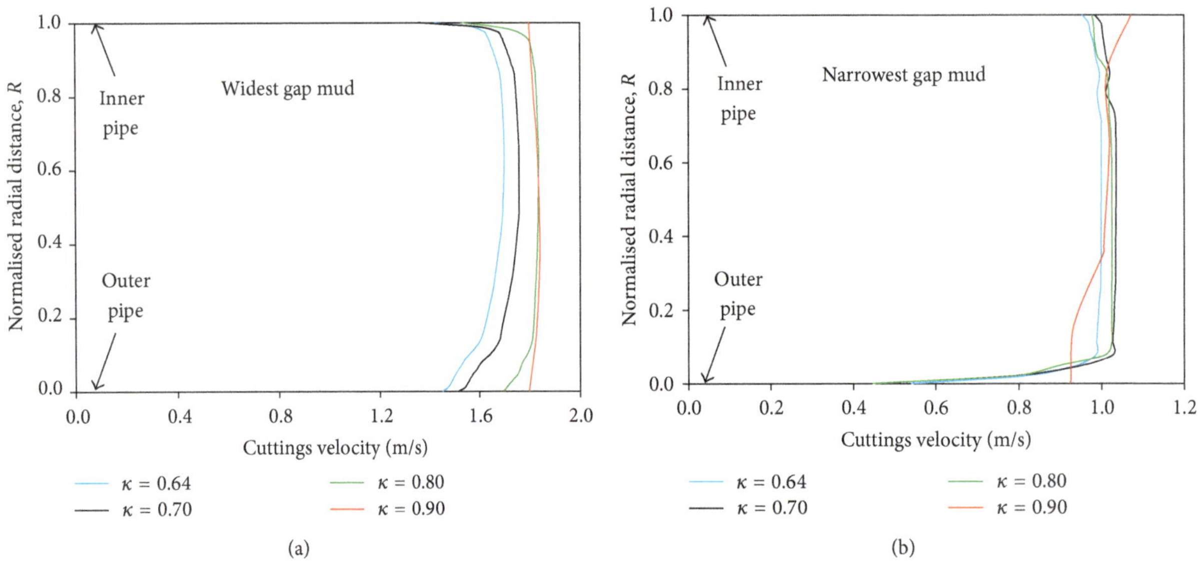

FIGURE 14: Cuttings velocity profiles with mud as carrier fluid for varying diameter ratio at 1.524 m/s and 120 rpm: (a) gap above inner pipe and (b) gap below inner pipe.

P_α: Phase pressure (Pa)
P_s: Solid particle pressure (Pa)
Q: Volumetric flow rate (m^3/s)
r: Radial distance (m)
R: Normalised radial distance ($(R_2 - r)/(R_2 - R_1)$)
R_1: Outer radius of inner pipe (m)
R_2: Inner radius of outer pipe (m)
ROP: Rate of penetration (m/s)
R_T: Transport ratio (—)
$T_{\alpha\beta}^{(\varepsilon)}$: Interphase transfer for ε
$T_{\alpha\beta}^{(k)}$: Interphase transfer for k
U_α: Phase velocity vector (m/s)
U_l: Fluid phase velocity vector (m/s)
U_s: Solid phase velocity vector (m/s).

Symbols

δ: Offset distance (m)
ε_α: Turbulence dissipation rate (m^2/s^3)
ρ_α: Phase density (kg/m^3)
ρ_l: Fluid phase density (kg/m^3)
ρ_s: Solid phase density (kg/m^3)
$\overline{\overline{\tau}}$: Viscous stress tensor (Pa)
κ: Diameter ratio (D_1/D_2)
σ_ε: (k-ε) turbulence model constant (1.3)
σ_k: (k-ε) turbulence model constant (1.0)
μ: Dynamic viscosity (Pa·s)
μ_a: Apparent viscosity (Pa·s)
μ_{eff}: Effective viscosity (Pa·s)
$\mu_{t\alpha}$: Phase turbulent viscosity (Pa·s)
υ: Specific volume (m^3/kg)
ω: Angular velocity (1/min)
$\dot{\gamma}$: Shear rate (1/s)
Ω: Rotation vector (1/min).

Unit Conversion Factors

ft × 0.3048:	$E + 00 = \text{m}$
inch × 25.4:	$E - 03 = \text{m}$
Gal (US) × 3.785:	$E + 00 = \text{liter}$
gal/min (gpm) × 6.309:	$E - 05 = \text{m}^3/\text{s}$
psi × 6.8948:	$E - 03 = \text{MPa}$
ppg × 1.198:	$E + 02 = \text{kg/m}^3$.

Conflict of Interests

The authors declare that there is no conflict of interests regarding the publication of this paper.

References

[1] P. H. Tomren, A. W. Iyoho, and J. J. Azar, "Experimental study of cuttings transport in directional wells," *SPE Drilling Engineering*, vol. 1, no. 1, pp. 43–56, 1986.

[2] T. E. Becker and J. J. Azar, *Mud-Weight and Hole-Geometry Effects on Cuttings Transport While Drilling Directionally*, Society of Petroleum Engineers, SPE-14711-MS, 1985.

[3] R. B. Adari, S. Miska, E. Kuru, P. Bern, and A. Saasen, "Selecting drilling fluid properties and flow rates for effective hole cleaning in high-angle and horizontal wells," in *Proceedings of the SPE Annual Technical Conference and Exhibition*, paper SPE-63050-MS, pp. 273–281, Dallas, Tex, USA, October 2000.

[4] T. R. Sifferman and T. E. Becker, "Hole cleaning in full-scale inclined wellbores," *SPE Drilling Engineering*, vol. 7, no. 2, pp. 115–120, 1992.

[5] R. Ahmed, M. Sagheer, N. Takach et al., "Experimental studies on the effect of mechanical cleaning devices on annular cuttings concentration and applications for optimizing ERD systems," in *Proceedings of the SPE Annual Technical Conference and Exhibition*, paper SPE-134269-MS, pp. 2016–2028, Florence, Italy, September 2010.

[6] M. E. Ozbayoglu, A. Saasen, M. Sorgun, and K. Svanes, "Critical fluid velocities for removing cuttings bed inside horizontal and deviated wells," *Petroleum Science and Technology*, vol. 28, no. 6, pp. 594–602, 2010.

[7] J. O. Ogunrinde and A. Dosunmu, "Hydraulic optimization for efficient hole cleaning in deviated and horizontal wells," in *Proceedings of the SPE Nigerian Annual Technical Conference and Exhibition*, paper SPE 162970, Abuja, Nigeria, August 2012.

[8] M. E. Ozbayoglu and M. Sorgun, "Frictional pressure loss estimation of water-based drilling fluids at horizontal and inclined drilling with pipe rotation and presence of cuttings," in *Proceedings of the SPE Oil and Gas India Conference and Exhibition*, paper SPE-127300-MS, Mumbai, India, January 2010.

[9] M. Sorgun, I. Aydin, and M. E. Ozbayoglu, "Friction factors for hydraulic calculations considering presence of cuttings and pipe rotation in horizontal/highly-inclined wellbores," *Journal of Petroleum Science and Engineering*, vol. 78, no. 2, pp. 407–414, 2011.

[10] O. M. Evren, E. Reza O, O. A. Murat, and Y. Ertan, "Estimation of "very-difficult-to-identify" data for hole cleaning, cuttings transport and pressure drop estimation in directional and horizontal drilling," in *Proceedings of the IADC/SPE Asia Pacific Drilling Technology Conference and Exhibition*, paper SPE-136304-MS, pp. 668–685, Ho Chi Minh City, Vietnam, November 2010.

[11] N. C. G. Markatos, R. Sala, and D. R. Spalding, "Flow in an annulus of non-uniform gap," *Transactions of the Institution of Chemical Engineers*, vol. 56, no. 1, pp. 28–35, 1978.

[12] S.-M. Han, Y.-K. Hwang, N.-S. Woo, and Y.-J. Kim, "Solid-liquid hydrodynamics in a slim hole drilling annulus," *Journal of Petroleum Science and Engineering*, vol. 70, no. 3-4, pp. 308–319, 2010.

[13] M. Mokhtari, M. Ermila, A. N. Tutuncu, and M. Karimi, "Computational modelling of drilling fluids dynamics in casing drilling," in *Proceedings of the SPE Eastern Regional Meeting*, paper SPE-161301-MS, Lexington, Ky, USA, October 2012.

[14] T. N. Ofei, S. Irawan, and W. Pao, "Modelling ofpressure drop in eccentric narrow horizontal annuli with the presence of cuttings and rotating drillpipe," *International Journal of Oil, Gas and Coal Technology*. In press.

[15] G. M. Faeth, "Mixing, transport and combustion in sprays," *Progress in Energy and Combustion Science*, vol. 13, no. 4, pp. 293–345, 1987.

[16] M. Eesa and M. Barigou, "Horizontal laminar flow of coarse nearly-neutrally buoyant particles in non-Newtonian conveying fluids: CFD and PEPT experiments compared," *International Journal of Multiphase Flow*, vol. 34, no. 11, pp. 997–1007, 2008.

[17] B. G. M. van Wachem and A. E. Almstedt, "Methods for multiphase computational fluid dynamics," *Chemical Engineering Journal*, vol. 96, no. 1–3, pp. 81–98, 2003.

[18] C. Y. Wen and Y. H. Yu, "Mechanics of fluidization," *Chemical Engineering Progress Symposium Series*, vol. 62, pp. 100–111, 1966.

[19] D. Gidaspow, *Multiphase Flow and Fluidization*, Academic Press, 1994.

[20] P. G. Saffman, "The lift on a small sphere in a slow shear flow," *Journal of Fluid Mechanics*, vol. 22, no. 2, pp. 385–400, 1965.

[21] P. G. Saffman, "The lift on a small sphere in a slow shear flow—corrigendum," *Journal of Fluid Mechanics*, vol. 31, no. 3, p. 624, 1968.

[22] R. Mei and J. F. Klausner, "Shear lift force on spherical bubbles," *International Journal of Heat and Fluid Flow*, vol. 15, no. 1, pp. 62–65, 1994.

[23] B. E. Launder and D. B. Spalding, "The numerical computation of turbulent flows," *Computer Methods in Applied Mechanics and Engineering*, vol. 3, no. 2, pp. 269–289, 1974.

[24] C. A. Shook and M. C. Roco, *Slurry Flow: Principles and Practice*, Butterworth-Heimemann, London, UK, 1991.

[25] R. E. Osgouei, *Determination of cuttings transport properties of gasified drilling fluids [Ph.D. thesis]*, Middle East Technical University, Ankara, Turkey, 2010.

[26] S. V. Patankar, *Numerical Heat Transfer and Fluid Flow*, Hemisphere Publishing Corp., 1980.

6

Study and Application of Polymer Film-Forming Shielding Plugging Drilling Fluid System

Yuxue Sun, Fei Teng, and Jingyuan Zhao

Northeast Petroleum University, Daqing, Heilongjiang 163318, China

Correspondence should be addressed to Fei Teng; 476122791@qq.com

Academic Editor: Alireza Bahadori

To solve the low cementing quality of acoustic logging and some distensible well diameters in high permeable layer of Daqing Lamadian oilfield, we have carried out the study and field application of polymer film-forming shielding plugging drilling fluid system. By adding this polymer film-forming shielding plugging agent to macromolecule emulsion drilling fluid, the shielding and plugging effect will be improved. Combined with the experiment and analysis, we optimize the adding amount of this agent to improve the performance of drilling fluid system. From the application effect, the cementing quality of acoustic logging has been improved obviously. This system has been applied in 20 wells and there are 15 high quality wells. The merit factor of well cementing reaches 75%.

1. Introduction

1.1. General Situation of Daqing Lamadian Oilfield. Daqing Lamadian oilfield is the central depression area of Songliao basin. It is an anticline sandstone reservoir in the extreme north of Daqing Changyuan area. This oilfield was developed in 1973. Through 30 years of development adjustment, it has experienced the adjustments of layer series, secondary infilling, and polymer flooding in main layer and so on. At present production has entered high water cut stage, and composite water cut has reached 94%. Due to many years of injection and production, the formation characteristics have certain changes. From the site statistics, we found that the permeability had a tendency to increase with the development of Lamadian oilfield. At the same time, the indoor test using formation water to flood the natural core of Lamadian area for long term showed that the clay volume decreased after long-term waterflooding. But both permeability and porosity increased. In addition, Lamadian oilfield often used produced water to recirculate, so the formation water became calcium bicarbonate water. The total mineralization was as high as 7150 mg/L. Therefore, the most prominent features of Lamadian oilfield are high porosity, high permeability, high water cut, and high salinity. Due to long-term waterflood development of Putaohua oil layer, the layer was serious water flush. Cementing of sandstone reservoir became loose, and rock strength decreased obviously. Conventional drilling fluid could not form a compact mud cake. It made the slurry fluid loss become large and seriously affected cementing quality. In order to solve the above problems, the zero permeability and low permeability plugging drilling fluid technology has been carried out in Daqing Lamadian oilfield. The field test applied in tertiary infilling adjustment well drilling has obtained good effect. For better improving cementing quality of Daqing Lamadian oilfield, our laboratory carried out the study and field application of polymer film-forming shielding plugging drilling fluid technology [1–6].

1.2. Outline of This Paper. Based on the current situation of Lamadian oilfield, we study the function mechanism of this polymer film-forming shielding plugging drilling fluid. In Section 3, we first use the new sand bed experimental method to test the plugging ability. According to the experimental results we optimize the adding amount of film-forming agent and analyze this new drilling fluid system. Finally we study the performance of this optimized drilling fluid system and extend it to the field application. In this paper, the novel point lies in the optimization of polymer film-forming shielding plugging drilling fluid system. This system, compatible with Daqing Lamadian strata, can improve the cementing quality of high permeable layer in Daqing Lamadian area obviously.

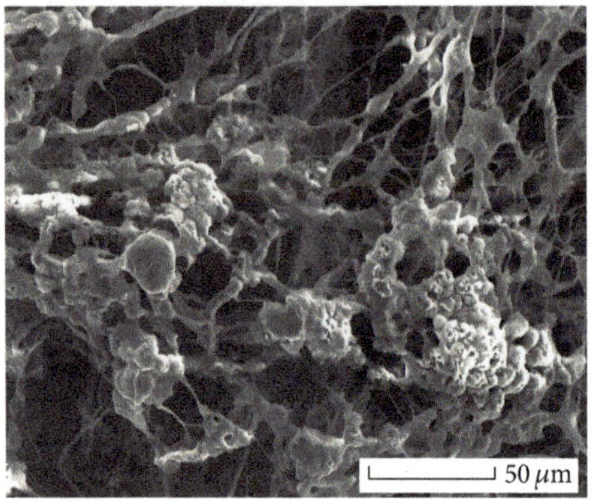

FIGURE 1: Water immersion of natural modified macromolecule compound for 1 hour.

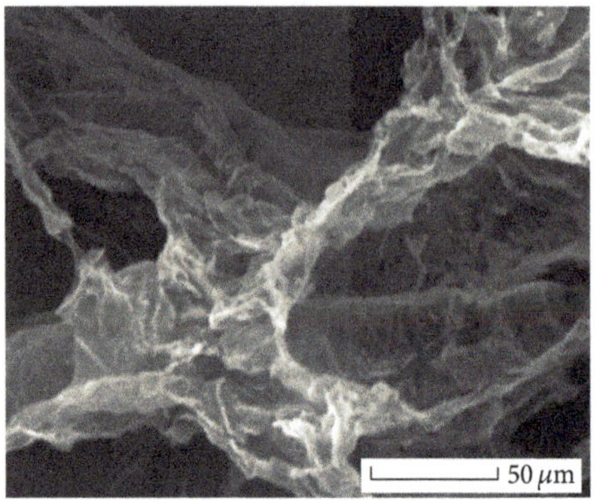

FIGURE 2: Water immersion of natural modified macromolecule compound for 24 hours.

FIGURE 3: Leakage of drilling fluid.

FIGURE 4: Plugging of drilling fluid.

2. Mechanism Study

This technology mainly uses special polymer addictive to concentrate into micelle in high permeable sandstone pore or microcrack port. We rely on the interface suction and deformability of polymer micelle or colloidal particle, and cooperate with inert solid particles or fibers. Under gradual swelling effect of the polymer particles, we form a dense ultra-low permeability sealing film on the wellbore surface [7–10].

2.1. Swelling Mechanism of Special Polymer Additive. This polymer addictive is a kind of natural modified macromolecular compound. The colloid contains lots of hydration groups. It can absorb water to form micelle. The micelle, macromolecule group, can swell by absorbing water, but it cannot dissolve completely. Because of the swelling characteristic of natural macromolecule compound, we can add hydrophobic molecules, which can prevent water molecules entering into micelle; thus water intake rate of natural macromolecule polymer will be delayed to achieve the goal of gradual water absorption and slow swelling. Due to the strong effect of cross-linked network, the polymer only swells not dissolves. Figures 1 and 2 are the water absorption swelling changes of natural macromolecule compound in 24 hours. From the two figures, we can see that both the swelling micelle density and grain diameter of micelle became large 24 hours later [11–14].

2.2. Plugging Mechanism. Using the characteristics of natural modified polymer and filling effect of inert particles and fibers will make the whole system have a good plugging effect. First, we match the inert materials reasonably by the optimization accumulation theory. Through the particle accumulation and bridge connection of fiber, we form a compacted mud cake. Then the entire sealing interface volume will increase gradually by water absorbing swelling effect of natural macromolecule. When the volume swells to a certain extent, the antipressure ability can be strengthened and particles in the sealing interface cannot divorce from plugged zone to the recesses of pore and fracture [15–17]. At the same time it can prevent drilling fluid and filtrate seeping into the formation. It makes near-zero filtration drilling come true and protects reservoir effectively.

3. Materials and Methods

3.1. Optimizing the Adding Amount of Polymer Film-Forming Shielding Plugging Agent. Using the new sand bed experimental method, we carried out the experiment to test the plugging ability and adding amount of this addictive. Figures 3 and 4 are the visual images of sand bed experiment. Experimental data is shown in Table 1.

TABLE 1: Sand bed leakage experiment.

Formula	ρ (g/cm^3)	FL (mL)	Experiment condition (0.7 MPa, 30 min, 20~40 mesh quartz sand, room temperature)
Emulsion slurry system	1.49	4	Instantaneous loss
Slurry + 1% MFDJ	1.49	3.6	Instantaneously soak to 6 cm and 10 minutes later soak to 8.5 cm. There is no increase. Forcing to 1.2 MPa, there are also no changes.
Slurry + 2% MFDJ	1.49	3	Instantaneously soak to 2.8 cm and 10 minutes later soak to 3.5 cm. There is no increase. Forcing to 1.2 MPa, there are also no changes.
Slurry + 3% MFDJ	1.50	2.8	Instantaneously soak to 2 cm and 10 minutes later soak to 2.5 cm. There is no increase. Forcing to 1.2 MPa, there are also no changes.
Slurry + 4% MFDJ	1.50	2.4	Instantaneously soak to 1.7 cm and 10 minutes later soak to 2 cm. There is no increase. Forcing to 1.2 MPa, there are also no changes.

MFDJ is a kind of polymer film-forming shielding plugging addictive.

TABLE 2: Compatible evaluation indoor experiment of various adding amounts.

Formula	ρ (g/cm^3)	AV (s)	PV (mPa·s)	τ_0/μ_p	Gel10″/Gel1′ (Pa)	FL (mL)	Mud cake (mm)
Emulsion slurry system	1.49	36	25.5	0.36	1/2.5	4.4	1.0
Slurry + 1% MFDJ	1.49	37	26.5	0.4	1/3	4.2	1.0
Slurry + 2% MFDJ	1.49	40	29	0.45	1.5/3.5	3.6	1.0
Slurry + 3% MFDJ	1.50	44	32.5	0.51	2/4	3	1.0
Slurry + 4% MFDJ	1.51	53	38	0.58	2.5/5	2.6	1.2

TABLE 3: Measured drilling fluid performance of La 5-PS1922.

Depth (m)	ρ (g/cm^3)	AV (s)	FL (mL)	Mud cake (mm)	YP (Pa)	PV (mPa·s)	τ_0/μ_p	Gel10″/Gel1′ (Pa)	R (Ω·M/18°C)
850	1.35	52	4.8	1.0	9.5	21	0.45	2/4	4.10
910	1.48	68	3.4	1.0	19	33	0.58	5/7.5	3.55
945	1.48	62	3.8	1.0	15	28	0.54	4.5/6.5	3.57
1035	1.48	60	3.6	1.0	14.5	27	0.54	4/6	3.68
1084	1.48	57	3.8	1.0	13	27	0.48	3/7	3.75
Electric log	1.48	55	3.6	1.0	10	25	0.40	3/5	3.55
Cementation	1.48	55	3.6	1.0	10	25	0.40	3/5	

3.2. Conventional Performance Experiment of Polymer Film-Forming Shielding Plugging Additive. We conduct the conventional performance comparison experiment by adding different amount of this addictive into the in situ macromolecule emulsion drilling fluid system. Data is shown in Table 2.

4. Results and Discussion

4.1. Optimizing the Adding Amount of Polymer Film-Forming Shielding Plugging Agent. From Table 1, sand bed leakage experiment data shows that MFDJ has good film-forming plugging ability. It can well prevent sandstone reservoir from seepage and improve the bearing capacity of borehole wall. Adding 1% of this addictive into the emulsion polymer drilling fluid can realize the zero filtration. The immersion depth into sand bed is 8.5 cm. When adding 2%, the depth will become 3.5 cm. Adding 3% and 4%, the depth is 2.5 cm and 2 cm, respectively. The adding amount of this addictive can be determined in the range of 2% to 4%.

4.2. Function Morphology and Microstate of Polymer Film-Forming Shielding Plugging Additive. We observe the surface of sand bed after the experiment and of mud cake after medium voltage loss experiment (shown in Figure 5). From the phenomenon of sand bed plugging experiment, we can see that modified high molecular polymer and inert particle material are embedded into the sand bed surface pore. Filter cake of drilling fluid presents dense reticular structure. Thus it can achieve the effect of blocking and strengthening borehole wall [10].

4.3. Conventional Performance of Polymer Film-Forming Shielding Plugging Additive. From Table 2, the experimental

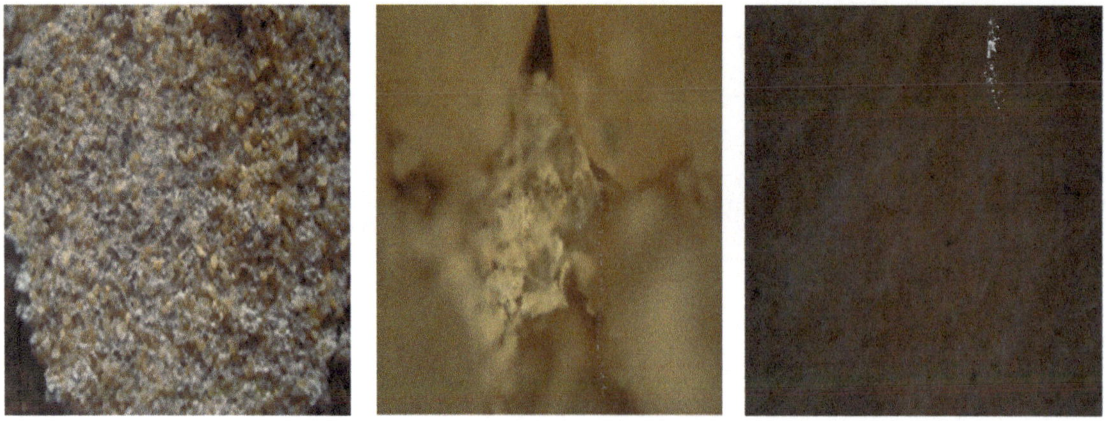

Surface state of sand bed after drilling fluid penetration | Drilling fluid microcrystalline state of embedding into sand bed pore | State of drilling fluid filter cake surface

Figure 5: Photos of sand bed surface after new sand bed experiment and mud cake surface of medium voltage loss experiment.

Lithology and quality situation of La 5-PS1821 conventional well | Lithology and quality situation of La 5-PS1922 test well

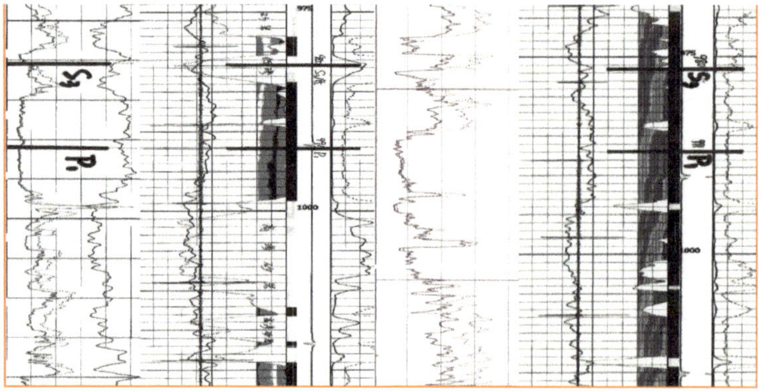

Figure 6: Contrast picture of cementing time-delay acoustic change detection.

data shows that when the slurry is normal formula of emulsion macromolecule drilling fluid system and the adding amount is less than 3%, the effect on density, viscosity, yield point and plastic viscosity ratio, and mud cake thickness of drilling fluid is smaller. When the adding amount is 4%, the parameter values of apparent viscosity, plastic viscosity, and shearing force increase obviously. The difficulty of protecting the drilling fluid rheological properties also increases correspondingly. According to the immersion depth into sand bed and rheological properties of drilling fluid, we optimize synthetically the adding amount of 2%~3%.

4.4. Field Application

4.4.1. Application Scheme. Based on indoor experimental data, we determine the amount of this additive to 3 tons per well. The applied interval is from weighting to finishing drilling. The specific addition method is to adjust drilling fluid properties after weighting and then to add this addictive 2 tons at a time. To maintain the effective adding amount of this addictive in the drilling fluid, after weighting, it needs to be supplied gradually in the drilling fluid. One ton addictive should be supplied before entering the P1 group.

4.4.2. Field Application. This drilling fluid technology has been applied into La 5-PS1922 well of 15556 drilling crew on September 6, 2010. During field test process, we adjust drilling fluid properties after weighting and we add 2 tons treated addictive at a time. During the adding process, the drilling fluid viscosity increased slightly. The drilling fluid viscosity returned to normal after adding some diluting addictive to maintain the viscosity. In the subsequent drilling process, 1 ton treatment addictive was supplied into drilling fluid successively. The electric log and cementing operation of this well was normal. The measured drilling fluid performance in the field is shown in Table 3.

4.4.3. Application Effect. Drilling fluid technology of polymer film-forming shielding plugging was applied to 20 wells of Lamadian oilfield. Site construction was smooth without any complicated engineering problems. Successful rate of electric log was 100%. The merit factor of cementing time-delay acoustic change detection was 75%, and qualification rate was 100%. Figure 6 is the contrast picture of cementing time-delay acoustic change detection of high permeable layer in a La 5-PS1922 test well.

From Figure 6, we can see that after applying this treatment addictive in the field, the cementation quality of high permeable layer S3 and P1 group in La 5-PS1922 is improved obviously compared with La 5-PS1821 well of conventional drilling fluid construction.

5. Conclusions

Indoor sand bed experiment shows that the polymer film-forming shielding plugging addictive can form a better compact and low permeable mud cake on the high permeable layer. The function mechanism in film-forming shielding plugging materials, which can effectively improve the low permeability property of drilling fluid, is the swelling effect of modified macromolecule polymer. In the optimized experiment, the best adding amount of polymer film-forming shielding plugging addictive in high permeable layer is 3%~5%. And field application effect shows that this drilling fluid technology can improve the time-delay acoustic change cementing quality of Lamadian high permeable layer obviously.

Conflict of Interests

The authors declare that there is no conflict of interests regarding the publication of this paper.

Acknowledgment

This study is supported by Scientific Research Fund of Heilongjiang Provincial Education Department (no. 12521045).

References

[1] J. Sun, J. Zheng, D. Huang, and B. Wang, "Study and application of ultra-low permeable drilling fluid: lost circulation prevention and control," *Drilling Fluid and Completion Fluid*, vol. 22, no. 4, pp. 21–24, 2005.

[2] J. Sun, X. Lin, B. Zhang et al., "Abroad technology summary of ultra-low permeable drilling fluid," *Drilling Fluid & Completion Fluid*, vol. 22, no. 1, pp. 57–59, 2005.

[3] J. Sun, Y. Su, P. Luo, and Y. Liu, "Mechanism research of ultra-low permeable drilling fluid to improve formation bearing capacity," *Drilling Fluid & Completion Fluid*, no. 5, pp. 1–3, 2005.

[4] B. Jing, W. Sui, and C. Pei, "Research and application of ultra-low permeable drilling fluid in H17 oilfield," *Drilling & Production Technology*, no. 5, pp. 36–39, 2008.

[5] Y. Peng, Y. Liu, Y. Song, P. Qing, and Y. Lu, "Application of ultra-low invasion fluid technology in Xiaolongwan block," *Drilling Fluid & Completion Fluid*, vol. 22, no. 5, pp. 31–34, 2005.

[6] F. Zhou, S. Tang, Y. Wang, and T. Huang, "The application of ultra-low filter loss drilling fluid in well Hua 3–8," *Drilling Fluid & Completion Fluid*, vol. 24, no. 6, pp. 29–31, 2007.

[7] J. Yan, *Drilling Fluid Technology*, 1st version in May, 2001, Petroleum University Press, Dongying, China, 2003.

[8] P. Reid and H. Santos, "Novel drilling, completion and workover fluids for depleted zones: avoiding losses, formation damage and stuck pipe," in *Proceedings of the SPE IADC*, October 2003.

[9] H. Santos, M. B. Villas-Boas, R. F. T. Lomba, C. H. M. Sa, S. F. Oliveira, and J. F. Costa, "API filtrate and drilling fluid invasion: is there any correlation," in *Proceedings of the Latin American and Caribbean Petroleum Engineering Conference*, Caracas, Venezuela, April 1999.

[10] H. Santos and R. Perez, "What have we been wrong in wellbore stability?" in *Proceedings of the SPE Latin American and Caribbean Petroleum Engineering Conference*, Buenos Aires, Argentina, March 2001.

[11] J. Xie, Z. Wang, B. Zhang, and X. Wang, "New drilling fluid system and plugging technology in Zhongyuan oilfield," *Advance in Fine Petrochemical Complex*, no. 1, pp. 1–5, 2010.

[12] M. Arabloo, M. P. Shahri, and M. Zamani, "Characterization of colloidal gas aphron drilling fluids produced from a new plant based surfactant," *Journal of Dispersion Science and Technology*, vol. 34, pp. 669–678, 2013.

[13] S. Luo, Y. Li, Y. Meng, and L. Zhang, "A new drilling fluid for formation damage control used in underbalanced drilling," in *Proceedings of the IADC/SPE Drilling Conference*, New Orleans, La, USA, February 2000.

[14] S. Zhang, G. Jiang, H. Guo, X. Tang, and H. Jin, "Study of new film-forming agent in drilling fluid system and application in Chenghai oilfield," *Petroleum Drilling Techniques*, vol. 41, no. 2, pp. 44–47, 2013.

[15] Y. Chun, S. Jinsheng, W. Pingquan et al., "Development of CMJ-1: a high temperature film-forming fluid loss additive and the properties," *Petroleum Drilling Techniques*, vol. 32, no. 2, pp. 30–32, 2004.

[16] B. Xiaodong and P. Xiaolin, "Evolution of membrane forming technology of water-based mud," *Natural Gas Industry*, vol. 26, no. 8, pp. 75–77, 2006.

[17] L. Haitao, Z. Xiutai, L. Qiulian et al., "Laboratory study on drilling fluid of mosaic agent-membrane agent synergistic interaction for reservoir protection," *Petroleum Drilling Techniques*, vol. 40, no. 4, pp. 65–71, 2012.

7

Optimization of Multiple Hydraulically Fractured Horizontal Wells in Unconventional Gas Reservoirs

Wei Yu and Kamy Sepehrnoori

Department of Petroleum and Geosystems Engineering, The University of Texas at Austin, Austin, TX 78712, USA

Correspondence should be addressed to Wei Yu; weiyu@utexas.edu

Academic Editor: Alireza Bahadori

Accurate placement of multiple horizontal wells drilled from the same well pad plays a critical role in the successful economical production from unconventional gas reservoirs. However, there are high cost and uncertainty due to many inestimable and uncertain parameters such as reservoir permeability, porosity, fracture spacing, fracture half-length, fracture conductivity, gas desorption, and well spacing. In this paper, we employ response surface methodology to optimize multiple horizontal well placement to maximize Net Present Value (NPV) with numerically modeling multistage hydraulic fractures in combination with economic analysis. This paper demonstrates the accuracy of numerical modeling of multistage hydraulic fractures for actual Barnett Shale production data by considering the gas desorption effect. Six uncertain parameters, such as permeability, porosity, fracture spacing, fracture half-length, fracture conductivity, and distance between two neighboring wells with a reasonable range based on Barnett Shale information, are used to fit a response surface of NPV as the objective function and to finally identify the optimum design under conditions of different gas prices based on NPV maximization. This integrated approach can contribute to obtaining the optimal drainage area around the wells by optimizing well placement and hydraulic fracturing treatment design and provide insight into hydraulic fracture interference between single well and neighboring wells.

1. Introduction

The combination of horizontal drilling and multistage hydraulic fracturing technology has made possible the current flourishing gas production from shale gas reservoirs in the United States, as well as the global fast growing investment in shale gas exploration and development. Multiple transverse hydraulic fractures are generated when all wellbores are drilled in the direction of the minimum horizontal stress. Maximizing the total stimulated reservoir volume (SRV) plays a major role in successful economic gas production. The unprecedented growth of shale reservoirs has brought a new perspective and focus to the optimization of multiwell placement in the same pad. Drilling multiple horizontal wells from a single pad has increasingly become a common approach for developing shale reservoirs due to significant cost, time, and environmental savings. The surface footprint is reduced greatly by drilling multi-well from the same pad due to minimizing the number of surface locations required while increasing the bottom hole contact of the shale resource [1]. Zipper fracturing (zipper-frac) and simultaneous fracturing (simul-frac) [2], where two adjacent horizontal wells are hydraulically fractured alternatingly and simultaneously, respectively, are two commonly used hydraulic fracturing techniques to stimulate multi-well from the same pad. Although hydraulic fractures improve gas production from shale gas wells, the cost of operation is expensive. Long laterals require greater volume of liquids and proppants which contribute to greater cost [3]. The well economics is also sensitive to well performance and natural gas price due to higher drilling and completion costs. Therefore, optimizing well parameters such as well number and well distance in conjunction with hydraulic fracture parameters, such as fracture spacing and fracture half-length based on economic analysis, are very important, especially in the current environment of low natural gas prices.

Optimization of multi-well placement is primarily valuable for overall project economic viability and minimizing

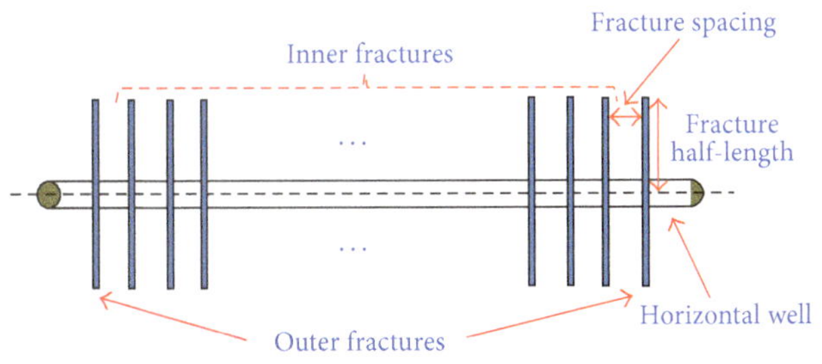

FIGURE 1: A sketch of multiple hydraulic fractured horizontal shale gas well.

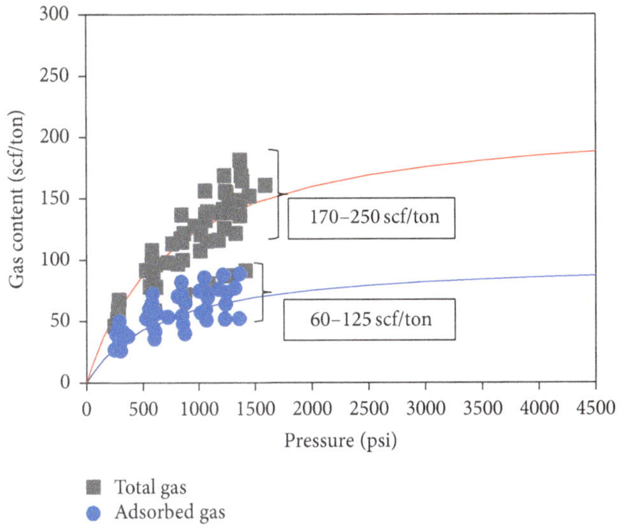

FIGURE 2: Adsorption isotherms for Barnett Shale core samples [4].

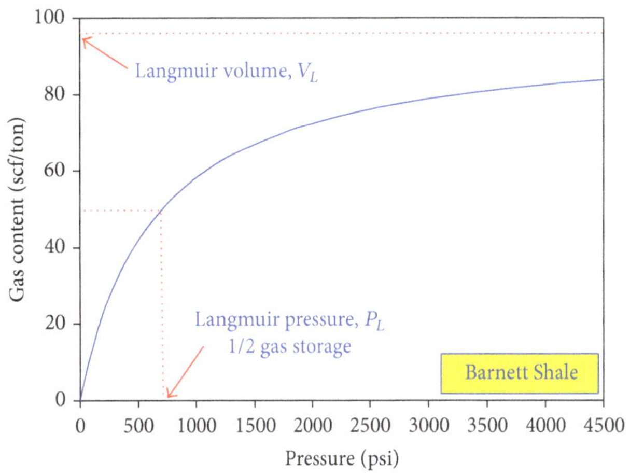

FIGURE 3: Langmuir isotherm curve for Barnett Shale (V_L = 96 scf/ton, P_L = 650 psi, and ρ_B = 2.58 g/cm^3).

the risks of well collision in shale gas reservoirs. Closer well placement will result in stimulated reservoir volume intersects, leading to well competition and penalizing overall production [5]. However, only recently there have been limited studies of optimizing fracturing design together with multi-well placement simultaneously. Esmaili et al. [6] defined three types of horizontal well based on the number of neighboring wells for sharing drainage area when drilling multiple wells from a pad. Rafiee et al. [7] proposed a new design for two horizontal wells by the modification of the traditional zipper-frac and demonstrated this new design maximized reservoir contact and improved well performance when compared to the original zipper-frac design from both rock mechanics and fluid production aspects. However, such a design assumed a larger fracture spacing of 500 ft and did not optimize well placement and fracture spacing simultaneously to obtain the optimal design for economic gas production. Díaz de Souza et al. [8] did sensitivity studies of three different wells placement with 2, 3, and 4 horizontal wells within the same stimulation volume in the Haynesville Shale to obtain the optimal well spacing. They stated that four horizontal wells with a well distance of 660 ft was a near-optimal solution for this reservoir. However, critical parameters for developing a play economically, such as fracture spacing and fracture half-length, have not been considered for optimization based on optimal well spacing. Harpel et al. [9] reported that well spacing becomes tighter in parts of the Fayetteville Shale from 600 ft to 400 ft or 300 ft, leading to fluid and proppant volume reductions, while further optimization of stimulation design, especially fracture half-length, is very much required for future development. Ramakrishnan et al. [10] suggested that it is particularly challenging to optimize the stimulation treatment in a multi-well design drilled with few hundreds feet of spacing between one another to maximize coverage around each horizontal well, because the optimization process involves perforation and stage placements, sequential stimulation of these wells, fluid and proppant schedules, treatment rates, and application of diversion technology in an effort to achieve effective stimulation along these wells and between the wells. Hence, a detailed study and a comprehensive approach for optimization of fracturing design and multi-well placement are still significantly necessary.

In this paper, we employed response surface methodology to build the response surface in terms of NPV with six parameters such as reservoir porosity, permeability, fracture half-length, fracture conductivity, fracture spacing, and well distance from Barnett Shale, to obtain the best economic scenario for a given range of these influential parameters. The effect of gas desorption is integrated in the numerical modeling of multistage hydraulic fractures. The impact of different gas prices is also taken into account for the optimization process. The goal of this work is to provide insights into the effective exploitation of shale gas reservoirs via optimization of the fracturing design and multiple wells placement simultaneously.

2. Shale Gas Reservoir Modeling

Given the complex nature of hydraulic fracture growth and the very low permeability of the matrix rock in shale gas reservoirs, coupled with the predominance of horizontal completions, reservoir simulation is the preferred method to predict and evaluate well performance [11–13]. Local grid

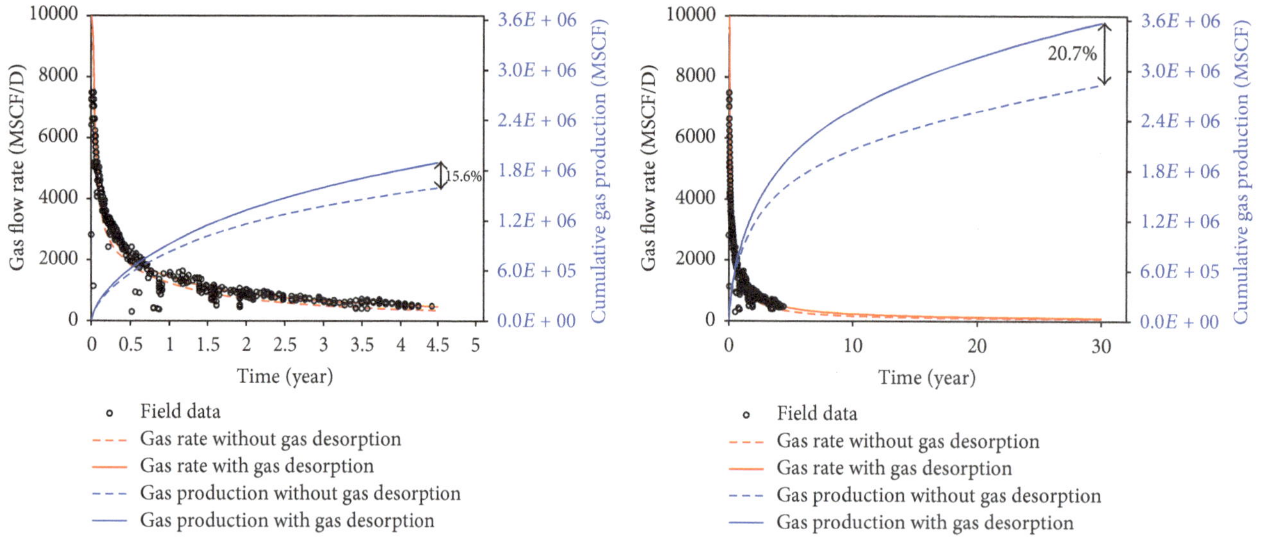

(a) Gas production for a 4.5-year period (b) Gas production for a 30-year period

FIGURE 4: History matching of Barnett Shale with and without gas desorption effect.

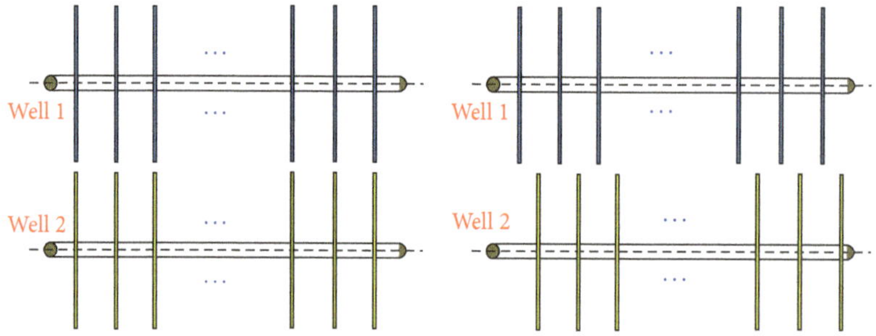

(a) Scenario 1: aligning fracturing (b) Scenario 2: alternating fracturing

FIGURE 5: Two scenarios of multiple horizontal well placement.

refinement with logarithmic cell spacing is used in the simulation to accurately model flow from the shale to the fracture, that is, properly incorporate the transient flow behavior from the matrix to the fracture. In a block, the hydraulic fracture is explicitly modeled; moreover, the matrix is described as some subcells whose size increases logarithmically, while moving away from the hydraulic fracture to properly simulate the large pressure drop between the matrix and the fracture. In addition, a dual permeability grid is used to allow simultaneous matrix-to-matrix and fracture-to-fracture flows. This method can accurately and efficiently model transient gas production from hydraulic fractures of the horizontal wells in shale gas reservoirs [14, 15]. The reservoir is assumed to be homogeneous and the fractures evenly spaced, with stress-independent porosity and permeability. It is assumed that there is no water flow in the reservoir modeling of shale gas. In our simulation, gas is only flowing into the wellbore through the hydraulic fractures, that is, no matrix-wellbore communication. The turbulent gas flow due to high gas flow rate in hydraulic fractures is modeled as non-Darcy flow. The non-Darcy Beta factor, used in the Forchheimer number, is determined using a correlation proposed by Evans and Civan [16] as follows:

$$\beta_{(f)} = \frac{1.485E9}{K^{1.021}}, \tag{1}$$

where the unit of K is md and the unit of β is ft^{-1}. The $\beta_{(f)}$ correlation was obtained using over 180 data points including those for propped fractures and was found to match the data very well with the correlation coefficient of 0.974 [14]. This equation is implemented into the numerical model and used for accounting for non-Darcy flow in hydraulic fractures. Figure 1 is a diagram of typical shale gas completion design with a multistage hydraulic fracture treatment, which illustrates several important geometric fracture parameters, such as outer fractures, inner fractures, fracture spacing, and fracture half-length.

3. Economic Model

NPV is one of the most common methods used to evaluate the economic viability of investing a project. It is referred to

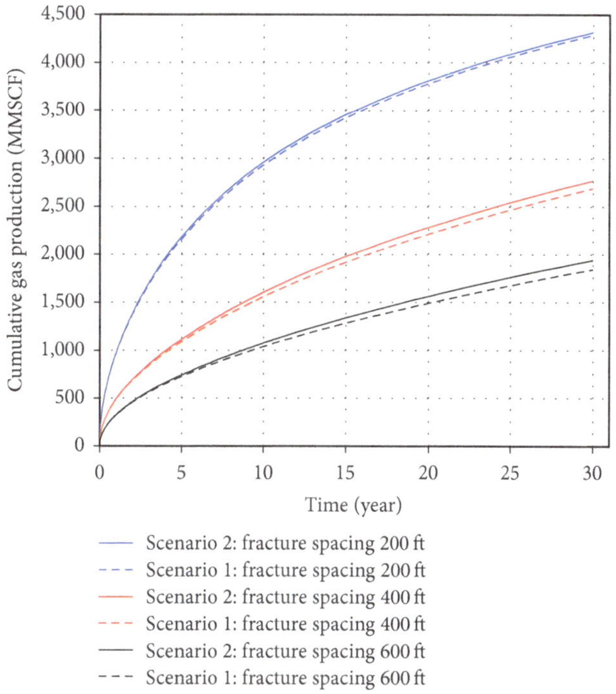

FIGURE 6: Comparison of cumulative gas production.

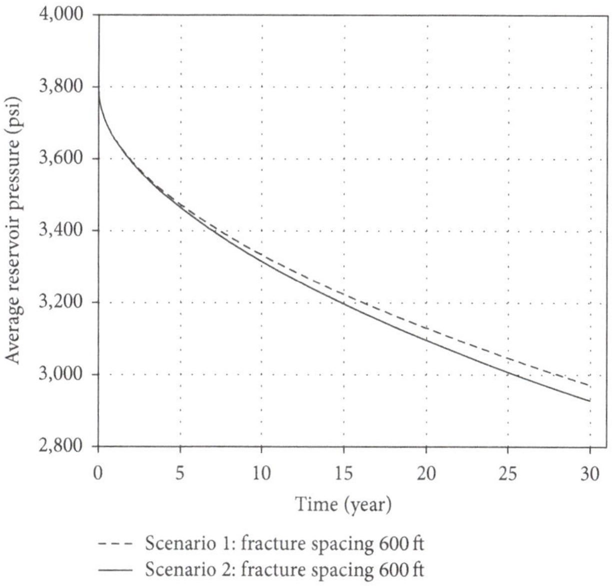

FIGURE 7: Comparison of average reservoir pressure.

the sum of all cash flows discounted to a specific point in time at the investor's minimum discount rate. The correlation between the present value P and the future value F is

$$P = \frac{F}{(1+i)^n}, \tag{2}$$

where i is the currency escalation rate or interest rate; n is the number of periods.

The NPV is calculated using the following expression:

$$\text{NPV} = \sum_{j=1}^{n} \frac{(V_F)_j}{(1+i)^j} - \sum_{j=1}^{n} \frac{(V_o)_j}{(1+i)^j} - \left(\text{FC} + \sum_{k=1}^{N} (C_{\text{well}} + C_{\text{fracture}}) \right), \tag{3}$$

where V_F is future value of production revenue for a fracture reservoir; V_o is future value of production revenue for an unfractured reservoir; FC is the total fixed cost; C_{well} is the cost of one horizontal well; C_{fracture} is the cost of hydraulic fracture in a horizontal well; N is the number of horizontal wells. The costs of well and fracture used in the economic analysis are based on the work of Schweitzer and Bilgesu [17], as shown in Table 1.

4. Langmuir Isotherm

Gas shales are organic-rich formations. Gas storage in the shale is mainly divided into free gas in natural fractures and matrix pore structure and adsorbed gas in organic materials. Langmuir isotherm is widely used to describe the gas adsorption phenomenon. The amount of gas stored in shale is often described by Langmuir equation:

$$G_s = \frac{V_L P}{P + P_L}, \tag{4}$$

where G_s is the gas content in scf/ton, V_L is the Langmuir volume in scf/ton, P_L is the Langmuir pressure in psi, and P is pressure in psi. The bulk density of shale (ρ_B) is needed to convert the typical gas content in scf/ft^3 to scf/ton. Langmuir pressure and Langmuir volume are two key parameters. Langmuir volume is referred to as the gas volume at the infinite pressure representing the maximum storage capacity for gas; Langmuir pressure is referred to as the pressure corresponding to one-half Langmuir volume. As the reservoir pressure is decreased, gas is desorbed from the surface of the matrix. Figure 2 shows a graph of gas content with pressure for the adsorbed gas and total gas used for Barnett Shale [4]. Both free gas and adsorbed gas add together to generate the total gas content. In Barnett Shale, the adsorbed gas is approximately 46% of the total gas. Contrary to conventional gas reservoirs, the amount of gas desorption in the matrix is commonly described by the Langmuir isotherm in a range of reservoir pressures. The Langmuir isotherm of the Barnett Shale used in this study is illustrated in Figure 3. It is clearly shown that higher Langmuir pressure releases more adsorbed gas and results in higher gas production. Generally, in early stage of production, when reservoir pressure is high, the gas desorption contribution to the gas production is insignificant; however, for long-term production, it is necessary to account for gas desorption, based on a laboratory measured isotherm due to the more substantial pressure depletion, resulting in more gas desorption. CMG [18] was used to model the effect of gas desorption from a shale gas reservoir in a black oil model with a technique developed by Seidle and Arri [19].

TABLE 1: Economic data for NPV calculation.

Horizontal well length (ft)	Cost ($)	Fracture half-length per stage (ft)	Cost ($)	Parameter	Value
1,000	2,000,000	250	100,000	Gas Price	$3, 4, 5/MSCF
2,000	2,100,000	500	125,000	Interest rate	10.0%
3,000	2,200,000	750	150,000	Royalty tax	12.5%
4,000	2,300,000	1,000	175,000		

TABLE 2: Parameters used in history matching.

Parameter	Value(s)	Unit
The model dimensions	3,000 (length) × 1,500 (width) × 300 (height)	ft
Initial reservoir pressure	2950	psi
Bottom hole pressure (BHP)	500	psi
Production time	30	year
Reservoir temperature	150	°F
Gas viscosity	0.0201	cp
Initial gas saturation	0.70	fraction
Total compressibility	3×10^{-6}	Psi^{-1}
Matrix permeability	0.00015	md
Matrix porosity	0.06	fraction
Fracture conductivity	1	md-ft
Fracture half-length	155	ft
Fracture spacing	100	ft
Fracture height	300	ft
Horizontal well length	2968	ft
Number of fractures	28	number

TABLE 3: Parameters used in multiwell modeling.

Parameter	Value(s)	Unit
The model dimensions	5,000 (length) × 1,600 (width) × 200 (height)	ft
Initial reservoir pressure	3800	psi
BHP	500	psi
Production time	30	year
Reservoir temperature	180	°F
Gas viscosity	0.0201	cp
Initial gas saturation	0.70	fraction
Total compressibility	3×10^{-6}	Psi^{-1}
Matrix permeability	0.0001	md
Matrix porosity	0.06	fraction
Fracture conductivity	50	md-ft
Fracture half-length	300	ft
Fracture spacing	200, 400, 600	ft
Fracture height	200	ft
Horizontal well length per well	3600	ft
Number of fractures per well	18, 9, 6	number
Number of wells	2	number
Well distance	620	ft

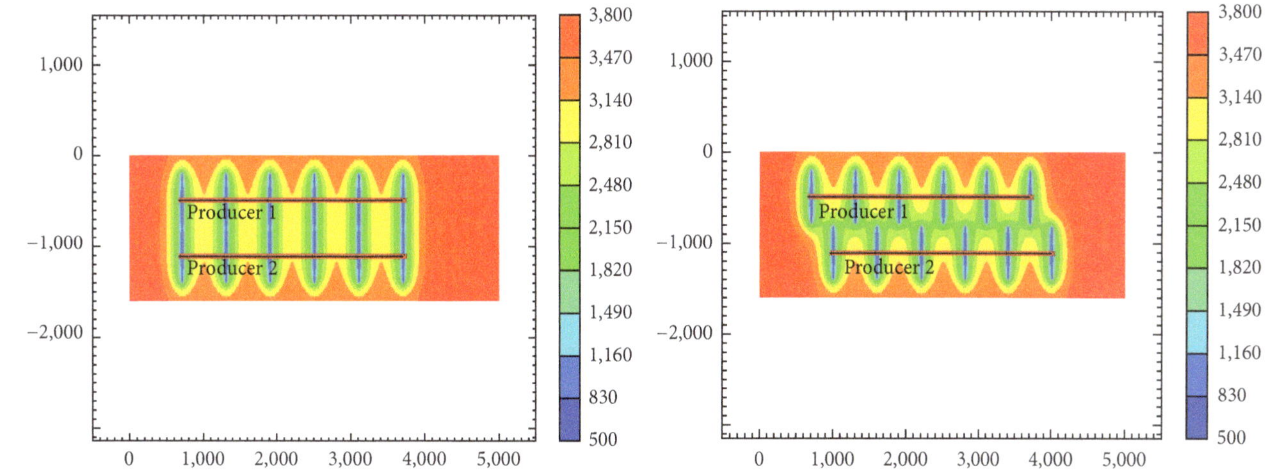

(a) Scenario 1: aligning fracturing

(b) Scenario 2: alternating fracturing

FIGURE 8: Pressure distribution at 30 years of gas production.

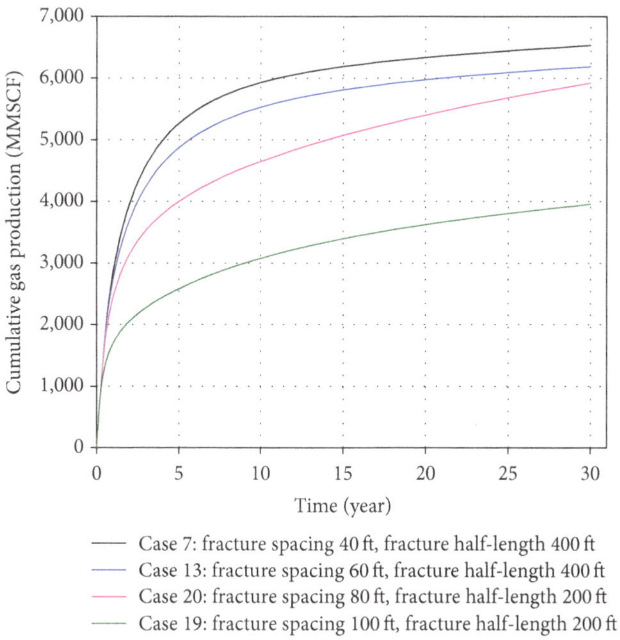

FIGURE 9: Cumulative gas production of 30 run cases for a 30-year period.

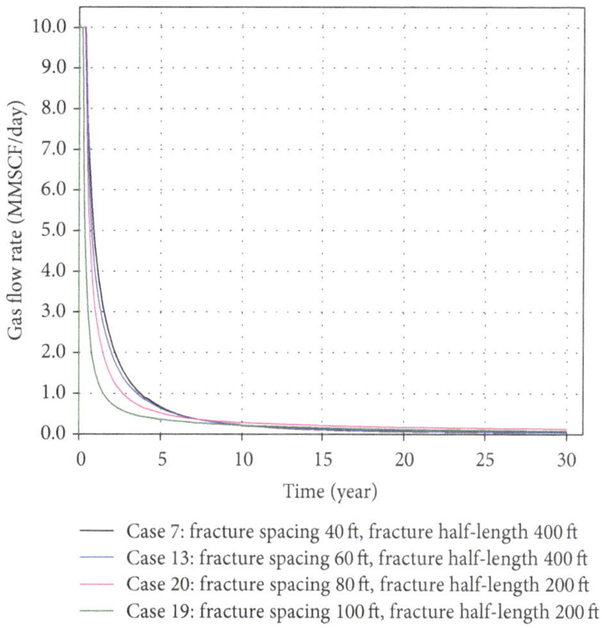

FIGURE 10: Gas flow rate of 30 run cases for a 30-year period.

A Langmuir isotherm is replicated by a black oil model's solution gas ratio to include the effect of gas desorption in shale.

5. History Matching for Barnett Shale

Published average reservoir data for a Barnett Shale well were used for history matching [20]. In this case, the well was stimulated by a multistage fracturing with a single, perforated interval for each stage. In this simulation study, we set up a reservoir with a volume of 3000 ft × 1500 ft × 300 ft. The fracture spacing and half-length are set at 100 ft and 150 ft, respectively, and the number of fractures is 28. Detailed reservoir information about this section of the Barnett Shale is listed in Table 2. The reservoir is assumed to be homogeneous and the fractures evenly spaced, with stress-independent porosity and permeability. Only gas is flowing in the reservoir, which is assumed to behave as non-Darcy flow. The history matching of field data is presented in Figure 4(a). It shows a more reasonable match between the numerical simulation results and the actual field gas flow data, considering the effect of gas desorption, contributing to 15.6% of total gas production at around 4.5 years of gas production. In addition, Figure 4(b) shows the forecasting of gas production for a 30-year period with and without considering gas desorption.

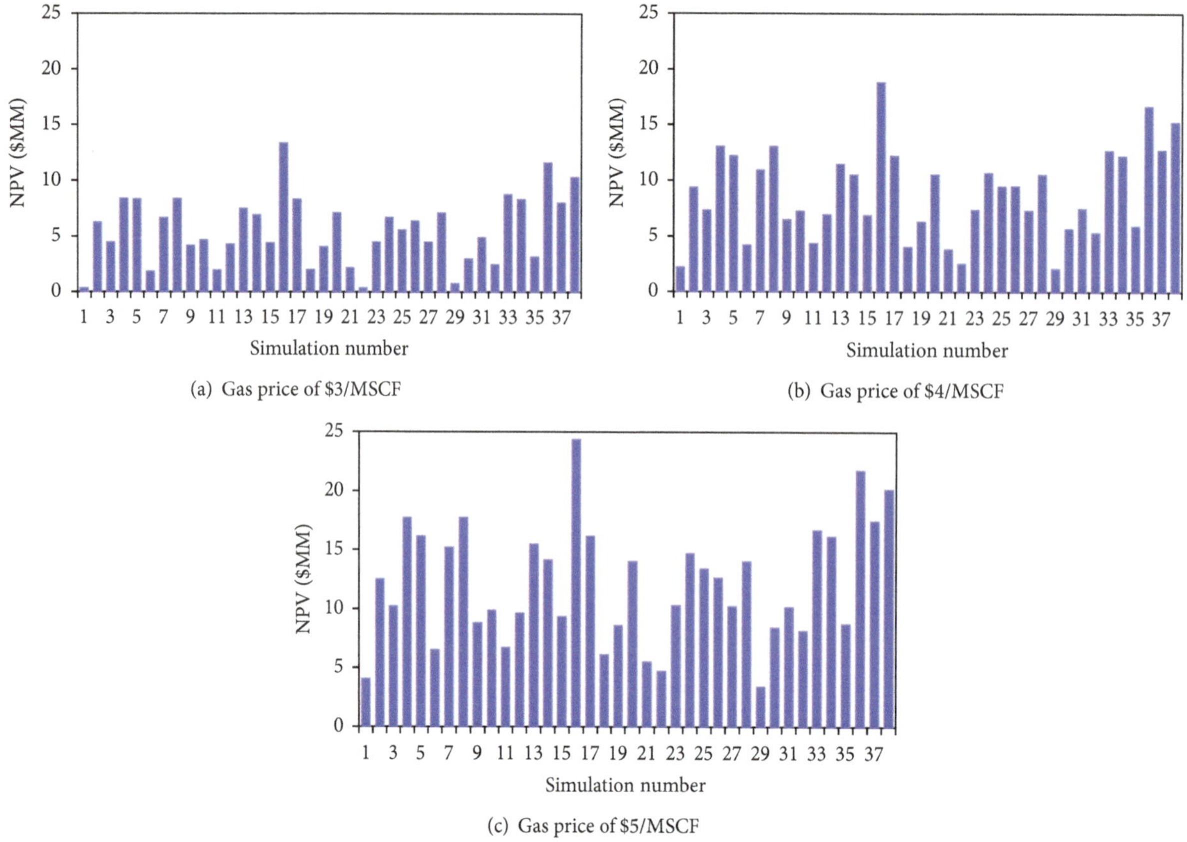

FIGURE 11: NPVs of 38 simulation cases at 30 years of gas production with different gas prices.

TABLE 4: Parameters used in multiwell optimization.

Parameter	Value(s)	Unit
The model dimensions	5,000 (length) × 2,000 (width) × 200 (height)	ft
Initial reservoir pressure	3800	psi
BHP	500	psi
Production time	30	year
Reservoir temperature	180	°F
Gas viscosity	0.0201	cp
Initial gas saturation	0.70	fraction
Total compressibility	3×10^{-6}	Psi^{-1}
Fracture height	200	ft
Horizontal well length per well	3500	ft
Number of wells	2	number

TABLE 5: Uncertainty parameters in this study.

Parameters	Coded symbol	Minimum (−1)	Maximum (+1)	Unit
Porosity	A	0.04	0.08	fraction
Permeability	B	0.00005	0.0005	md
Fracture half-length	C	200	400	ft
Fracture conductivity	D	1	50	md-ft
Fracture spacing	E	40	100	ft
Well distance	F	500	1000	ft

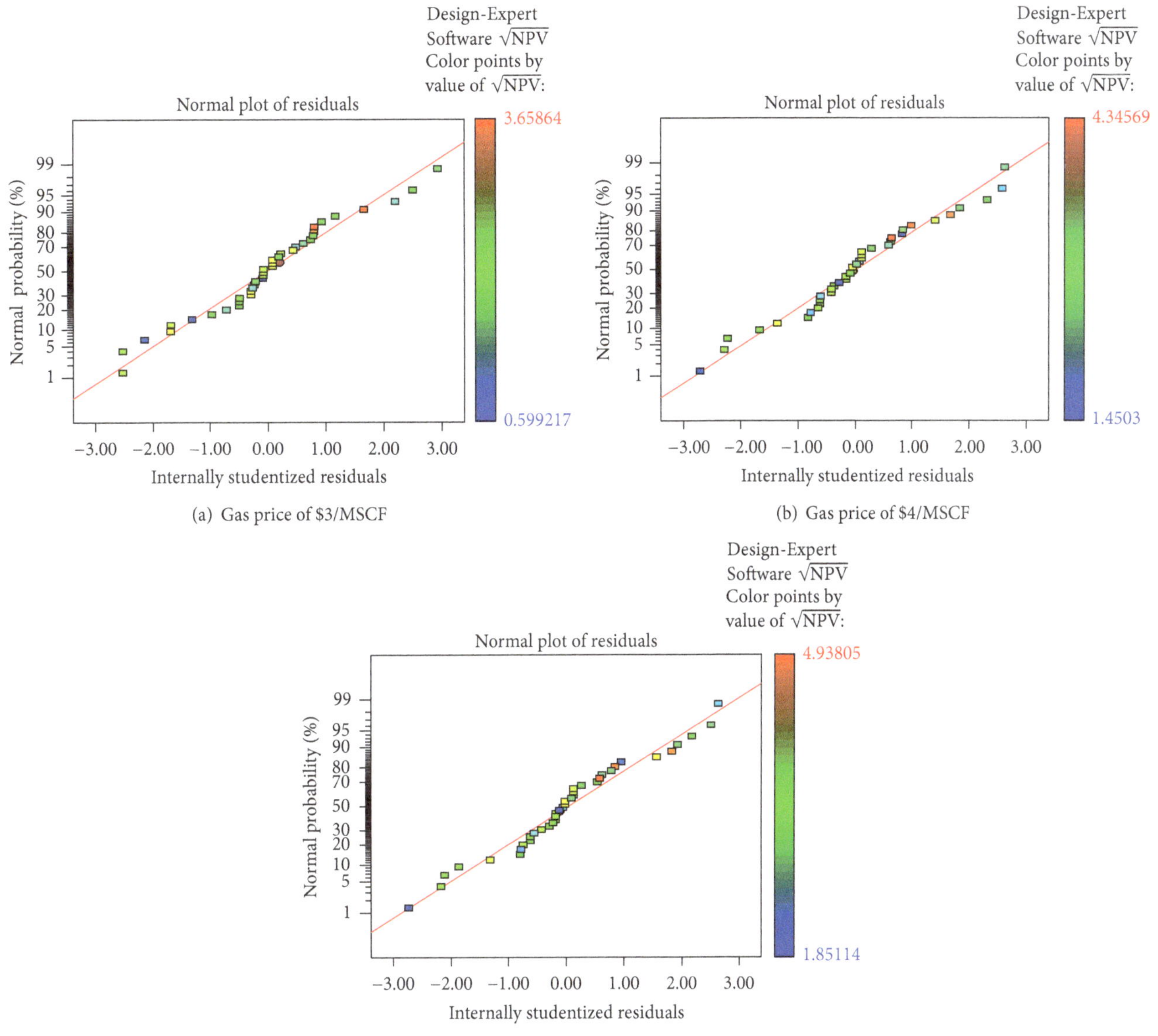

(a) Gas price of $3/MSCF

(b) Gas price of $4/MSCF

(c) Gas price of $5/MSCF

FIGURE 12: Normal plot of residuals.

As shown, with the gas production, the gas desorption contributes more due to substantial pressure depletion and larger gas drainage area and finally contributes to 20.7% of the total gas production at 30 years of gas production. Thus, the impact of gas desorption cannot be ignored when performing history matching and assessing production forecast of gas production in Barnett Shale formation. Hence, this study takes into account gas desorption effect for the subsequent optimization of multiwell placement in Barnett Shale.

6. Multiwell Modeling

Two scenarios describing multiple horizontal well placement were studied, as illustrated in Figure 5. Scenario 1 is referred to as aligning fracturing, where hydraulic fracturing is between two wells in an aligned pattern, and Scenario 2 is referred to as alternating fracturing, where hydraulic fracturing is between two wells in a staggered pattern. We investigate the effect of fracture spacing towards comparison of gas production between these two scenarios. We set up a shale gas reservoir model with a volume of 5000 ft × 1600 ft × 200 ft. Detailed reservoir information used for Barnett Shale is listed in Table 3. Comparison of cumulative gas production is shown in Figure 6. It shows that there is almost no difference of gas production for these two scenarios when fracture spacing is below 400 ft; however, Scenario 2 yields higher gas production than Scenario 1 when the fracture spacing is 400 ft or above. Figure 7 presents the comparison of average reservoir pressure with fracture spacing of 600 ft. It can be seen that for Scenario 2, it has a larger average reservoir pressure drawdown, leading to higher cumulative gas production. Pressure distribution at

Table 6: D-Optimal design table.

Run	A—Porosity	B—Permeability (md)	C—Fracture half-length (ft)	D—Fracture conductivity (md-ft)	E—Fracture spacing (ft)	F—Well distance (ft)
1	0.05	$5.00E-05$	200	1	40	900
2	0.04	$5.00E-04$	400	41	100	900
3	0.06	$2.52E-04$	300	1	60	500
4	0.08	$3.38E-04$	300	27	40	700
5	0.06	$5.00E-04$	300	25	80	800
6	0.05	$2.37E-04$	200	50	40	600
7	0.08	$5.00E-05$	400	1	40	700
8	0.08	$3.38E-04$	300	27	40	700
9	0.05	$5.00E-04$	200	3	100	500
10	0.05	$7.25E-05$	400	12	100	700
11	0.04	$4.71E-04$	200	28	40	1000
12	0.04	$4.55E-04$	400	41	80	500
13	0.08	$5.00E-05$	400	50	60	600
14	0.08	$5.00E-04$	200	50	60	500
15	0.04	$3.76E-04$	300	1	100	1000
16	0.08	$3.20E-04$	400	50	100	1000
17	0.06	$5.00E-04$	300	25	80	800
18	0.04	$5.00E-05$	300	50	60	900
19	0.04	$5.00E-04$	200	50	100	900
20	0.08	$2.71E-04$	200	15	80	800
21	0.06	$5.00E-05$	200	33	100	1000
22	0.04	$5.00E-05$	400	26	40	500
23	0.04	$5.00E-04$	300	9	60	1000
24	0.08	$5.00E-04$	200	1	40	1000
25	0.06	$5.00E-05$	400	43	40	900
26	0.07	$1.92E-04$	300	45	100	500
27	0.06	$2.52E-04$	300	1	60	500
28	0.08	$2.71E-04$	200	15	80	800
29	0.04	$5.00E-05$	200	17	80	500
30	0.08	$5.00E-05$	200	14	40	500
31	0.08	$5.00E-05$	300	1	100	1000
32	0.04	$5.00E-04$	400	1	40	700
33	0.08	$5.00E-04$	400	1	100	500
34	0.06	$5.00E-04$	300	25	80	800
35	0.08	$5.00E-05$	200	50	40	1000
36	0.08	$5.00E-04$	400	50	80	600
37	0.06	$5.00E-04$	400	50	40	1000
38	0.07	$1.98E-04$	400	14	60	1000

Table 7: Statistical approach to select the RSM model with gas price of \$3/MSCF.

Source	Std. Dev.	R-Squared	Adjusted R-Squared	Predicted R-Squared	Press	
Linear	0.27	0.8812	0.8582	0.8094	3.72	—
2FI	0.26	0.9433	0.8688	0.2155	15.31	—
Quadratic	0.064	0.9979	0.9921	0.9032	1.89	Suggested
Cubic	0	1	1	—	—	Aliased

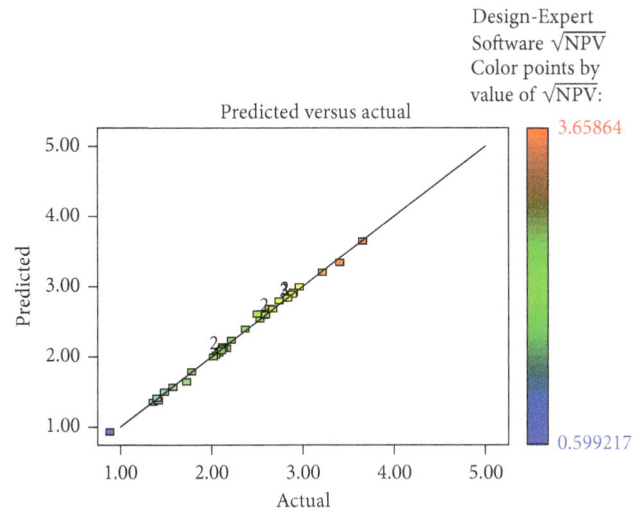

(a) Gas price of $3/MSCF

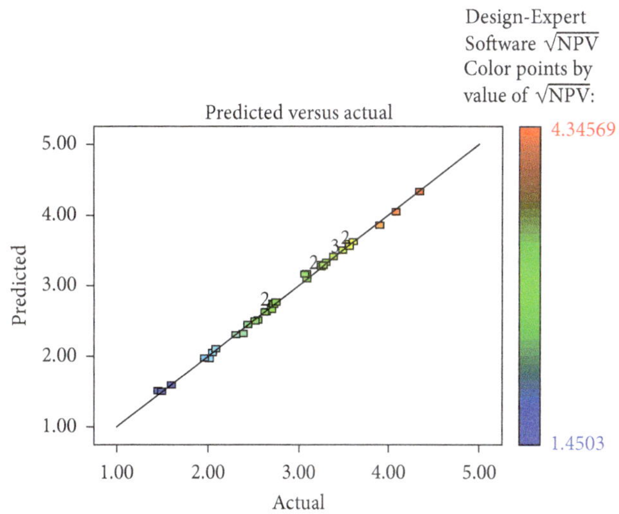

(b) Gas price of $4/MSCF

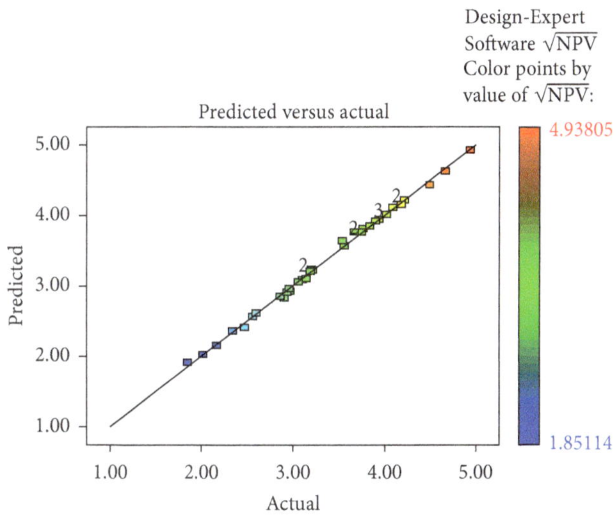

(c) Gas price of $5/MSCF

FIGURE 13: Predicted NPV versus the actual NPV plot.

TABLE 8: Statistical approach to select the RSM model with gas price of $4/MSCF.

Source	Std. Dev.	*R*-Squared	Adjusted *R*-Squared	Predicted *R*-Squared	Press	
Linear	0.25	0.8950	0.8747	0.8314	3.16	—
2FI	0.24	0.9497	0.8837	0.2936	13.26	—
Quadratic	0.065	0.9978	0.9917	0.8851	2.16	Suggested
Cubic	0	1	1	—	—	Aliased

TABLE 9: Statistical approach to select the RSM model with gas price of $5/MSCF.

Source	Std. Dev.	*R*-Squared	Adjusted *R*-Squared	Predicted *R*-Squared	Press	
Linear	0.26	0.8985	0.8788	0.8368	3.37	—
2FI	0.25	0.9525	0.8901	0.3213	14.01	—
Quadratic	0.071	0.9975	0.9908	0.8712	2.66	Suggested
Cubic	0	1	1	—	—	Aliased

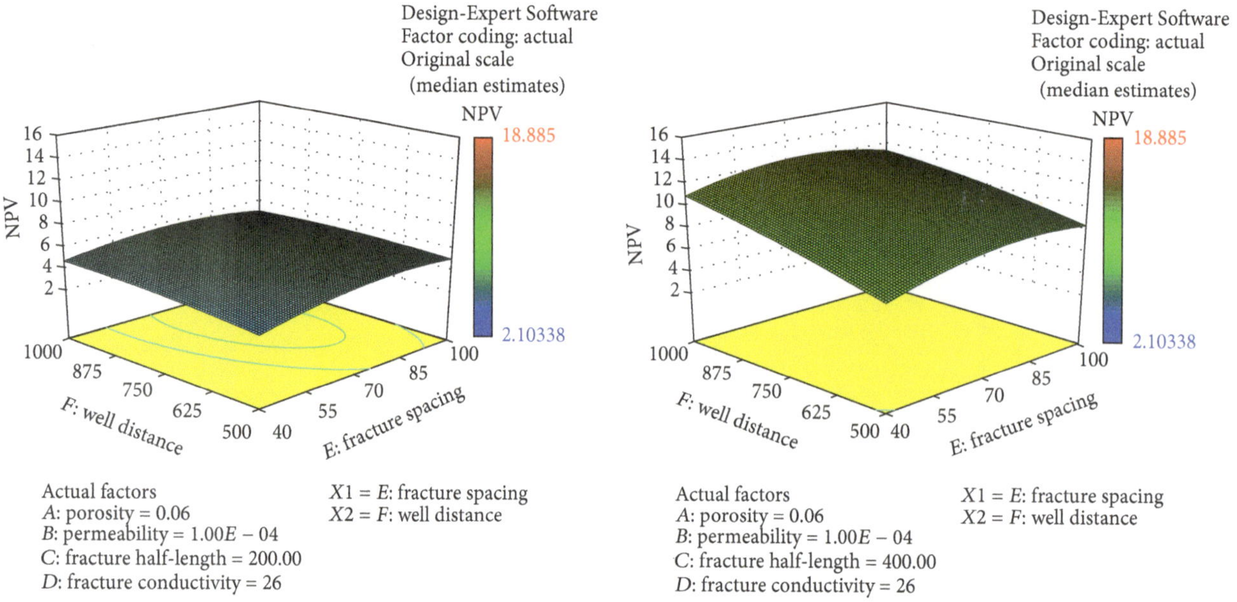

(a) Fracture half-length of 200 ft (b) Fracture half-length of 400 ft

FIGURE 14: 3D surface of NPV at varied values of well distance and fracture spacing with gas price of $4/MSCF.

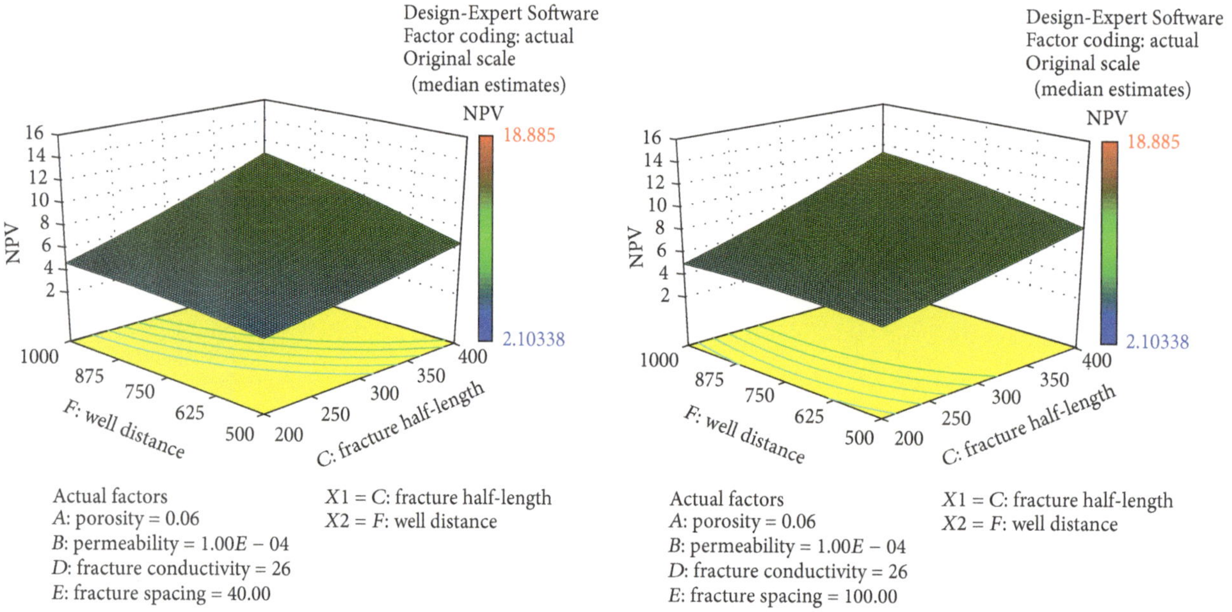

(a) Fracture spacing of 40 ft (b) Fracture spacing of 100 ft

FIGURE 15: 3D surface of NPV at varied values of well distance and fracture half-length with gas price of $4/MSCF.

TABLE 10: Optimal combinations and optimization validation.

Porosity	Permeability (md)	Fracture half-length (ft)	Fracture conductivity (md-ft)	Fracture spacing (ft)	Well distance (ft)	Gas price ($/MSCF)	Optimal NPV ($MM)	Validated NPV ($MM)	Relative error
0.06	0.0001	400	26	80	1000	3	8.70	7.91	0.091
0.06	0.0001	400	26	70	1000	4	12.40	11.81	0.047
0.06	0.0001	400	26	60	1000	5	16.34	15.75	0.036

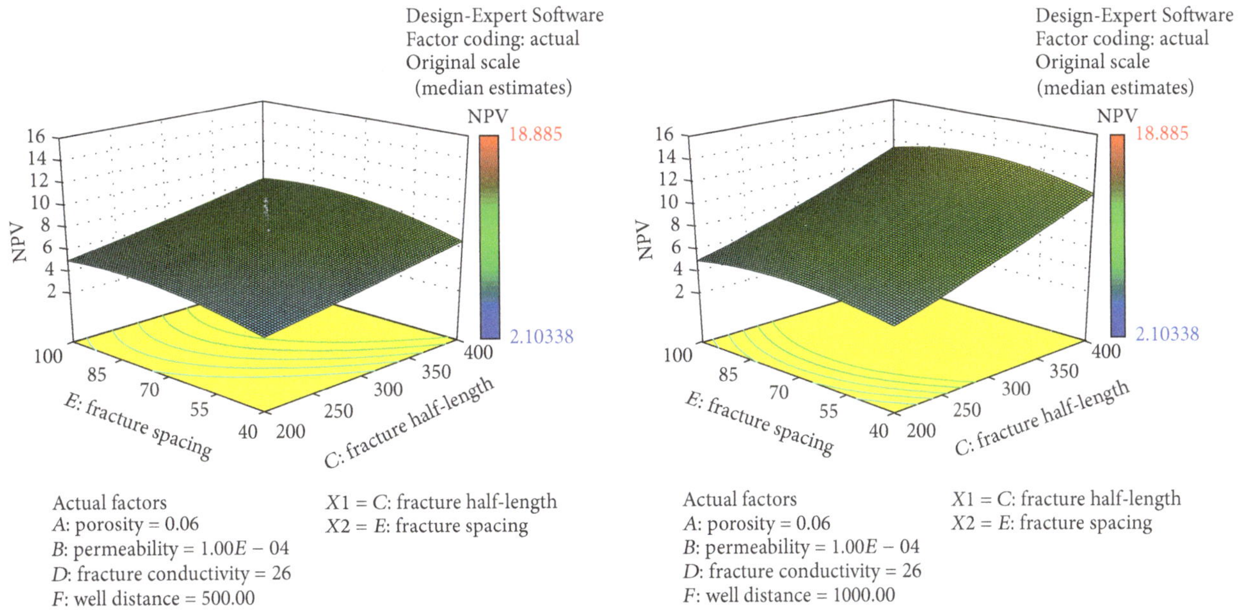

(a) Well distance of 500 ft

(b) Well distance of 1000 ft

FIGURE 16: 3D surface of NPV at varied values of fracture spacing and fracture half-length with gas price of $4/MSCF.

30 years of gas production for these two scenarios is shown in Figure 8. It shows that more contact reservoir area is drained effectively by Scenario 2, compared to Scenario 1. In addition, Rafiee et al. [7] reported that Scenario 2 design can increase the stress interference between fractures and create more effective stimulated reservoir volume to improve gas production. Therefore, Scenario 2 is used for optimization of multi-well placement in the subsequent study of this paper.

7. Multiwell Optimization

Response surface methodology (RSM) approach is applied to the optimization of two horizontal well placement based on the Barnett Shale reservoir information. RSM is utilized to approximate a response, in terms of maximum NPV in this work, over a region of interest specified by the range of variability of input factors based on the least squares criterion. The RSM model can be linear or fully quadratic. It can offer a cost-effective and efficient way to manage the uncertainties for shale gas reservoir development. More detailed mathematical and statistical theories of RSM can be found in the work by Myers and Montgomery [21]. We set up a shale gas reservoir model with a volume of 5000 ft × 2000 ft × 200 ft. Detailed reservoir information used for the Barnett Shale is listed in Table 4. Six uncertain parameters such as porosity (A), permeability (B), fracture half-length (C), fracture conductivity (D), fracture spacing (E), and well distance (F) are given a reasonable range with the actual maximum and minimum values or coded symbol of "+1" and "−1," respectively, as shown in Table 5. The number of hydraulic fractures is 87 and 35, corresponding to fracture spacing of 40 ft and 100 ft, respectively. These uncertainty ranges come from field data, analogues, and history matching of the Barnett Shale reservoirs. According to 6 variables, 38 cases were required, based on the approach of D-Optimal design, which was originated from the optimal design theory [22]. Table 6 shows the 38 combinations of these uncertain parameters generated by the D-Optimal design. After numerical simulation of each case, cumulative gas production and gas flow rate are obtained and shown in Figures 9 and 10 for selected cases with various fracture spacing and fracture half-length. It clearly shows that the cumulative gas production at 30 years of gas production is in the range of 3934–6529 MMSCF and gas flow rate has a large uncertainty. This means further optimization is needed.

Once the cumulative gas production of 38 run cases obtained, the economic Excel spreadsheet is used to calculate the corresponding NPVs. In this work, the impact of different gas price is considered for calculation of NPVs and optimization of fracturing design in multi-well placement to determine the optimal stimulation design. The price of natural gas plays a critical role in economic success of shale gas development. The activity in shale gas exploitation increases with the increasing gas price. Negative NPV of the project suggests that an unoptimized shale gas production can result in not only no income but also loss of money. For gas price of $3/MSCF, $4/MSCF, and $5/MSCF, the positive NPV is in the range of $0.36–13.39 million, $2.10–18.88 million, and $3.42–24.38 million, respectively, as shown in Figure 11.

Once NPVs of 38 run cases obtained, the Design-Expert Software is used to build the NPV response surface model in this study. To select the appropriate model, the statistical approach was used to determine which polynomial fits the equation among linear model, two-factor interaction model (2FI), quadratic model, and cubic model, as shown

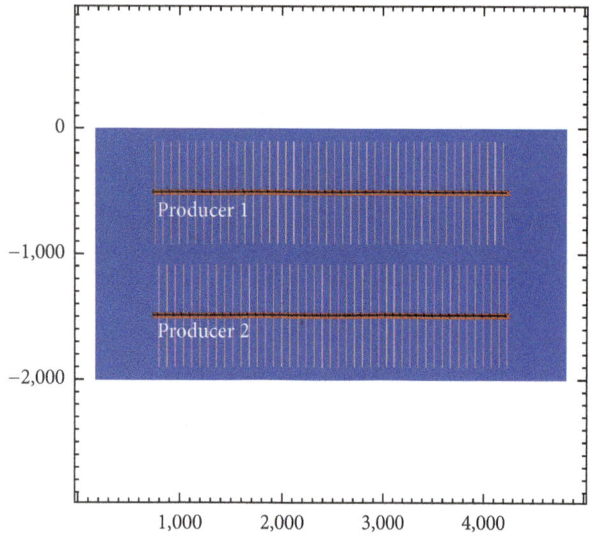

(a) Fracture spacing of 80 ft for gas price of \$3/MSCF

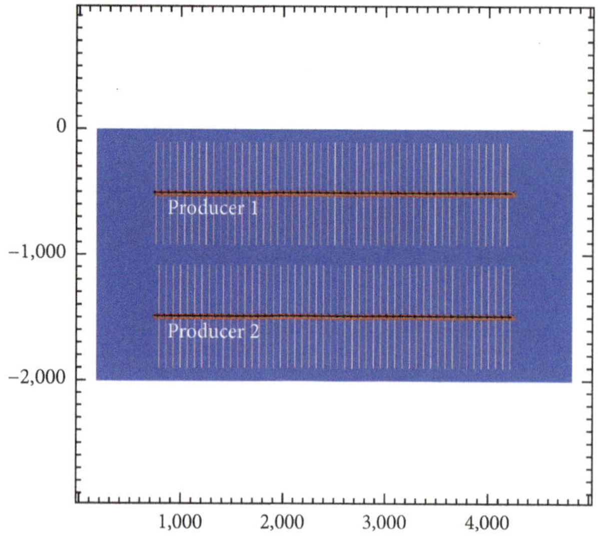

(b) Fracture spacing of 70 ft for gas price of \$4/MSCF

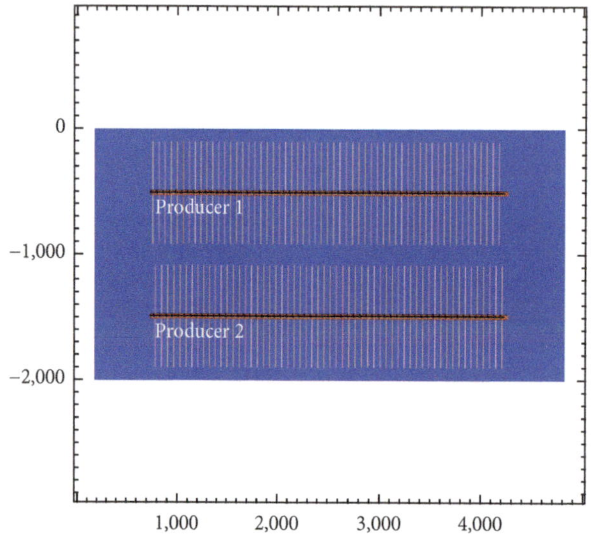

(c) Fracture spacing of 60 ft for gas price of \$5/MSCF

FIGURE 17: The optimum fracture spacing with different gas prices.

in Tables 7, 8, and 9, for different gas prices. The criterion for selecting the appropriate model is choosing the highest polynomial model, where the additional terms are significant and the model is not aliased. Although the cubic model is the highest polynomial model, it is not selected because it is aliased. Aliasing is a result of reducing the number of experimental runs. When it occurs, several groups of effects are combined into one group and the most significant effect in the group is used to represent the effect of the group. Essentially, it is important that the model is not aliased. In addition, other criteria are to select the model that has the maximum "Adjusted R-Squared" and "Predicted R-Squared". Thus, the fully quadratic model is selected to build the NPV response surface in the subsequent optimization process.

The equation fitted to the NPV response surface with the coded symbol for different gas prices is presented below.

For gas price of \$3/MSCF,

$$\begin{aligned}\sqrt{\text{NPV}} = {} & 2.67 + 0.59 \times A + 0.35 \times B + 0.35 \times C \\ & + 0.056 \times D + 0.21 \times E + 0.170 \times F - 0.025 \times AB \\ & + 0.058 \times AC - 0.011 \times AD - 0.160 \times AE \\ & - 0.037 \times AF - 0.090 \times BC - 0.026 \times BD \\ & - 0.028 \times BE - 0.0056 \times BF + 0.15 \times CD \\ & + 0.0098 \times CE + 0.13 \times CF + 0.046 \times DE \\ & + 0.044 \times DF - 0.064 \times EF - 0.086 \times A^2\end{aligned}$$

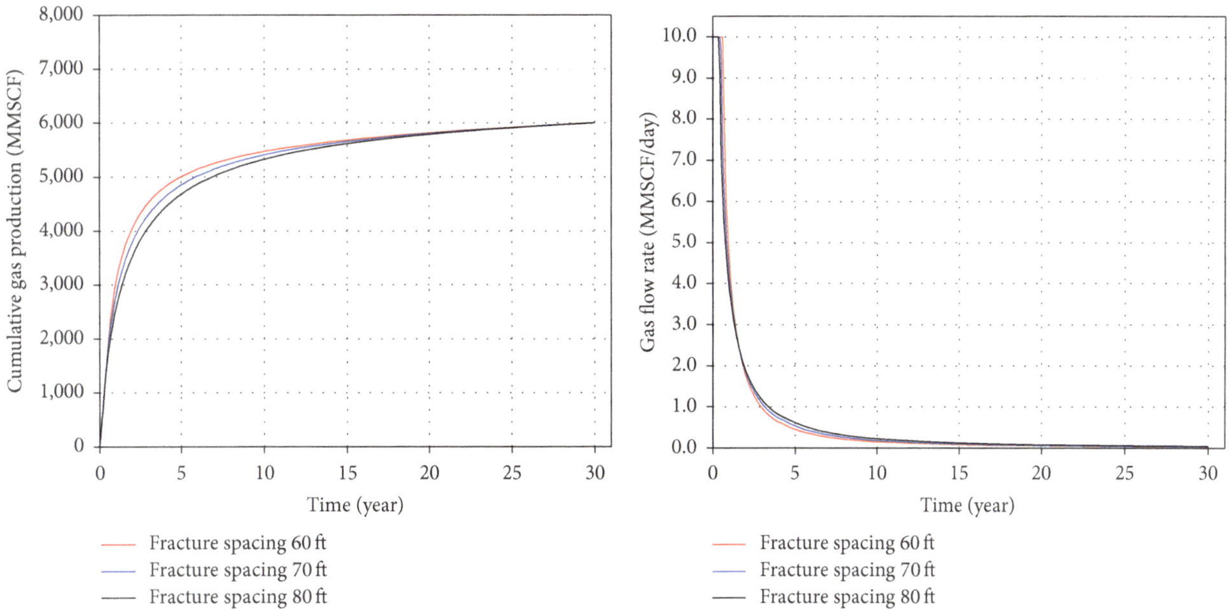

(a) Cumulative gas production

(b) Gas flow rate

FIGURE 18: Three optimum cases with different fracture spacing for different gas prices.

$$- 0.24 \times B^2 - 0.013 \times C^2 - 0.069 \times D^2 - 0.25 \times E^2 - 0.057 \times F^2. \quad (5)$$

For gas price of \$4/MSCF,

$$\begin{aligned}\sqrt{\text{NPV}} = {} & 3.27 + 0.60 \times A + 0.35 \times B + 0.38 \times C \\ & + 0.050 \times D + 0.12 \times E + 0.190 \times F + 0.014 \times AB \\ & + 0.055 \times AC + 0.018 \times AD - 0.11 \times AE \\ & - 0.018 \times AF - 0.088 \times BC - 0.0012 \times BD \\ & + 0.028 \times BE + 0.011 \times BF + 0.020 \times CD \\ & + 0.0014 \times CE + 0.12 \times CF + 0.055 \times DE \\ & + 0.027 \times DF - 0.062 \times EF - 0.082 \times A^2 \\ & - 0.25 \times B^2 - 0.0083 \times C^2 - 0.094 \times D^2 \\ & - 0.19 \times E^2 - 0.062 \times F^2.\end{aligned} \quad (6)$$

For gas price of \$5/MSCF,

$$\begin{aligned}\sqrt{\text{NPV}} = {} & 3.78 + 0.63 \times A + 0.37 \times B + 0.410 \times C \\ & + 0.053 \times D + 0.070 \times E + 0.20 \times F \\ & + 0.027 \times AB + 0.060 \times AC + 0.027 \times AD \\ & - 0.096 \times AE - 0.011 \times AF - 0.091 \times BC \\ & + 0.0042 \times BD + 0.046 \times BE + 0.017 \times BF \\ & + 0.021 \times CD - 0.00061 \times CE + 0.13 \times CF \\ & + 0.062 \times DE + 0.024 \times DF - 0.063 \times EF \\ & - 0.083 \times A^2 - 0.26 \times B^2 - 0.0052 \times C^2 \\ & - 0.11 \times D^2 - 0.17 \times E^2 - 0.07 \times F^2,\end{aligned} \quad (7)$$

where A is formation porosity; B is formation permeability, md; C is fracture half-length, ft; D is fracture conductivity, md-ft; E is fracture spacing, ft; F is well distance, ft.

The normal plot of residuals, reflecting the distribution of the residuals, for different gas prices is shown in Figure 12. All the points in the "Normal Plot of Residuals" fall on the straight line, meaning the residuals are normally distributed. Figure 13 shows the plot of "Predicted versus Actual" for different gas prices, illustrating whether the generated equation of NPV response surface accurately predicts the actual NPV values. It can be seen that generated NPV response surface models provide such reliable predicted values of NPV, as compared with the actual values of NPV. This means that the generated NPV response surface models are reliable.

Figure 14 shows the 3D surface of the well distance and fracture spacing with gas price of \$4/MSCF. It shows that there exists an optimal combination between well distance and fracture spacing. For fracture half-length of 200 ft, the NPV first increases and then decrease with increasing well distance. For fracture half-length of 400 ft, the NPV increases with increasing well distance within the range of this study. With the increasing fracture half-length, the optimal point moves to larger well distance. Figure 15 presents the 3D surface of the well distance and fracture half-length with gas price of \$4/MSCF. It can be seen that larger well distance and fracture half-length will lead to higher NPV. Similarly, Figure 16 presents the 3D surface of the fracture spacing and fracture half-length with gas price of $4/MSCF. It shows that there exists an optimal value for fracture spacing. With the increasing well distance, the optimal point moves to larger fracture half-length and smaller fracture spacing. Therefore, it can provide some insights into optimization of stimulation designs and completion strategies to obtain the maximum economic viability of the field.

The numerical optimization option selects the set of variables that leads to the maximum NPV value. In this study for the Barnett Shale, the optimal designs with porosity of 0.06 and permeability of 0.0001 md and different gas prices are listed in Table 10. The optimal NPV value is $8.70 MM, $12.40 MM, and $16.34 MM, for gas price of $3/MSCF, $4/MSCF, and $5/MSCF, respectively. What is more, the optimal fracture spacing is 80 ft for gas price of $3/MSCF, 70 ft for gas price of $4/MSCF, and 60 ft for gas price of $5/MSCF, as shown in Figure 17. It is extremely vital to validate the optimal results. Verification is performed by running the simulation with the best design condition. Figure 18 shows the cumulative gas production and gas flow rate for these three optimum cases with different gas prices, respectively. The NPV is calculated as $7.91 MM, $11.81 MM, and $15.75 MM, for gas price of $3/MSCF, $4/MSCF, and $5/MSCF, respectively. The absolute NPV difference between the NPV from the response surface and the real NPV is small. As indicated by the small relative error, all three solutions show a very good agreement between the calculated NPV and the real NPV.

8. Summary and Conclusions

The economic success of shale gas reservoirs depends on optimization of the number of treatment stages and number of fractures and horizontal wells. In this paper, response surface methodology was used to obtain the optimal design for shale gas production by optimizing reservoir and fracture uncertainty parameters in two horizontal wells. We applied this method to optimize 6 uncertain parameters, such as permeability, porosity, fracture spacing, fracture half-length, fracture conductivity, and well distance for the Barnett Shale development. Also, the gas desorption effect is considered for modeling shale gas production because of its large contribution to the estimated ultimate recovery for the Barnett Shale. The impact of varying natural gas price is taken into account during the optimization process. The following conclusions can be drawn from this study.

(1) With a porosity of 0.06, a permeability of 0.0001 md, and a fracture conductivity of 26 md-ft, the optimal design combinations for Barnett Shale is fracture half-length of 400 ft, well distance of 1000 ft, and fracture spacing of 80 ft, 70 ft, and 60 ft for gas price of $3/MSCF, $4/MSCF, and $5/MSCF, respectively.

(2) The gas desorption contributes to 20.7% of EUR at 30 years of gas production for an actual Barnett Shale horizontal well.

(3) The proposed approach is practical and efficient for the design and optimization of hydraulic fracturing for multiple horizontal wells in shale gas reservoirs.

Nomenclature

DoE:	Design of experiment
RSM:	Response surface methodology
2FI:	Two-factor interaction
EUR:	Estimated ultimate recovery
SCF:	Standard cubic feet, ft^3
NPV:	Net present value
MSCF:	10^3 standard cubic feet, ft^3
MMSCF:	10^6 standard cubic feet, ft^3
β:	Non-Darcy Beta factor, ft^{-1}
K:	Permeability, md
P:	Present value, $
F:	Future value, $
i:	Currency escalation rate
n:	Number of periods, year
V_F:	Future value of production revenue for a fracture reservoir, $
V_o:	Future value of production revenue for an unfractured reservoir, $
FC:	Total fixed cost, $
C_{well}:	Cost of one horizontal well, $
C_{fracture}:	Cost of hydraulic fracture in a horizontal well, $
N:	Number of horizontal wells
V_L:	Langmuir pressure, scf/ton
P_L:	Langmuir volume, psi
G_s:	Gas content, scf/ton
ρ_B:	Bulk density of shale, g/cm^3.

Acknowledgments

Funding for this project is provided by the Hilcorp Energy Company. The authors would like to acknowledge Computer Modeling Group Ltd. for providing with usage of CMG-IMEX software. The authors would also like to express their gratitude for reviewers for their careful review of this paper.

References

[1] J. L. Miskimins, "Design and life-cycle considerations for unconventional-reservoir wells," *SPE Production and Operations*, vol. 24, no. 2, pp. 353–359, 2009.

[2] G. Waters, B. Dean, R. Downie, K. Kerrihard, L. Austbo, and B. McPherson, "Simultaneous hydraulic fracturing of adjacent horizontal wells in the woodford shale," in *Proceedings of the SPE Hydraulic Fracturing Technology Conference*, pp. 694–715, Woodlands, Tex, USA, January 2009.

[3] M. J. Kaiser, "Haynesville shale play economic analysis," *Journal of Petroleum Science and Engineering*, vol. 82-83, pp. 75–89, 2012.

[4] S. L. Montgomery, D. M. Jarvie, K. A. Bowker, and R. M. Pollastro, "Mississippian Barnett Shale, Fort Worth basin, north-central Texas: gas-shale play with multi-trillion cubic foot potential," *AAPG Bulletin*, vol. 89, no. 2, pp. 155–175, 2005.

[5] M. Segatto and I. Colombo, "Use of reservoir simulation to help gas shale reservoir estimation," in *Proceedings of the International Petroleum Technology Conference (IPTC '11)*, Bangkok, Thailand, 2011.

[6] S. Esmaili, A. Kalantari-Dahaghi, and S. D. Mohaghegh, "Modeling and history matching of hydrocarbon production from Marcellus shale using data mining and pattern recognition technologies," in *Proceedings of the SPE Eastern Regional Meeting (SPE '12)*, Lexington, Ky, USA, 2012.

[7] M. Rafiee, M. Y. Soliman, and E. Pirayesh, "Hydraulic fracturing design and optimization: a modification to zipper frac," in *Proceedings of the SPE Eastern Regional Meeting (SPE '12)*, Lexington, Ky, USA, 2012.

[8] O. C. Díaz de Souza, A. J. Sharp, R. C. Martinez et al., "Integrated unconventional shale gas reservoir modeling: a worked example from the Haynesville Shale, De Soto Parish, North Lousiana," in *Proceedings of the Americas Unconventional Resources Conference (SPE '12)*, Pittsburgh, Pa, USA, 2012.

[9] J. Harpel, L. Barker, J. Fontenot, C. Carroll, S. Thomson, and K. Olson, "Case history of the Fayetteville shale completions," in *Proceedings of the SPE Hydraulic Fracturing Technology Conference (SPE '12)*, The Woodlands, Tex, USA, 2012.

[10] H. Ramakrishnan, R. Yuyan, and J. Belhadi, "Real-time completion optimization of multiple laterals in gas shale reservoirs: Integration of geology, log, surface seismic, and microseismic information," in *Proceedings of the SPE Hydraulic Fracturing Technology Conference*, pp. 691–705, The Woodlands, Tex, USA, January 2011.

[11] S. Tavassoli, W. Yu, F. Javadpour, and K. Sepehrnoori, "Well screen and optimal time of refracturing: a Barnett shale well," *Journal of Petroleum Engineering*, vol. 2013, pp. 1–10, 2013.

[12] W. Yu and K. Sepehrnoori, "Simulation of gas desorption and geomechanics effects for unconventional gas reservoirs," in *Proceedings of the SPE Western Regional and AAPG Pacific Section Meeting (SPE '13)*, Monterey, Calif, USA,, 2013.

[13] W. Yu and K. Sepehrnoori, "An efficient reservoir simulation approach to design and optimize unconventional gas production," in *Proceedings of the SPE Western Regional and AAPG Pacific Section Meeting (SPE '13)*, Monterey, Calif, USA,, 2013.

[14] B. Rubin, "Accurate simulation of non-darcy flow in stimulated fractured shale reservoirs," in *Proceedings of the SPE Western Regional Meeting (SPE '10)*, pp. 19–34, Anaheim, Calif, USA, May 2010.

[15] C. L. Cipolla, E. P. Lolon, J. C. Erdle, and B. Rubin, "Reservoir modeling in shale-gas reservoirs," *SPE Reservoir Evaluation and Engineering*, vol. 13, no. 4, pp. 638–653, 2010.

[16] R. D. Evans and F. Civan, "Characterization of non-Darcy multiphase flow in petroleum bearing formations," U.S. DOE Contract DE-AC22-90BC14659, School of Petroleum and Geological Engineering, University of Oklahoma, 1994.

[17] R. Schweitzer and H. I. Bilgesu, "The role of economics on well and fracture design completions of marcellus shale wells," in *Proceedings of the SPE Eastern Regional Meeting (SPE '09)*, pp. 423–428, September 2009.

[18] CMG, *IMEX User's Guide*, Computer Modeling Group, 2011.

[19] J. P. Seidle and L. E. Arri, "Use of conventional reservoir models for coalbed methane simulation," in *Proceedings of the CIM/SPE International Technical Meeting (SPE '90)*, Calgary, Canada, 1990.

[20] H. A. Al-Ahmadi, S. Aramco, and R. A. Wattenbarger, "Triple-porosity models: one further step towards capturing fractured reservoirs heterogeneity," in *Proceedings of the SPE/DGS Saudi Arabia Section Technical Symposium and Exhibition (SPE '11)*, Al-Khobar, Saudi Arabia, 2011.

[21] R. H. Myers and D. C. Montgomery, *Response Surface Methodology: Process and Product Optimization Using Designed Experiments*, John Wiley and Sons, Hoboken, NJ, USA, 2002.

[22] J. Kiefer and J. Wolfowitz, "Optimum designs in regression problems," *The Annals of Mathematical Statistics*, vol. 30, no. 2, pp. 271–294, 1959.

8

Dispersing of Petroleum Asphaltenes by Acidic Ionic Liquid and Determination by UV-Visible Spectroscopy

Eshagh Rezaee Nezhad,[1,2] Fariba Heidarizadeh,[1] Sami Sajjadifar,[2] and Zahra Abbasi[2]

[1] *Department of Chemistry, Faculty of Sciences, Shahid Chamran University, Ahvaz 61357-43337, Iran*
[2] *Department of Chemistry, Payame Noor University, P.O. BOX 19395-4697, Tehran, Iran*

Correspondence should be addressed to Eshagh Rezaee Nezhad; e.rezaee66@yahoo.com

Academic Editor: Jorge Ancheyta

Nowadays, constructing a mechanism to prevent the aggregation petroleum asphaltenes by the use of new acidic ionic liquids has become of fundamental importance. In this research, 3-(2-carboxybenzoyl)-1-methyl-1H-imidazol-3-ium chloride ([CbMIM][Cl]) and other ionic liquids such as [CbMIM]BF_4, [HMIM]Cl, [BMIM]Br, and [HMIM]HSO_4 were tested. It should be noted that during the experiment the presence of the acidic ionic liquid moiety enhanced interactions between asphaltenes and acidic ionic liquids and it greatly limited asphaltene aggregation. We considered parameters such as temperature, amount of dispersant, effect of water: toluene ratio, the stirring time and effects of other ionic liquids, and determination of concentration of petroleum asphaltenes after dispersing by acidic ionic liquid under various parameters using UV-Visible spectroscopy.

1. Introduction

Asphaltene is defined as a component of the crude oil which is precipitated by adding low molecular weight solvents (n-alkanes) [1]. According to recent studies, asphaltenes are defined as solid soluble asphaltenes in aromatic solvents (e.g., benzene/toluene) and are also regarded as insoluble asphaltenes in paraffinic solvents (n-heptane/n-pentane/petroleum ether, etc.) [2]. Asphaltenes are petroleum hydrocarbons with extremely complex molecular structure containing sulfur (0.3–10.3%), oxygen (0.3–4.8%), nitrogen (0.6–3.3%), and metal elements, such as Fe, Ni, and V in a small amount [3–6]. One of the most important problems in oil industries is the precipitation of heavy components of crude oil such as asphaltenes. The precipitation of these compounds causes different problems like the blockage of crude oil extraction in transport pipes and pollution of ecosystems [7, 8]. The aromatic moiety, through π-π^* and dipole interactions, was thought to be one of the dominant contributors to asphaltene self-association. Hence, proper characterization of the asphaltene molecular structure and its aromaticity is fundamental in understanding of the self-association phenomenon [9]. The efficiency of dispersants is dependent on their polarity and structure. Therefore, dispersants should have an affinity to asphaltenes and should either be absorbed on the surface of the asphaltene aggregation or become bond with the polar group of asphaltene molecules [10]. Recent studies show that the petroleum asphaltenes are strong hydrogen bond acceptors and weak hydrogen bond donors [11]. However, strong acids become effective asphaltene dispersants if their alkyl tails are long enough to provide the necessary steric-stabilization layers around the asphaltenes. Besides, organic salts may be strong donors of the hydrogen bonding and can be used to stabilize asphaltenes [12].

Hu an Guo reported the alkylpyridinium-based and alkylisoquinolinium-based ionic liquids they claimed that these cations of ionic liquids cannot have a stable bond with asphaltene and form complexes because of a low density of the cation which are delocalized from the whole head of aromatic rings [13]. Association of asphaltenes in solutions has been the subject of a number of studies, Evdokimov et al. [14–16], Sheu [17] in which The reported concentrations of aggregation were typically in the range 2–18 g/L, depending on a solvent.

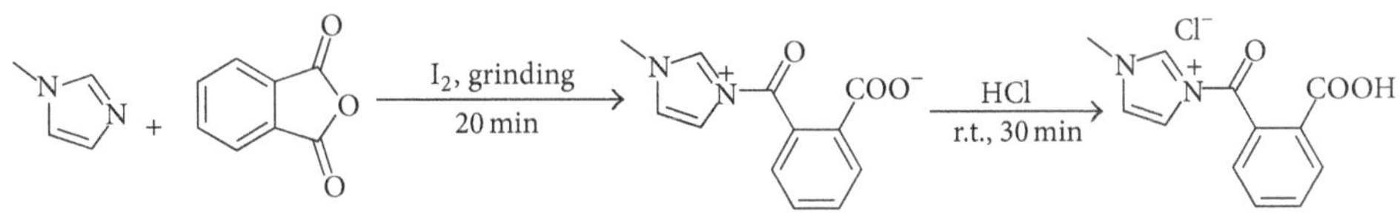

SCHEME 1: Synthesis of (3-(2-carboxybenzoyl)-1-methyl-1H-imidazol-3-ium chloride).

TABLE 1: Elemental analysis of crude oil studied.

Element	Crude oil %
C	83.15
H	12.37
O	0.68
N	0.36
S	0.83
Total	97.32
Residue	2.61

TABLE 2: Results of the SARA analysis.

Family	W/W %
Saturated	33.28
Aromatics	20.52
Resins	34.69
Asphaltene	11.51
Total	100
Ratio resins/asphaltene	3.01

Asphaltenes precipitation is one of the problems affecting crude oil production and refining processes. asphaltenes change a very high temperature and pressure where asphaltenes are expected to exist in a molecular form [18].

In this paper, we synthesized the acidic ionic liquid and used a great deal of petroleum asphaltenes' dispersant. In present study, we observed that the acidic ionic liquid, with the help of its hydrogen bond and electrostatic interaction, prevents asphaltene deposition.

2. Experimental

2.1. Chemicals and Instruments. Phthalic anhydride was provided from Merck company, and toluene was purchased from Fluka company. All solvents and reagents were used without further purification. UV-Visible spectrophotometer (Jasco model 7850), quartz cell with 1 cm. The crude oils were from the Maroon petroleum field. The chemical composition of crude oil obtained analysis is given in Table 1. Results of SARA analysis are presented in Table 2.

2.2. Preparation of the Acidic Ionic Liquid (AIL). Acidic ionic liquid [(3-(2-carboxybenzoyl)-1-methyl-1H-imidazol-3-ium chloride ([CbMIM] [Cl]) synthesized by our group in the chemical laboratory of the shahid chamran university.

Phthalic anhydride (14.8 g, 0.1 mmol), 1-methyl imidazole (8.2 g, 0.1 mmol), and a catalytic amount of iodine as a catalyst were grinded for 20 min. After completion of the reaction, a viscous liquid, which turned into a yellow solid after drying, was obtained. Then, crude product was washed with diethyl ether (3 × 10 mL) to remove iodine. The obtained salt (5 mmol, 1.15 gr) was added to HCl (5 mmol, 37%) and stirred for 1 h. After that, the obtained colorless ionic liquid was dried under low pressure for 4 h (Scheme 1). Yield: 96%. IR (KBr): 2400–3400, 1721, 1582, 1491, 1284. ^{1}H-NMR (CD_3CN):14.2(s, acidic H), 7.3–8.5(m, 7H), 3.8 (s, 3H)^{13}CNMR (CD_3CN): 35.7, 117.44, 119.534, 123.038, 128.924, 131.04, 132.661, 135.361, 169.07, Anal. Calc. for $C_{12}H_{11}O_3N_2Cl$ ($3H_2O$): C, 44.9; H, 5.3; N, 8.73%. Found: C, 44.15; H, 4.82; N, 9.61%.

2.3. Analysis of the Asphaltene Dispersing. UV-Vis Measurements. Determination of the asphaltene dispersing. The Dispersing of asphaltenes was determined with the use of UV-Vis spectroscopy. The spectrophotometer (Jasco model 7850) was operated at the wavelength of 346 nm. The quartz cell with 1 cm optical path was used. Solutions of asphaltenes in toluene (1 gL^{-1}) were dispersed with acidic ionic liquid in water. it should be mentioned that we used asphaltene concentration with 30 ppm.

2.4. Determination of the Asphaltene Concentration in Solution after Dispersion. The absorbance of light depends on the concentration of asphaltenes in toluene solutions. The calibration curve at a wavelength of 346 nm was established and then used to measure the asphaltene concentration in toluene solutions. Experiments were conducted to quantify the effect of dispersants on the amount of asphaltenes remained in the solution after dispersion with acidic ionic liquid. Toluene solutions containing 30 ppm of asphaltenes and the amount of dispersant between 1 and 5 mmol were mixed with water at the water/toluene ratio 2 : 1. Samples were stirred during several periods of time, 1–20 h, in the presence of several ranges of temperature 273–353°K, and then the mixture was centrifuged and filtered on the 0.22 μm Millipore filter. The concentration of asphaltenes in the filtrate was measured and compared with the initial concentration. After calculating the optimal conditions such as dispersant, time of stirring, ratio of water/toluene, and temperature. Afterwards, ionic liquids as dispersant of asphaltenes were observed/tested in the presences of optimal conditions.

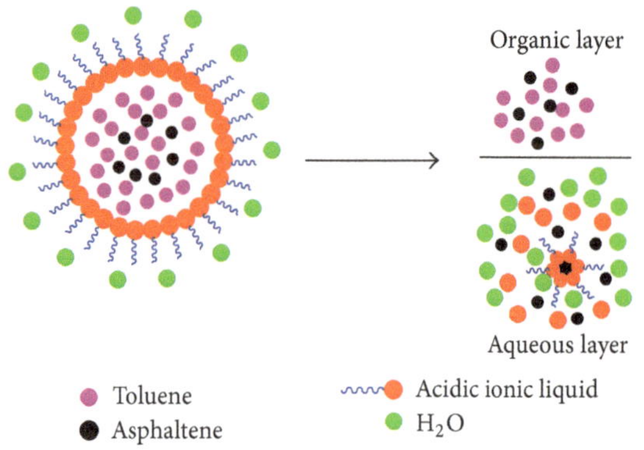

SCHEME 2: Asphaltene dispersing.

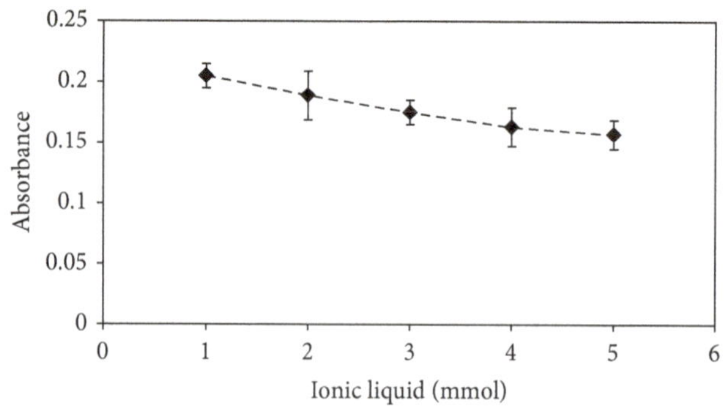

FIGURE 2: Effect amounts of acidic ionic liquid as dispersant.

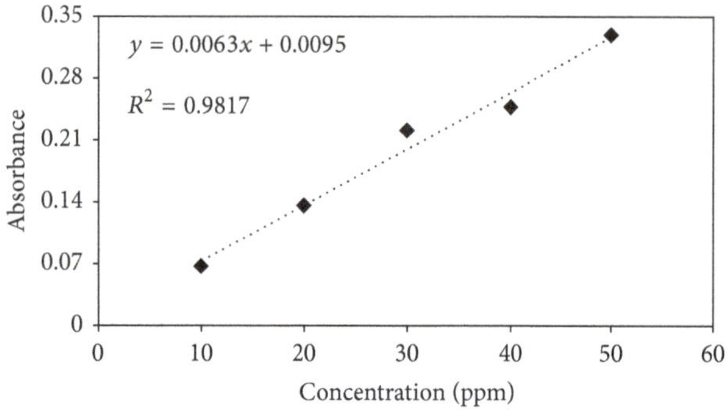

FIGURE 1: Standard curve.

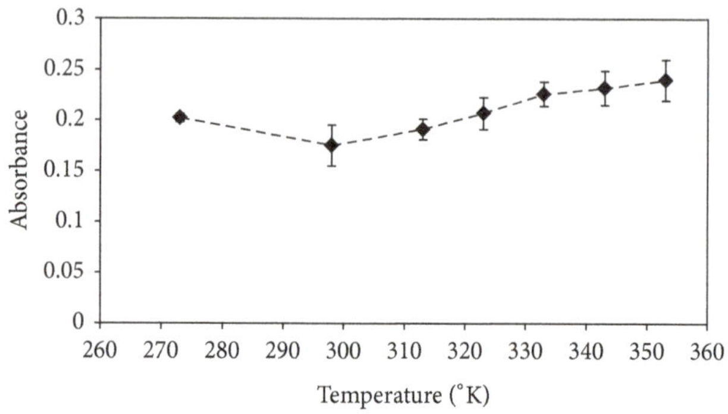

FIGURE 3: Effect of the temperature on the aggregation kinetics.

3. Results and Discussion

Ionic liquids are also regarded as the organic salts, but because of their specific properties such as capability to dissolve various organic and inorganic compounds and potential recyclable properties, they have attracted considerable attention as ecofriendly reaction media in the green organic synthesis and may be used as asphaltene dispersants (Scheme 2).

3.1. Preparation of Standard Solutions. A 1000 ppm asphaltene solution was prepared by dissolving 1 g of asphaltenes in 1000 mL toluene (1 gL^{-1}). For the preparation of standard solutions (10–50 ppm) 1–5 mL of primary solution (1000 ppm) was diluted in 100 mL of toluene, respectively (Figure 1).

3.2. Effect Amounts of Acidic Ionic Liquid as Dispersant. Best dispersion of asphaltene depends on the amount of acidic ionic liquid. According to Figure 2 with the increase in the amount of acidic ionic liquid we have the decreases in the concentration of asphaltene in organic phase, it is so because interaction of dispersant with asphaltene increases in organic and aqueous phases. [CbMIM] Cl as a novel acidic ionic liquid contains carboxyl groups as well as electrostatic forces and π-π interaction with polar and π-π groups of asphaltene. The ionic liquid due to asphaltene deposition is from these interactions.

3.3. Effect of Temperature. The amounts of asphaltene precipitation have been measured at temperatures of 273°K, 298°K, 313°K, 323°K, 333°K, 343°K, and 353°K. At each temperature the [CbMIM] [Cl] solution in water was used as a dispersant agent. Figure 3 illustrates the effect of temperature on the amount of asphaltene precipitation. As it is shown, the amount of asphaltene precipitation increases as temperature decreases. An increase in temperature results in the decrease of asphaltene precipitation and also less interaction between [CbMIM] [Cl] and asphaltene in aqueous phase because by increasing the temperature ionic liquids dissolve more likely in the organic phase. This data was determined by UV-Visible spectroscopy.

3.4. Effect of the Stirring Time. Samples were stirred during several periods of time, 1–20 h, in the presence of optimal temperature, and then were centrifuged and filtered on the 0.22 μm Millipore filter. The concentration of asphaltenes in the filtrate was measured and compared with the initial concentration. Figure 4 illustrates the effect of stirring time on the amount of asphaltene precipitation. As it is shown, the amount of optimal stirring time is 5 h. the amount of asphaltene precipitation increases as stirring time increases for 5 h and after that it decreases.

3.5. Effect of Water: Toluene Ratio. Figure 5 reported results of the determination of the asphaltene after dispersing by [CbMIM] [Cl] using UV-Vis technique. The best dispersing

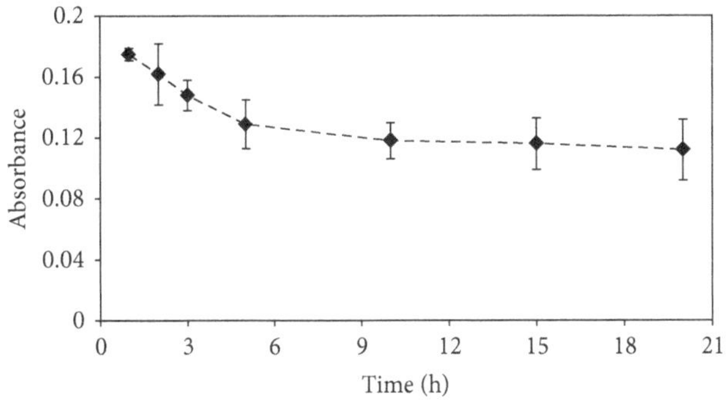

FIGURE 4: Effect stirring time as dispersant.

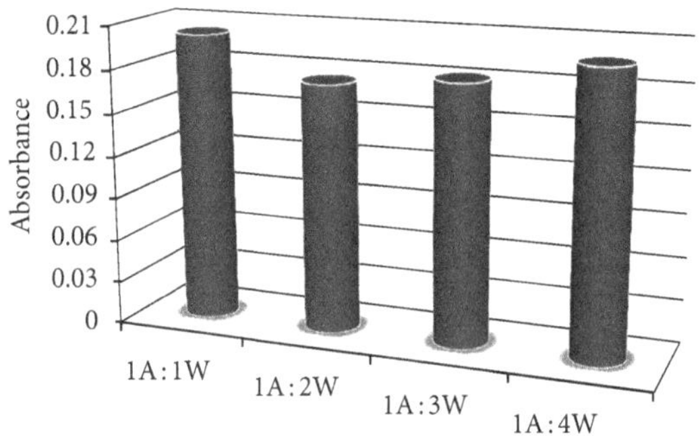

FIGURE 5: Effect of water : toluene ratio as dispersant (W = water, A = asphaltenes).

of asphaltene occurs after using a value of about 2 (ratio of water and toluene volumes). In this ratio of water and toluene, high interaction between asphaltene and dispersing agent occurs.

3.6. Effect of Other Ionic Liquids. The effect of the ionic liquids (amphiphiles) is to prevent asphaltene precipitation and during this research were studied, using [CbMIM]Cl, [CbMIM]BF_4, [HMIM]Cl, [BMIM]Br, and [HMIM]HSO_4. Results were compared of [CbMIM]Cl and [CbMIM]BF_4 with other ionic liquids (i.e., [ILX], [HMIM][Cl], [BMIM][Br] and [HMIM]HSO_4). Ionic liquids are effective in most types of interactions, including dispersive, π-π, acid-base (hydrogen bonding), dipolar, and Coulomb interactions. After determination of optimal conditions: 3 mmol of [CbMIM][Cl] as dispersant, room temperature, stirring time 5 h, and ration of water/toluene 2, a comparison between [CbMIM][Cl] and other ionic liquids such as [CbMIM][BF_4], [HMIM][Cl], [BMIM][Br], and [HMIM]HSO_4 (Figure 6) was done.

4. Conclusions

In this research, a new method for dispersing of asphaltene flocculation by the use of novel acidic ionic liquid was proposed. Acidic ionic liquid with their hydrogen bond and electrostatic interaction can be used to prevent asphaltene deposition. Finally, it should also be mentioned that we

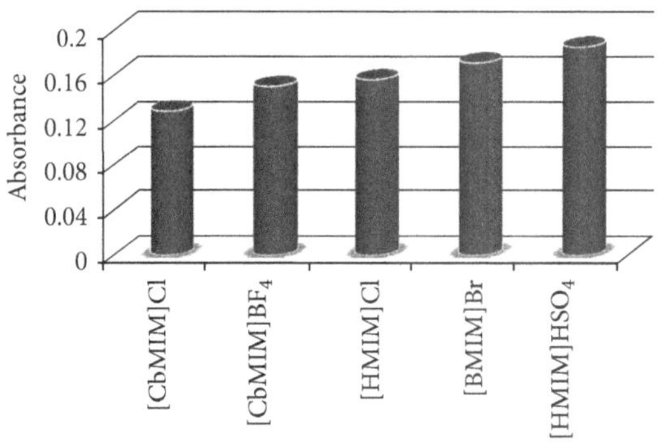

FIGURE 6: Effect of ionic liquids as dispersant.

observed and tested various parameters effective on the inhibition of the asphaltene precipitation such as temperature, amount of dispersant, the stirring time, and effects of other ionic liquids.

Acknowledgments

The authors gratefully acknowledge the financial support of this work by Shahid Chamran (Ahvaz) University Research Council and Payame Noor University of Ilam, Iran.

References

[1] M. Nikookar, G. R. Pazuki, M. R. Omidkhah, and L. Sahranavard, "Modification of a thermodynamic model and an equation of state for accurate calculation of asphaltene precipitation phase behavior," *Fuel*, vol. 87, no. 1, pp. 85–91, 2008.

[2] A. K. Ghosh, "Spectrophotometric study of molecular complex formation of asphaltene with two isomeric chloranils," *Fuel*, vol. 84, no. 2-3, pp. 153–157, 2005.

[3] L. Buch, H. Groenzin, E. Buenrostro-Gonzalez, S. I. Andersen, C. Lira-Galeana, and O. C. Mullins, "Molecular size of asphaltene fractions obtained from residuum hydrotreatment," *Fuel*, vol. 82, no. 9, pp. 1075–1084, 2003.

[4] H. Groenzin and O. C. Mullins, "Molecular size and structure of asphaltenes from various sources," *Energy and Fuels*, vol. 14, no. 3, pp. 677–684, 2000.

[5] J. T. Miller, R. B. Fisher, P. Thiyagarajan, R. E. Winans, and J. E. Hunt, "Subfractionation and characterization of mayan asphaltene," *Energy and Fuels*, vol. 12, no. 6, pp. 1290–1298, 1998.

[6] T. Tavassoli, S. M. Mousavi, S. A. Shojaosadati, and H. Salehizadeh, "Asphaltene biodegradation using microorganisms isolated from oil samples," *Fuel*, vol. 93, pp. 142–148, 2012.

[7] E. Hong and P. Watkinson, "A study of asphaltene solubility and precipitation," *Fuel*, vol. 83, no. 14-15, pp. 1881–1887, 2004.

[8] T. J. Kaminski, H. S. Fogler, N. Wolf, P. Wattana, and A. Mairal, "Classification of asphaltenes via fractionation and the effect of heteroatom content on dissolution kinetics," *Energy and Fuels*, vol. 14, no. 1, pp. 25–30, 2000.

[9] Y. Bouhadda, P. Florian, D. Bendedouch, T. Fergoug, and D. Bormann, "Determination of Algerian Hassi-Messaoud asphaltene aromaticity with different solid-state NMR sequences," *Fuel*, vol. 89, no. 2, pp. 522–526, 2010.

[10] M. Boukherissa, F. Mutelet, A. Modarressi, A. Dicko, D. Dafri, and M. Rogalski, "Ionic liquids as dispersants of petroleum

asphaltenes," *Energy and Fuels*, vol. 23, no. 5, pp. 2557–2564, 2009.

[11] F. Mutelet, G. Ekulu, R. Solimando, and M. Rogalski, "Solubility parameters of crude oils and asphaltenes," *Energy and Fuels*, vol. 18, no. 3, pp. 667–673, 2004.

[12] O. León, E. Rogel, A. Urbina, A. Andújar, and A. Lucas, "Study of the adsorption of alkyl benzene-derived amphiphiles on asphaltene particles," *Langmuir*, vol. 15, no. 22, pp. 7653–7657, 1999.

[13] Y. F. Hu and T. M. Guo, "Effect of the structures of ionic liquids and alkylbenzene-derived amphiphiles on the inhibition of asphaltene precipitation from CO_2-injected reservoir oils," *Langmuir*, vol. 21, no. 18, pp. 8168–8174, 2005.

[14] I. N. Evdokimov, N. Y. Eliseev, and B. R. Akhmetov, "Assembly of asphaltene molecular as studied by near-UV/visible spectroscopy aggregates, I. Structure of the absorbance spectrum," *Journal of Petroleum Science and Engineering*, vol. 37, no. 3-4, pp. 135–143, 2003.

[15] I. N. Evdokimov, N. Y. Eliseev, and B. R. Akhmetov, "Assembly of asphaltene molecular as studied by near-UV/visible spectroscopy aggregates, II. Concentration dependencies of absorptivities," *Journal of Petroleum Science and Engineering*, vol. 37, no. 3-4, pp. 145–152, 2003.

[16] I. N. Evdokimov, N. Y. Eliseev, and B. R. Akhmetov, "Initial stages of asphaltene aggregation in dilute crude oil solutions: studies of viscosity and NMR relaxation," *Fuel*, vol. 82, no. 7, pp. 817–823, 2003.

[17] E. Y. Sheu, "Petroleum asphaltene—properties, characterization, and issues," *Energy and Fuels*, vol. 16, no. 1, pp. 74–82, 2002.

[18] M. Idris and L. N. Okoro, "A review on the effects of asphaltene on petroleum processing," *European Chemical Bulletin*, vol. 2, no. 6, pp. 393–396, 2013.

9

Corrosion Inhibition of Tubing Steel during Acidization of Oil and Gas Wells

M. Yadav,[1] Sumit Kumar,[1] and P. N. Yadav[2]

[1] *Department of Applied Chemistry, Indian School of Mines, Dhanbad 826004, India*
[2] *Department of Physics, Post Graduate College, Ghazipur 233001, India*

Correspondence should be addressed to M. Yadav; yadav_drmahendra@yahoo.co.in

Academic Editor: Jorge Ancheyta

Acidization is an oil reservoir stimulation technique for increasing oil well productivity. Hydrochloric acid is used in oil and gas production to stimulate the formation. The acid treatment occurs through N80 steel tubes. The process requires a high degree of corrosion inhibition of tubing material (N80 steel). In the present investigation effect of synthesized amino acid compounds, namely, acetamidoleucine (AAL) and benzamidoleucine (BAL) as corrosion inhibitors for N80 steel in 15% HCl solution was studied by polarization, AC impedance (EIS), and weight loss measurements. It was found that both the inhibitors were effective inhibitors and their inhibition efficiency was significantly increased with increasing concentration of inhibitors. Polarization curves revealed that the studied inhibitors represent mixed type inhibitors. AC impedance studies revealed that charge transfer resistance increases and double layer capacitance decreases in presence of inhibitors. Adsorption of inhibitors at the surface of N80 steel was found to obey Langmuir isotherm.

1. Introduction

Acidization of a petroleum oil well is one of the important stimulation techniques for enhancing oil production. It is commonly brought about by forcing a solution of 15% to 28% hydrochloric acid into the well through N80 tubing to open up near bore channels in the formation and hence to increase the flow of oil. To reduce the aggressive attack of the acid on tubing and casing materials (N80 steel), inhibitors are added to the acid solution during the acidifying process. The effective acidizing inhibitors that are usually found in commercial formulations are acetylenic alcohols, alkenyl phenones, aromatic aldehydes, nitrogen-containing heterocyclics, quaternary salts, and condensation products of carbonyls and amines [1–6]. However, these inhibitors are effective only at high concentrations, and they are harmful to the environment due to their toxicity, so it is important to search for new nontoxic and effective organic corrosion inhibitors for N80 steel—15% HCl system. In this regard, amino acids have a promising alternative for the design of ecofriendly corrosion inhibitors [7–10] which satisfy the environmental requirements.

In the present investigation acetamidoleucine (AAL) and benzamidoleucine (BAL) were synthesized, and their corrosion inhibiting properties on N80 (oil-well tubular steel) steel was studied in 15% HCl by using weight loss, polarization, and AC impedance techniques. Influence of temperature (from 298 to 333 K) on the inhibition behavior was studied. Both the activation and thermodynamic parameters governing the corrosion process were calculated and discussed.

2. Material and Methods

2.1. Materials. The working electrode and specimens for weight loss experiments were prepared from N80 steel sheets (supplied by ONGC) having the following percentage by weight (%wt) composition: C: 0.31, Mn: 0.92, Si: 0.19, P: 0.01, S: 0.008, Cr: 0.20, and Fe: remainder.

2.2. Weight Measurements. The specimens for the weight loss experiments were of the size 3 cm × 3 cm × .1 cm, and for electrochemical studies the exposed surface area of the working electrode was 1 cm^2. For weight loss experiments

300 mL of 15% HCl (v/v) was taken in 500 mL glass beakers. The inhibition efficiencies (IE) were evaluated after a preoptimized time interval of 6 h using 10, 20, 50, 100, and 150 ppm by weight of inhibitors. The specimens were removed from the electrolyte, washed thoroughly with distilled water, dried, and weighed. The inhibition efficiencies (IE) were calculated using the following equation:

$$\%\mathrm{IE} = \left[\frac{\mathrm{CR}_0 - \mathrm{CR}_i}{\mathrm{CR}_0}\right] \times 100, \quad (1)$$

where CR_0 is the weight loss in absence of inhibitor, and CR_i is the weight loss in presence of inhibitor.

Corrosion rate (CR) for the specimen was calculated with the help of the following equation:

$$\mathrm{CR}\,(\mathrm{mpy}) = \frac{534W}{DAT}, \quad (2)$$

where W = weight loss (mg); D = density of specimen ($\mathrm{g/cm^3}$); A = area of specimen ($\mathrm{cm^2}$); T = exposure time (hours).

2.3. Adsorption Isotherms. The information about the electrochemical interaction of inhibitors with metal surface can be provided by the nature of adsorption isotherm. Different adsorption isotherm such as Langmuir, Temkin, and Frumkin, were tested graphically by ploting surface coverages (θ) and concentration of studied (AAL and BAL) on N80 steel in HCl solution. The surface coverage was calculated using the following equation:

$$\theta = \left[\frac{\mathrm{CR}_0 - \mathrm{CR}_i}{\mathrm{CR}_0}\right] \times 100, \quad (3)$$

where CR_0 is the weight loss in absence of inhibitor, and CR_i is the weight loss in presence of inhibitor.

2.4. Electrochemical Polarization Studies. The electrochemical studies were performed in a three necked glass assembly containing 150 mL of the electrolyte with different concentrations of inhibitors (10, 50, and 150 ppm by weight) dissolved in it. The potentiodynamic polarization studies were carried out with N80 steel strips having an exposed area of 1 $\mathrm{cm^2}$. A conventional three electrode cell consisting of N80 steel as working electrode, platinum as counter electrode, and a saturated calomel electrode as reference electrode were used. Polarisation studies were carried out using VoltaLab 10 electrochemical analyser, and data was analysed using Voltamaster 4.0 software. The potential sweep rate was 10 $\mathrm{mVs^{-1}}$. All experiments were performed at 25 ± 0.2°C in an electronically controlled air thermostat. For calculating inhibition efficiency by electrochemical polarization method, the following formula was used:

$$\%\mathrm{IE} = \left[\frac{I_0 - I_{\mathrm{inh}}}{I_0}\right] \times 100, \quad (4)$$

where I_0 is corrosion current in absence of inhibitor, and I_{inh} is the corrosion current in presence of inhibitor.

2.5. AC Impedance Studies. AC impedance studies were carried out in a three electrode cell assembly using computer-controlled VoltaLab 10 electrochemical analyzer, as well as N80 steel as the working electrode, platinum as counter electrode, and saturated calomel as reference electrode. The data were analysed using Voltamaster 4.0 software. The electrochemical impedance spectra (EIS) were taken in the frequency range from 10 kHz to 1 mHz at the rest potential by applying 5 mV sine wave AC voltage. The charge transfer resistance (R_{ct}) and double layer capacitance (C_{dl}) were determined from Nyquist plots. The inhibition efficiencies were calculated from charge transfer resistance values by using the following formula:

$$\%\mathrm{IE} = \left[\frac{R_{\mathrm{ct(inh)}} - R_{\mathrm{ct}}}{R_{\mathrm{ct(inh)}}}\right] \times 100, \quad (5)$$

where R_{ct} is the charge transfer resistance in absence of inhibitor, and $R_{\mathrm{ct(inh)}}$ is the charge transfer resistance in presence of inhibitor.

2.6. Synthesis of Inhibitors. One gram of leucine was dissolved in 25 mL of 5% NaOH solution in a conical flask. To this mixture benzoyl chloride/acetyl chloride (2.25 mL) was added in five portions in (0.49 mL increments) and shaken vigorously until all the chloride has reacted. Acidified with diluted hydrochloric acid the crude product was washed with cold ether. Finally, the desired product was recrystallized from ethanol [11]. The purity of the compounds was checked by TLC. The structure of the inhibitors are given in Scheme 1.

3. Results

3.1. Weight-Loss Study. Weight-loss studies were performed in accordance with ASTM method. Test was conducted in 15% HCl (v/v) solution for 6 h at different temperatures (from 298 to 333 K) containing different concentration of inhibitors (10–150 ppm).

3.1.1. Effect of Concentration. The percentage inhibition efficiencies (%IEs) in presence of 10, 20, 50, 100, and 150 ppm of AAL and BAL have been evaluated by weight-loss technique at 25°C, and the results are summarized in Table 1. It is evident from these values that the inhibitor is significantly effective even at low concentrations like 10 ppm and there is a linear increase in %IE in the whole range of concentrations studied. The corrosion rate decreases with an increase in concentration of each inhibitor and reached the maximum value in the range of 150 ppm concentration (Figure 1).

3.1.2. Effect of Temperature. The effect of temperature on the performance of AAL and BAL as corrosion inhibitors for N80 steel in 15% HCl was investigated by weight loss measurements in the temperature range 298–333 K in absence and presence of both the inhibitors at optimum concentration (150 ppm) in a thermostat. The variation of inhibition efficiency and corrosion rate with temperature are shown

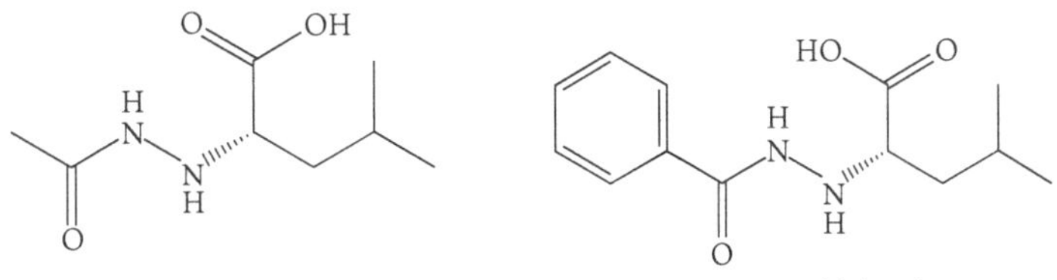

SCHEME 1

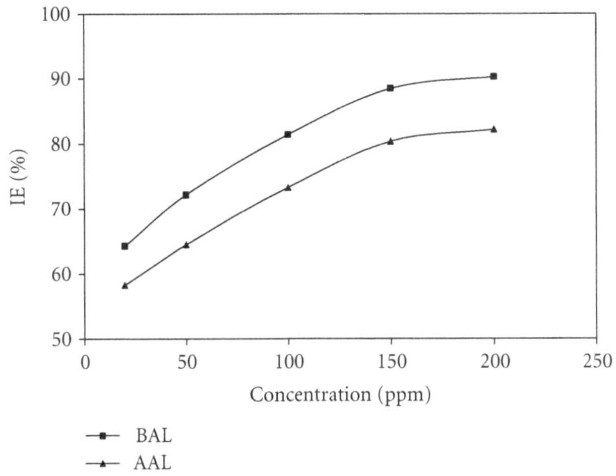

FIGURE 1: Variation of inhibition efficiency with concentration in presence of AAL and BAL.

TABLE 1: Corrosion inhibition of N80 steel in 15% HCl in presence and absence of inhibitors AAL and BAL at different concentration.

Conc. (ppm)	AAL CR (mpy)	AAL %IE	BAL CR (mpy)	BAL %IE
0	9.55	—	9.55	—
10	3.98	58.2	3.41	64.2
20	3.39	64.4	2.66	72.1
50	2.55	73.3	1.78	81.3
100	1.88	80.3	1.10	88.4
150	1.70	82.2	0.93	90.2

in Table 2. The observation depicts that rate of corrosion increases with increase in temperature. This may be due to the fact that at higher temperature the metal organic complex layer dissociates leaving a porous diffused film, which is responsible for corrosion [12].

3.1.3. Adsorption Isotherms. To determine the adsorption mode, various isotherms were tested, and Langmuir adsorption isotherm was found to be best, which gives a straight line graph for the plot of $\log(\theta/1-\theta)$ verses logarithmic concentration of inhibitors (Figure 2).

3.1.4. Kinetic Study. The apparent activation energy (E_a) was calculated by using Arrhenius equation:

$$\log k = -\frac{E_a}{2.303RT} + \log A, \quad (6)$$

where k is rate of corrosion, E_a is the apparent activation energy, R is the universal gas constant, T is absolute temperature, and A is the Arrhenius preexponential factor. By plotting $\log k$ against $1/T$ the values of activation energy (E_a) have been calculated ($E_a = -(\text{Slope}) \times 2.303 \times R$) (Figure 3). Activation energy for N80 steel in 15% HCl increases in presence of inhibitors. The E_a values for AAL and BAL were found to be 57.6 kJ/mol, 68.7 kJ/mol, respectively, (Table 3), which are higher than the E_a value for blank solution (38.4 kJ/mol).

The values of entropy of activation (ΔS^*) and enthalpy of activation (ΔH^*) were calculated by using the formula:

$$k = \left(\frac{RT}{Nh}\right)\exp\left(\frac{\Delta S^*}{R}\right)\exp\left(\frac{\Delta H^*}{RT}\right), \quad (7)$$

where k is rate of corrosion, h is Planks constant, N is Avogadro's number, ΔS^* is the entropy of activation, and ΔH^* is the enthalpy of activation. A plot of $\log(k/T)$ verses $1/T$ (Figure 4) should give a straight line, with a slope of $(-\Delta H^*/2.303R)$ and a intercept of $[\log(R/Nh) + \Delta S^*/2.303R]$, from which the values of ΔS^* and ΔH^* can be calculated (Table 3). The ΔH^* values for AAL and BAL were found to be as −62.1 kJ/mol and −69.4 kJ/mol, respectively, that revealed the exothermic nature of corrosion reaction. ΔS^* values were found to be −32.5 J/mol and −22.5 J/mol for AAL and BAL, respectively (Table 3).

The average value for free energy of adsorption (ΔG_{ads}) were calculated using the following equation:

$$K = \frac{1}{55.5}\exp\left(\frac{-\Delta G_{ads}}{RT}\right), \quad (8)$$

where θ is degree of coverage on metal surface, C is concentration of inhibitors in $\text{mol}\,\text{l}^{-1}$, R is molar gas constant in $\text{J}\,\text{K}^{-1}\,\text{moL}^{-1}$, and T is temperature. The value of 55.5 in the above equation is the concentration of water in the solution in mole/litre. The equilibrium constant (K) has been replaced by the equation

$$K = \frac{\theta}{(1-\theta)C}. \quad (9)$$

By plotting $\log K$ against $1/T$ the value of ΔG_{ads} can be calculated ($\Delta G_{ads} = -2.303 \times R \times \text{Slope}$) from the slope of

Table 2: Corrosion parameters in absence and in presence of AAL and BAL at different temperatures.

Temperature (K)	Blank		AAL		BAL	
	CR (mpy)	IE%	CR (mpy)	IE%	CR (mpy)	IE%
298	9.55	—	1.70	82.2	0.93	90.2
303	12.09	—	2.63	78.2	1.79	85.1
313	19.27	—	4.87	74.7	4.21	78.1
323	30.42	—	10.51	65.4	9.01	70.3
333	48.94	—	20.09	58.9	18.31	62.5

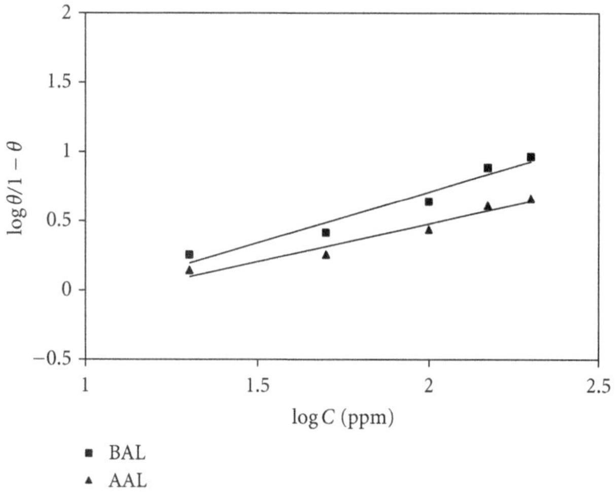

Figure 2: Langmuir adsorption isotherm in presence of AAL and BAL.

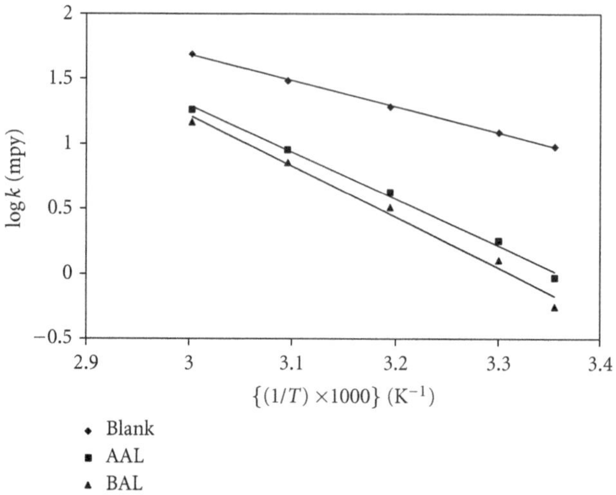

Figure 3: Arrhenius plot for the corrosion of N80 steel in 15% HCl in presence and in absence of AAL and BAL.

the straight line obtained (Figure 5). The values of free energy of adsorption (ΔG_{ads}) for AAL and BAL were found to be −32.2 kJ/mol and −36.7 kJ/mol, respectively (Table 3). Values of ΔG_{ads} around −20 kJ/mol or lower are consistent with the electrostatic interaction between charged organic molecules and the charged metal surface (physisorption); those around −40 kJ/mol or higher involve charge sharing or transfer from the organic molecules to the metal surface to form a coordinate type of bond (chemisorptions). The value of ΔG_{ads} for AAL and BAL that was found to be −32.2 kJ/mol and −36.7 kJ/mol which is less than −40 kJ/mol indicated that in addition to electrostatic interaction, there may be some other interactions [13]. Lebrini et al. [14] studied some triazole derivatives as corrosion inhibitors of mild steel in 1 M $HClO_4$. The Gibbs free energy of adsorption of these molecules were reported to be around −34 kJ/mol. They concluded that the adsorption mechanism of these molecules on steel involved two types of interactions, chemisorptions, and physisorption. Similar conclusion was found by Ozcan [15], who studied the use of cystine as a corrosion inhibitor on mild steel in sulfuric acid. Thus, adsorption of studied inhibitors at the surface of N80 steel is not pure physisorption but it is combination of physisorption as well as chemisorptions.

3.2. Potentiodynamic Polarization Study. Figures 6 and 7 show the polarization curves of N80 steel in 15% HCl solution in absence and presence of 10, 50, and 150 ppm of inhibitors AAL and BAL, respectively. The nature of the polarization curves remains almost same in absence and presence of both the inhibitors but in presence of inhibitors the curves shifted towards the lower current density as compared to the blank. The shift in current density towards lower current density in presence of inhibitors increases on increasing the concentration of the inhibitors. The electrochemical corrosion parameters including corrosion current densities (I_{corr}), corrosion potential (E_{corr}), cathodic Tafel slope (β_c), anodic Tafel slope (β_a), and inhibition efficiency obtained from the polarization curves of AAL and BAL are presented in Table 4. It is clear from the Figures 6 and 7 that with the increase of inhibitors concentration both anodic and cathodic currents were inhibited.

It is apparent from the Table 4 that I_{corr} decreases considerably in the presence of both the inhibitors, and %IE increases with the increase in the inhibitor concentration, due to the increase in the blocked fraction of the electrode surface by adsorption. The value of anodic and cathodic Tafel slopes increases in presence of both the inhibitors as compared to the blank solution. A minor shift in E_{corr} values towards the negative direction was obtained in presence of both the inhibitors, indicating mixed nature of the inhibitors.

3.3. Electrochemical Impedance Spectroscopy. Nyquist plots of N80 steel in 15% HCl in the presence and absence of

Table 3: Thermodynamic parameter in absence and in presence of AAL and BAL.

Inhibitor	ΔH^* (kJ/mol)	E_a (kJ/mol)	ΔG_{ads} (kJ/mol)	ΔS^* (J/mol)
Blank	—	38.4	—	—
AAL	−62.1	57.6	−32.2	−32.5
BAL	−69.4	68.7	−36.7	−22.5

Table 4: Electrochemical corrosion parameters in absence and in presence of AAL and BAL at different concentration.

Inhibitors	Conc of inhibitors (ppm)	Tafel slopes β_a (mV dec^{-1})	Tafel slopes β_c (mV dec^{-1})	I_{corr} (μA/cm^2)	E_{corr} (mV)	%IE	%IE from wt. loss
Blank	—	109	153	471	−468	—	—
	10	113	162	193	460	59.0	58.2
AAL	50	118	170	108	−462	77.0	73.3
	150	121	178	74	−478	84.1	82.2
	10	116	171	172	−476	63.3	64.2
BAL	50	119	180	84	−480	82.1	81.3
	150	125	187	35	−484	92.5	90.2

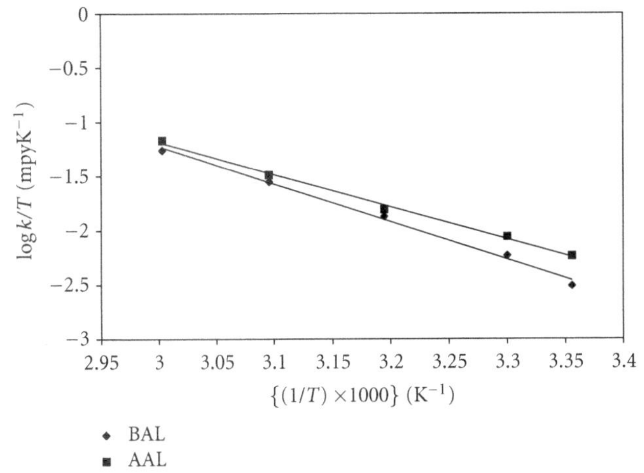

Figure 4: A plot of $\log k/T$ versus $1/T$ in presence of AAL and BAL.

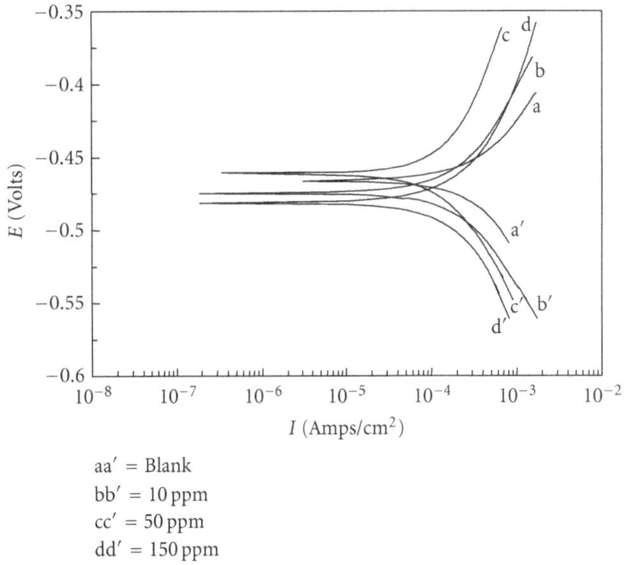

Figure 6: Potentiodynamic polarization curves for N80 steel in 15% HCl in absence and in presence of AAL.

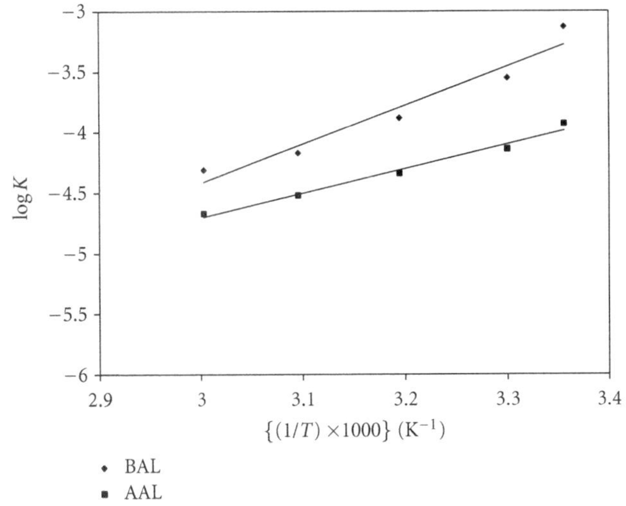

Figure 5: A plot of $\log K$ versus $1/T$ in presence of AAL and BAL.

different concentrations (10, 50, and 150 ppm) of AAL and BAL at 25 ± 1°C are shown in the Figures 8 and 9. All the Nyquist plots obtained were semicircle in nature, the diameter of the semicircles increases with increase in inhibitor concentration, and shape is maintained throughout the tested concentration, indicating that almost no change in the corrosion mechanism occurs due to inhibitor action. The electrochemical parameters (R_{ct}, C_{dl}) calculated from the Nyquist plots by using the equivalent circuit (Figure 10) in presence and absence of inhibitors are presented in Table 5. Inspection of Table 5 reveals that R_{ct} values increase prominently while C_{dl} values tend to decrease with the increasing concentration of inhibitors, which suggest that inhibitors molecule function by adsorption on metal/solution

TABLE 5: Electrochemical impedance parameters in absence and in presence of AAL and BAL.

Inhibitors	Conc. (ppm)	R_{ct} ($\Omega\,cm^2$)	C_{dl} ($\mu F\,cm^{-2}$)	IE%
Blank	—	176	662	—
AAL	10	439	369	59.9
	50	745	173	76.3
	150	1006	145	82.5
BAL	10	505	286	65.1
	50	998	165	82.3
	150	2301	55	92.3

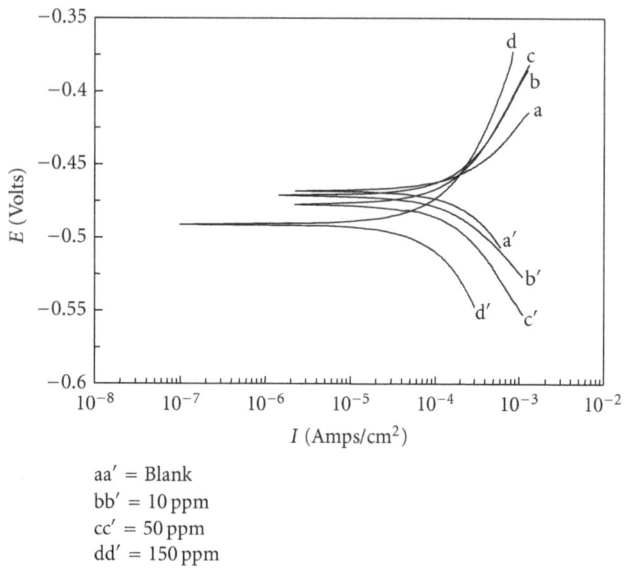

FIGURE 7: Potentiodynamic polarization curves for N80 steel in 15% HCl in absence and in presence of BAL.

interface. The values of electrochemical double layer capacitance (C_{dl}) were calculated at the frequency f_{max}, at which the imaginary component of the impedance is maximal ($-Z_i$) by the following equation:

$$C_{dl} = \frac{1}{2\pi f_{max} R_{ct}}. \quad (10)$$

3.4. SEM Study of Metal Surface Product. SEM microphotographs (Figures 11(a), 11(b), 11(c), and 11(d)) of the film formed on the metal surface at 1000x magnifications were studied to analyze the change in the morphology of the metal surface after corrosion tests in presence and absence of the inhibitors. Figure 11(a) shows the SEM of polished sample (magnification 1000x). Figure 11(b) shows the SEM microphotographs of N80 steel (magnification 1000x) when exposed to 15% HCl solution at room temperature in absence of inhibitors. Figures 11(c) and 11(d) are the microphotographs of the metal surface when exposed to the acid medium in presence of BAL and AAL, respectively, at the same magnifications.

4. Discussion

The experimental results in the (Table 1) reveal that inhibition efficiency offered by AAL and BAL increases on increasing their concentration indicating that adsorption of inhibitors increases as concentration increases resulting in reduction of corrosion rate. It is clear for the Table 2 that the inhibition efficiency of both the inhibitors decreases on increasing the temperature. The decrease in efficiency at higher temperature may be due to desorption of inhibitors from the N80 steel surface at higher temperatures. The surface coverage data of AAL and BAL shows that both the inhibitors followed Langmuir adsorption isotherm indicating that adsorption is the root cause of corrosion inhibition.

The values of apparent activation energy (E_a) in the inhibited solution were found to be higher as compared to that in the blank acid (Table 3). The increase in activation energy E_a indicates the retardation in corrosion rate which could have occurred because of adsorption of the inhibitors at the surface of the metal. The higher value of E_a in presence of both the inhibitors as compared to the blank solution also indicated that the dissolution of N80 steel was decreased by making a barrier by their adsorption on metal surface. The ΔH^* values for AAL and BAL were found to be as −62.1 kJ/mol and −69.4 kJ/mol, respectively, that revealed the exothermic nature of corrosion reaction. ΔS^* values were found to be −32.5 J/mol and −22.5 J/mol for AAL and BAL, respectively (Table 3).

The value of ΔG_{ads} for AAL and BAL was found to be −32.2 kJ/mol and −36.7 kJ/mol, respectively (Table 3). The negative values of ΔG_{ads} are consistent with the spontaneity of the adsorption process. Generally, the energy values of −20 kJ/mol or less negative are associated with an electrostatic interaction between charged molecules and charged metal surface, physisorption; those of −40 kJ/mol or more negative involve charge sharing or transfer from the inhibitor molecules to the metal surface to form a coordinate covalent bond, chemisorptions [16]. In present study, the ΔG_{ads} value obtained for both inhibitors on N80 steel in 15% HCl solution was higher than −20 kJ/mol but less than −40 kJ/mol (Table 3); this indicates that the adsorption is neither typical physisorption nor typical hemisorptions but it is complex mixed type; that, is the adsorption of inhibitor molecules on the N80 steel surface in present study involves both physisorption and chemisorptions. The negative values

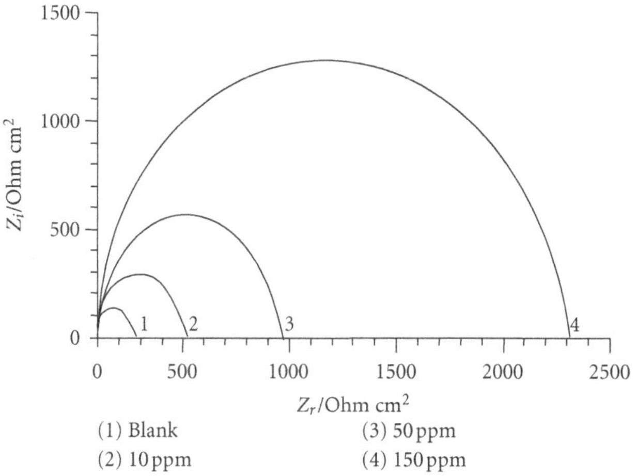

Figure 8: Nyquist plots of the corrosion of N80 steel in 15% HCl without and with different concentration of BAL.

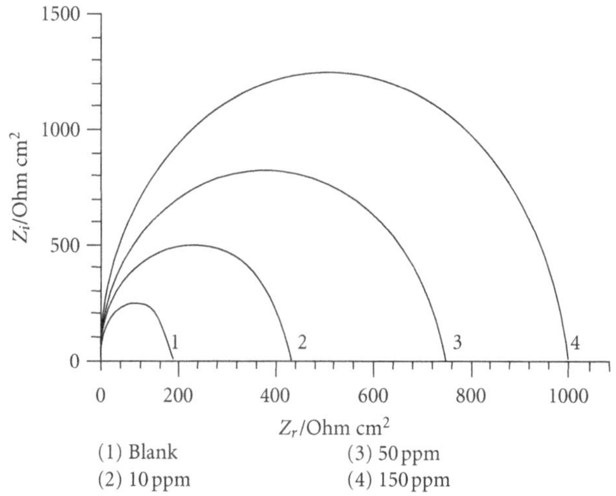

Figure 9: Nyquist plots of the corrosion of N80 steel in 15% HCl without and with different concentration of AAL.

of ΔS^* obtained in the inhibited and uninhibited solutions suggest that the activated complex is the stage that determines the corrosion rate [17]. The negative values of the entropy change of activation (ΔS^*) for AAL and BAL also imply that the activated complex in the rate determining step represents an association rather than a dissociation step, indicating that a decrease in disordering takes place on going from reactants to the activated complex [18]. The negative value of enthalpy change of activation (ΔH^*) reflects the exothermic nature of activation of corrosion process.

The variation in the values of β_a and β_c in presence of both the inhibitors (Table 4) indicates that both the anodic and cathodic processes are controlled. The significant reduction in I_{corr} at higher concentration level (150 ppm) indicated better inhibition performance at higher concentration. In 15% HCl solution, the presence of AAL and BAL cause a minor shift in E_{corr}, which indicates that inhibitor molecules are mixed type inhibitors. Generally, if the displacement in E_{corr} is >85 mV with respect to E_{corr} in uninhibited solution,

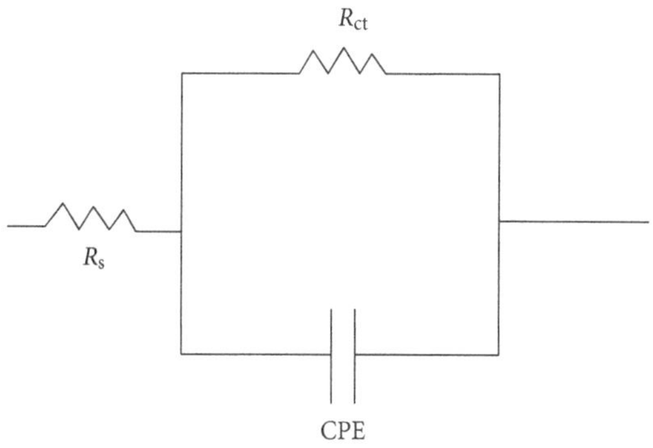

Figure 10: Equivalent circuit diagram.

the inhibitor can be seen as a cathodic or anodic type [19, 20]. In our study the maximum displacement is 12 mV in BAL and 10 mV in AAL, which indicates that both inhibitors can be arranged as a mixed type inhibitor.

In EIS study an increase in R_{ct} value is observed with increasing the inhibitors concentration, suggesting that the charge transfer process is retarded due to decrease in the uncovered surface available for corrosion reaction. Due to the nonhomogeneity or roughness of the metal surface the observed semicircles of capacitive loops were depressed into the Zi which is often referred to as frequency dispersion [21]. It is worth mentioning that the double layer capacitance (C_{dl}) value is affected by imperfections of the surface [22]. The C_{dl} values were found to decrease with increase in concentration of inhibitors solutions. This behavior is generally seen for system where inhibition occurred due to the formation of a surface film by the adsorption of inhibitor on the metal surface [23]. Decrease in C_{dl}, which can result from a decrease in local dielectric constant and/or an increase in the thickness of the electrical double layer, suggests that the inhibitor molecules act by adsorption at the metal/solution interface [24].

Figure 11(b) shows the SEM microphotographs of N80 steel (magnification 1000x) when exposed to 15% HCl solution at room temperature in absence of inhibitors. Figure 11(b) shows that the steel surface appears to be very rough in absence of inhibitors. This is due to formation of uniform flake type corrosion products on the metal surface. No pitting and other separate phase are visible in microphotograph. Figures 11(c) and 11(d) are the microphotographs of the metal surface when exposed to the acid medium in presence of BAL and AAL, respectively, at the same magnifications. In comparison to the microphotograph of the steel surface without inhibitor the photographs of the exposed surface in presence of inhibitors are found to be covered with a semiglobular type protective film of compounds uniformly spread over the surface. The protective film is formed due to adsorption of the inhibitor molecules on the steel surface.

The inhibition efficiency afforded by AAL and BAL may be attributed to the presence of electron-rich N atoms of amino group. The inhibition efficiency of BAL is higher as compared to AAL due to presence of delocalized π-electrons and its bigger size. In the aqueous acidic solutions AAL

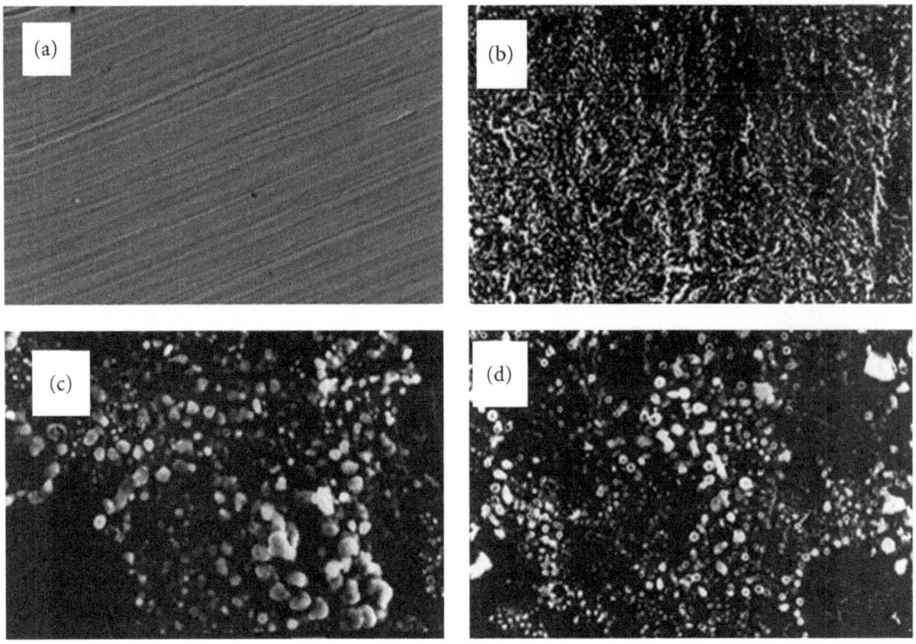

FIGURE 11: SEM of (a) polished sample; (b) sample in presence of 15% hydrochloric acid; (c) sample in presence of 150 ppm of BAL; (d) sample in presence of 500 ppm of AAL.

and BAL exist either as neutral molecules or in the form of protonated molecules (cations). These inhibitors may adsorb on the metal/acid solution interface by one and/or more of the following ways:

(i) electrostatic interaction of protonated inhibitors with already adsorbed chloride ions,

(ii) interaction between unshared electron pairs of heteroatoms and vacant d orbital of iron surface atoms.

Generally, two modes of adsorption could be considered. In one mode, the neutral inhibitors may be adsorbed on the surface of N80 steel through the chemisorptions mechanism, involving the displacement of water molecules from the N80 steel surface and the sharing electrons between the heteroatoms and iron. In another mode, since it is well known that the steel surface bears positive charge in acid solution [25], so it is difficult for the protonated inhibitors to approach the positively charged mild steel surface (H_3O^+/metal interface) due to the electrostatic repulsion. Since chloride ions have a smaller degree of hydration, thus they could bring excess negative charges in the vicinity of the interface and favour more adsorption of the positively charged inhibitor molecules, and the protonated inhibitors adsorb through electrostatic interactions between the positively charged molecules and the negatively charged metal surface. Thus there is a synergism between adsorbed Cl^- ions and protonated inhibitors. Experimental data reveals that %IE inhibitors are in the order BAL > AAL.

5. Conclusion

AAL and BAL both act as good corrosion inhibitors for corrosion of N80 steel in 15% HCl solution. The inhibition efficiency values increase with the increase in inhibitors concentration but decrease with increasing temperature for corrosion of N80 steel in 15% HCl. The adsorption of the AAL and BAL on N80 steel surface obeys Langmuir adsorption isotherm. Variation in the values of β_a and β_c (Tafel slopes) and shift in the values of corrosion potential (E_{corr}) indicate that both the tested inhibitors are mixed type inhibitors; EIS measurements show that charge transfer resistance (R_{ct}) increases and double layer capacitance (C_{dl}) decreases in presence of inhibitors. The SEM of the inhibited metal surface suggested the adsorption of the inhibitor molecules on the surface of N80 steel.

Acknowledgments

Financial assistance from Indian School of Mines Dhanbad under the "Faculty Research Scheme" to M. Yadav is gratefully acknowledged.

References

[1] G. W. Poling, "Infrared studies of protective films formed by acetylenic corrosion inhibitors," *Journal of The Electrochemical Society*, vol. 114, pp. 1209–1214, 1967.

[2] K. D. Neemla, R. C. Saxena, and A. Jayaraman, "Corrosion inhibitor studies on steels in hydrochloric acid," *Corrosion Prevention and Control*, vol. 6, pp. 69–76, 1992.

[3] W. W. Frenier, F. B. Growcock, and V. R. Lopp, "α-alkenylphenones—a new class of acid corrosion inhibitors," *Corrosion*, vol. 44, no. 9, pp. 590–598, 1988.

[4] W. W. Frenier, Eur. Patent, 047400, 1972.

[5] A. 1. Cizek, U.S. Patent 4997040, 1991.

[6] R. F. Monroe, C. H. Kuchera, and B. D. Oates, U.S. Patent, 3007454,1963.

[7] S. Ghareba and S. Omanovic, "Interaction of 12-aminododecanoic acid with a carbon steel surface: towards the development of "green" corrosion inhibitors," *Corrosion Science*, vol. 52, no. 6, pp. 2104–2113, 2010.

[8] A. R. Sathiya Priya, V. S. Muralidharan, and A. Subramania, "Development of novel acidizing inhibitors for carbon steel corrosion in 15% boiling hydrochloric acid," *Corrosion*, vol. 64, no. 6, pp. 541–552, 2008.

[9] P. Bommersbach, C. Alemany-Dumont, J. P. Millet, and B. Normand, "Hydrodynamic effect on the behaviour of a corrosion inhibitor film: characterization by electrochemical impedance spectroscopy," *Electrochimica Acta*, vol. 51, no. 19, pp. 4011–4018, 2006.

[10] Q. Qu, Z. Hao, L. Li, W. Bai, Y. Liu, and Z. Ding, "Synthesis and evaluation of Tris-hydroxymethyl-(2-hydroxybenzylidenamino)-methane as a corrosion inhibitor for cold rolled steel in hydrochloric acid," *Corrosion Science*, vol. 51, no. 3, pp. 569–574, 2009.

[11] M. Bodanszky, *Peptide Chemistry: A Practical Text Book*, Springer, Berlin, Germany, 1988.

[12] I. N. Putolova, S. A. Balezin, and V. P. Barannik, *Metallic Corrosion Inhibitors*, Pergamon Press, New York, NY, USA, 1960.

[13] M. Behpour, S. M. Ghoreishi, N. Soltani, M. Salavati-Niasari, M. Hamadanian, and A. Gandomi, "Electrochemical and theoretical investigation on the corrosion inhibition of mild steel by thiosalicylaldehyde derivatives in hydrochloric acid solution," *Corrosion Science*, vol. 50, no. 8, pp. 2172–2181, 2008.

[14] M. Lebrini, M. Traisnel, M. Lagrenée, B. Mernari, and F. Bentiss, "Inhibitive properties, adsorption and a theoretical study of 3,5-bis(*n*-pyridyl)-4-amino-1,2,4-triazoles as corrosion inhibitors for mild steel in perchloric acid," *Corrosion Science*, vol. 50, no. 2, pp. 473–479, 2008.

[15] M. Ozcan, "AC impedance measurement of cystine adsorption at mild steel/sulfuric acid interface as corrosion inhibitor," *Journal of Solid State Electrochemistry*, vol. 12, pp. 1653–1661, 2008.

[16] I. Dehri and M. Özcan, "The effect of temperature on the corrosion of mild steel in acidic media in the presence of some sulphur-containing organic compounds," *Materials Chemistry and Physics*, vol. 98, no. 2-3, pp. 316–323, 2006.

[17] S. S. Al-Juaid, "Inhibition of corrosion of carbon steel 1018 in acid medium with ethoxylated aliphatic alcohols," *Chemistry and Technology of Fuels and Oils*, vol. 47, no. 1, pp. 58–65, 2011.

[18] O. Olivares, N. V. Likhanova, B. Gómez et al., "Electrochemical and XPS studies of decylamides of α-amino acids adsorption on carbon steel in acidic environment," *Applied Surface Science*, vol. 252, no. 8, pp. 2894–2909, 2006.

[19] H. Ashassi-Sorkhabi, M. R. Majidi, and K. Seyyedi, "Investigation of inhibition effect of some amino acids against steel corrosion in HCl solution," *Applied Surface Science*, vol. 225, no. 1–4, pp. 176–185, 2004.

[20] X. Li, S. Deng, and H. Fu, "Synergism between red tetrazolium and uracil on the corrosion of cold rolled steel in H_2SO_4 solution," *Corrosion Science*, vol. 51, no. 6, pp. 1344–1355, 2009.

[21] M. Lebrini, M. Lagrenée, H. Vezin, M. Traisnel, and F. Bentiss, "Experimental and theoretical study for corrosion inhibition of mild steel in normal hydrochloric acid solution by some new macrocyclic polyether compounds," *Corrosion Science*, vol. 49, no. 5, pp. 2254–2269, 2007.

[22] J. R. Macdonald, W. B. Johnson, and J. R. Macdonald, Eds., *Impedance Spectroscopy*, John Wiley & Sons, New York, NY, USA, 1987.

[23] I. L. Rosenfield, *Corrosion Inhibitors*, McGraw-Hill, New York, NY, USA, 1981.

[24] M. MaCafferty and N. Hackerman, "Double layer capacitance of iron and corrosion inhibition with polymethylene diamines," *Journal of The Electrochemical Society*, vol. 119, pp. 146–154, 1972.

[25] G. N. Mu, T. P. Zhao, M. Liu, and T. Gu, "Effect of metallic cations on corrosion inhibition of an anionic surfactant for mild steel," *Corrosion*, vol. 52, no. 11, pp. 853–856, 1996.

10

Electromagnetic Heating of Heavy Oil and Bitumen: A Review of Experimental Studies and Field Applications

Albina Mukhametshina[1,2] and Elena Martynova[1,2]

[1] *Harold Vance Department of Petroleum Engineering, Texas A&M University, 3116 TAMU-407 Richardson Building, College Station, TX, USA*
[2] *Gubkin Russian State University of Oil and Gas, 65 Leninsky Prospekt, Moscow, Russia*

Correspondence should be addressed to Elena Martynova; elena.martynova@pe.tamu.edu

Academic Editor: Jorge Ancheyta

Viscosity is a major obstacle in the recovery of low API gravity oil resources from heavy oil and bitumen reservoirs. While thermal recovery is usually considered the most effective method for lowering viscosity, for some reservoirs introducing heat with commonly implemented thermal methods is not recommended. For these types of reservoirs, electromagnetic heating is the recommended solution. Electromagnetic heating targets part of the reservoir instead of heating the bulk of the reservoir, which means that the targeted area can be heated up more effectively and with lower heat losses than with other thermal methods. Electromagnetic heating is still relatively new and is not widely used as an alternate or addition to traditional thermal recovery methods. However, studies are being conducted and new technologies proposed that could help increase its use. Therefore, the objective of this study is to investigate the recovery of heavy oil and bitumen reservoirs by electromagnetic heating through the review of existing laboratory studies and field trials.

1. Introduction

High-frequency electromagnetic radiation is a relatively new technique for use in enhanced oil recovery methods. It has been tested by theoretic, laboratories and field trial research in Russia [1–10], the United States [11–17], Canada [18–21], and other countries [22–34]. Traditional thermal recovery and well stimulation techniques using hot steam or fluid are not effective in some cases [7, 35] due to prohibitive heat losses from injection wells and reservoirs, low reservoir injectivity (especially for bitumen deposits), steam leakage, large overburden heat loss at thin pay zones, permafrost conditions, and so forth. Furthermore, commonly used thermal recovery methods are not considered environmentally friendly, damaging the hydrogeologic environment and contributing to the greenhouse effect.

The most important thing in electromagnetic heating is that the heat is developed within the material rather than being brought from outside, which means the material is heated more uniformly throughout the medium [27]. Therefore, instead of heating the bulk reservoir volume, part of the reservoir can be targeted and heated more effectively with lower heat loss than other thermal methods. Unlike traditional thermal recovery methods, microwave heating causes friction by vibration of molecules, which results in dielectric heating of the reservoir. Heat and mass transfer in different environments under microwave influence was studied by a number of scientists around the globe, but its application as an EOR method is not yet fully understood. Microwave heating is not used productively because of the lack of reliable information about the processes of heat and mass transfer in a multiphase system in porous media under the influence of electromagnetic radiation, which does not allow effective control. Therefore, current research studies use modeling to discover optimum design parameters for the use of microwave heating in field applications.

2. Review of Experimental Electromagnetic Heat Studies

The success of near wellbore heating with electromagnetic waves has been proven experimentally [1]. To represent reservoir rock, quartz sand with a 7.7 dielectric constant and a 0.083 tangent was used in the laboratory. A 20% initial water saturation and an 80% initial oil saturation with 16.61 cps ($16.61 * 10^{-3}$ Pa·s) gravity and 0.86 g/cm^3 (860 kg/m^3) density oil were maintained. The dielectric constant and the loss tangent of the oil sample were 2.23 and 0.019, respectively. In this study of electric and high-frequency electromagnetic heating of a reservoir model, the temperature of the medium was measured by thermometers located at different points of the experimental setup. In another case, a linear radiator with a length of 87 cm and a diameter of 19 mm was placed in the center of the setup. The linear radiator was connected via acoaxial cable to the generator supplying electromagnetic waves with a frequency of 13.56 MHz. In the experiment, it was discovered that when fluid temperature was exposed to a high-frequency electromagnetic field at the same distance from the radiation source it was greater than one initiated by electric heating. In this case, the thermal conductivity of the medium is affected only slightly. When heat is induced throughout the whole volume of the medium, the amount of heat introduced depends largely on the electrical properties of the medium. The study summarized the advantages of microwave heating over electrical heating as having deeper penetration, quicker heating, and lower heat losses.

Sayakhov [2] discussed the physical foundations of fluid filtration in high-frequency fields. Radial fluid filtration through porous media in a heterogeneous high-frequency electromagnetic field was studied experimentally. A specially designed core holder, filled with a representative porous medium, was located in a coaxial resonator. A high-frequency wave generator was used to generate mw energy, which was directed to the resonator through cables. Studied liquid came into the porous media through the hose first, then through the lumen of the inner conductor of the resonator dripped into a graduated cylinder. Experiments were started with a 200–300 watts power microwave oven at 2400 MHz frequency and 500 watts vibrational power. Temperatures were recorded throughout the experiments via thermocouples inserted in the core holder. Kerosene was used as a representative reservoir fluid. The measurements were performed with and without the influence of a high-frequency electromagnetic field. It was established that exposure to high-frequency electromagnetic fields leads to a sharp increase in flow rate per unit of time and fluid temperature at the outlet. In addition, the flow rate increases dramatically once exposed to the electromagnetic field, while the temperature increases after 10 seconds. After the field is discontinued, a sharp decrease in the flow rate is observed and a gradual cooling of the porous medium takes place.

Experimental studies [3] on the influence of electromagnetic fields with a frequency of $3 * 10^5$ Hz to $6 * 10^5$ Hz on thermal conductivity of dielectric liquids showed that the thermal conductivity of the liquids increased when exposed. The thermal conductivity increases as the magnitude of the dipole moment of the liquid used in the experiment parameters increases with the frequency and intensity of field.

Fatikhov [5] conducted experimental research on the flow of bitumen oil at different pressure gradients in a high-frequency electromagnetic field. The experiment focused on changes in the volumetric flow rate of filtered oil under different pressures at different temperatures. The initial pressure drop for bitumen from the Mordovo-Karmalskoye deposit in reservoir conditions was 0.003 MPa/m. During the experiment, the pressure drop decreased rapidly as the temperatures were increased, resulting in bituminous oil becoming a Newtonian liquid. Therefore, it was established that the application of electromagnetic heating improves fluid flow behavior and the non-Newtonian properties of bitumen decrease rapidly with increasing temperature.

In 1992, Kasevich et al. [17] studied electromagnetic heating of rock samples at 1 kW (frequency 50.55 MHz) and 200 W (144 MHz) and found that a particular type of rock could be heated to 423 K when exposed to an RF electromagnetic field. The rock, which had low thermal conductivity, heats poorly when hot steam is pumped into the reservoir. They conducted experiments at both normal and formation pressures.

Ovalles et al. [30] used a microwave with 650 watts power to heat core samples saturated with oil of 25 API gravity and 7.7 API gravity (a sample from the Orinoco River Basin). Medium API oil temperatures were measured in 0.5, 1, and 1.5 minute intervals and heavy oil intervals were increased to 1, 5, and 10 minutes. The experimental results were used to test mathematical models and predict the production of three abstract oil reservoirs in Venezuela.

Chakma and Jha [23] conducted laboratory experiments using electromagnetic heating on a scaled thin heavy oil reservoir pay zone model. Gas injection with horizontal wells during electromagnetic heating was achieved. The aim was to decrease oil viscosity with electromagnetic heating and obtain a gas drive with the injected gas. Using nitrogen for the injected gas, they were able to prove that for thin pay zones heating of the wellbore vicinity is sufficient, by achieving oil recoveries as high as 45% of original oil in place compared to estimated primary recovery rates of less than 5%. Recovery achieved by use of the combined method was higher than that of nitrogen injection or electromagnetic heating alone. Chakma and Jha also discussed a number of parameters affecting the results of the combined method, including

(i) gas injection pressure (when no gas was injected, oil was produced only due to gravity drainage and no significant convective transport occurred; therefore, gas injection provided an oil rate increase with the increasing injection pressure);
(ii) temperature (the initial production rate was not significantly affected by temperature, but later there was an increase in the production rate, meaning that overall recovery increased with temperature);
(iii) electromagnetic frequency (the higher the frequency, the greater the recovery);
(iv) oil viscosity (as expected, a higher oil viscosity leads to a lower recovery for a given electromagnetic frequency, temperature, and gas injection pressure);

(v) salinity (higher salinity provides higher recovery due to the higher conductivity of saline water compared with distilled water);

(vi) electrode distance (recovery is similar, but closer electrode spacing provides faster production rates) [23].

Hascakir et al. [24] conducted a laboratory study of microwave-assisted gravity drainage on heavy oil samples from reservoirs in Turkey (Bati Raman, 9.5 API; Garzan, 12 API; and Camurlu, 18 API) using a specially designed novel graphite core holder packed with crushed limestone. Their study described the effects of operational parameters like heating time, waiting period and rock, and fluid properties on the effectiveness of microwave heating. Some of the conclusions made are confirming ones found in Chakma and Jha [23], like the positive effect of high water salinity and water saturation. Hascakir et al. [24] also concluded that water wet conditions are preferable for obtaining higher oil recoveries and that large porosity and permeability are also favorable. When microwave heating is applied to oil samples continuously, higher temperatures are reached, which allows better results to be achieved than with periodic heating when microwave heating is applied for a limited time in periodic intervals. Therefore, higher temperatures allow for better results in continuous heating.

Jha et al. [27] proposed using microwave-assisted gravity drainage (MWAGD) in the Mehsana oil field in India. They heated specially prepared samples with the required characteristics from that field in the laboratory using a microwave with variable power up to 1000 watts operating at 3 GHz frequency, which allowed them to obtain temperature and viscosity profiles of the gravity-drained oil. They described effects of initial oil and water saturations, wettability, porosity, and permeability similar to those found by Hascakir et al. [24] and Chakma and Jha [23].

Jha et al. [27] suggested using MWAGD commercially by drilling one horizontal well and multiple vertical ones with downhole microwave antennas. However, this might not allow deep enough heat penetration, so other options are also proposed such as a combination of two horizontal wells and installing antenna inside the horizontal production well. Vertical separation of the horizontal well pair is approximately 15 meters, which is far more efficient than SAGD in which the separation is around 5 meters. Because crude oil absorbs microwave heat weakly, Jha et al. also proposed increasing thermal conductivity by injecting powdered metallic oxides, chlorides, or activated carbon through a fracture operation. The working principle and description of laboratory applications of such additives to heavy oil can be found in various studies by Hascakir et al., Kershaw et al., and Odenbach [25, 28, 29].

Technical principles of the SAGD method assisted by electromagnetic heating (EM-SAGD process) were reported by Koolman et al. [26]. Inductive heating was initiated in the laboratory using an EM source with a working frequency of 142 kHz. The sample was heated for 10 minutes at a power of 7.2 kW, achieving a rise in the temperature of 7.5 K. Laboratory

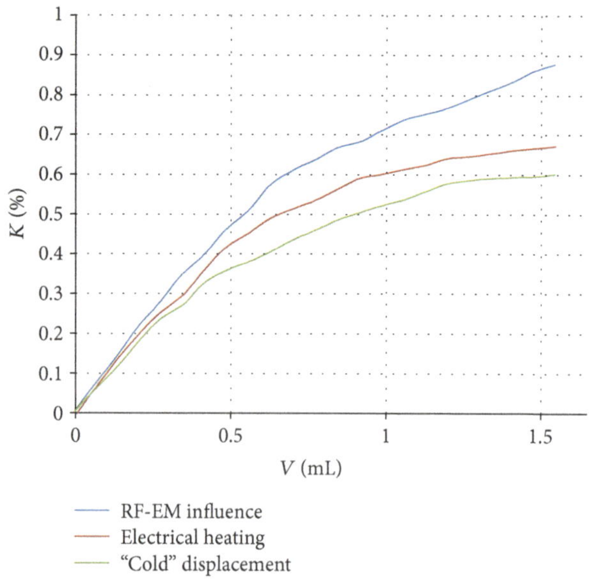

FIGURE 1: Dependence of oil recovery (K) on the relative volume of solvent injected. Adapted from—[6].

and field processes were modeled using a numerical simulator, combining electromagnetic and thermal modules. It was specially built and can be applied to field-scale simulations. According to simulation results, a 38% increase in bitumen production was predicted compared to conventional SAGD.

Kovaleva et al. investigated the effects of radio frequency electromagnetic (RF-EM) fields and electrical heating on the mass- and heat-transfer processes in a multicomponent hydrocarbon system flowing in porous media [4, 6]. Three different types of experiments were carried out: solvent (kerosene) flooding under the RF-EM field, solvent (kerosene) flooding under electrical heating, and cold solvent (kerosene) flooding. In all three experiments, the physical characteristics of the model and heating conditions (temperatures) were identically maintained. Two series of experiments on models with different granulometric composition of formation were also carried out.

Figure 1 shows the dependence of oil recovery on the volume of solvent injected. This figure demonstrates that the highest oil recovery was obtained by applying an RF-EM field. Kovaleva et al. [4, 6] concluded that following RF-EM influence on the oil-saturated samples the quantity of the received oil is more than the quantity received under electrical (and thermal) processing at identical temperatures of heating of the media. It confirms the additional "nonthermal" action of the electromagnetic field.

3. Technologies of In Situ Electromagnetic Heating of Heavy Oil and Bitumen

The first production method applying microwave heating to well production was patented in 1956 [14]. Electromagnetic waves were transferred to the well bottom from the surface through a coaxial system of internal and external pipes (tubing and casing). Interaction of electromagnetic waves with

Table 1: Characteristics of Well 40/19 of the Ishimbayskoye Oil Field.

Parameter	Value
Depth, m	830
Casing diameter, in.	6
Tubing diameter, in.	2
Flow rate, ton/day	3
Well temperature, K	287–289
Paraffin content, %	2.3
Resins content, %	11
Density, kg/m^3	890
Viscosity at 20°C, m^2/c	$20 * 10^{-6}$

Sayakhov et al. [10].

the formation causes the emergence of distributed volumetric heat sources and reduces the viscosity of the reservoir fluid. In 1965, Haagensen [13] described a device for generating high-frequency electromagnetic waves at the mouth of the well and a method of delivering electromagnetic energy through coaxial lines and waveguides to the bottom hole. In 1987, Wilson [16] described a similar device in his work, with some modifications of the radiating element of EM waves.

A huge drawback to the methods described by Haagensen and Wilson [13, 16] is the shallow penetration of electromagnetic waves, and, hence, a low sweep efficiency of heating. When the method described by Ritchey [14] was implemented, there were large losses of electromagnetic energy. Due to the finite conductivity of tubing, they are heated and electromagnetic energy is dissipated in rocks surrounding the well, resulting in large wellbore heat losses, especially if a permafrost layer is present.

Sayakhov et al. [8] proposed a method of recovery that included creating a combustion front with a simultaneous electromagnetic current influence. The environment is heated when exposed to an electromagnetic field, which decreases the viscosity and increases the mobility of crude oil. It is assumed in this method that the EM field continues influencing the reservoir after the combustion is initiated.

3.1. Review of Electromagnetic Heating Field Studies. Electromagnetic heating field trials have been carried out in Russia (Bashkortostan and Tatarstan) [2, 9, 10], the United States (California and Utah) [11, 17], and in Canada (Alberta and Saskatchewan) [18–21].

3.1.1. Russia. In Russia, field tests of radio frequency electromagnetic heating of the near-wellbore zone were first launched in 1969 at Well 40/19 in the Ishimbayskoye Oil Field in Bashkortostan and continued in the Yultimirovskoye Bitumen Field in Tatarstan according to Sayakhov et al. [2, 9, 10]. Characteristics of Well 40/19 from the Ishimbayskoye Oil Field are represented in Table 1.

The source of high-frequency electromagnetic energy was a generator providing an optimum oscillation output power of 63 kW at a frequency of 13.56 MHz. Standard RF coaxial cable was used to supply high-frequency electromagnetic energy from the generator to the well. Temperature measurements were carried out using a thermograph at the bottom hole. Temperature was recorded continuously at a fixed depth of 650–655 m (in the open hole, where the radiating element was working) while the generator was operating (Table 2).

Table 2: Dynamics of temperature growth.

Heating time, days	Temperature increase at the bottom hole, K
0.5	283
1	290
2	300
3	306
4	311
5	313

Sayakhov et al. [9].

Table 3: Characteristics of the Yultimirovskoye reservoir.

Parameter	Value
Porosity, %	25
Bitumen saturation, %	3.6
Permeability, micm^2	0–0.183

Sayakhov [2].

3.1.2. Yultimirovskoye Bitumen Field. RF electromagnetic heating was conducted in the Yultimirovskoye Bitumen Field by Sayakhov in 1980 [2] (Table 3). Two wells spaced 5 meters apart were studied, Well 150 and Well 1.

Electromagnetic heating of the bitumen reservoir was conducted in several stages at different conditions. Initially, the RF-EM installation was set to about of 20 kW power. After 36.5 hours, the temperature at the bottom of Well 150 has increased from 282 to 389 K. No temperature change was observed in Well 1. In the next phase, RF-EM installation was reset to approximately 30 kW of power. As a result, the temperature in Well 150 reached 423 K after six hours. It should be noted that the growth rate of temperature increased. In the third phase, the RF-EM installation was set to a maximum of 60 kW, which caused the heating intensity in Well 150 to increase greatly. After 5.5 hours, the temperature in Well 150 increased from 417 to 463 K. Next, the RF-EM unit was turned off for 2 hours, resulting in a temperature drop to 423 K.

During the next 32 hours, the RF-EM installation operated at maximum output, causing the heating of the bottom hole of Well 150 up to 583 K and the bottom hole of Well 1 up to 318 K. In Sayakhov's experiment [2], the fluoroplastic collars that centered the tubing in Well 150 melted (their maximum operating temperature was 573 K). As a result, a short circuit between the casing and tubing occurred and the RF-EM unit broke and was disabled. Before this disruption, the RF-EM installation worked steadily throughout the field experiment.

Deep heat penetration (up to 5 m in the reservoir) was demonstrated by temperature measurements done during the cooling of Well 150's drain zone after the exposure. The temperature decreased from 373 K to 343 K in three days.

After a few days of electromagnetic heating, wellbore heat distribution revealed low heat losses to both the overburden and underburden.

3.2. The United States

3.2.1. Bakersfield, California, United States. In 1992, Kasevich et al. [17] conducted field tests of RF-EM heating in the United States in the Bakersfield, California, field. The goal was to prove the concept that controlled RF-EM radiation could be used as a thermal EOR method. The production of reservoir fluids was not measured because it was a quality, not quantity, study designed to gain a better understanding of underground processes.

A high-frequency electromagnetic wave generator with a capacity of 25 kW and frequency of 13.56 MHz was used to heat the reservoir at Well 100D. Heat penetration was determined by temperature measurements in the surrounding observation wells T10, T20, and T30, which were located 3, 6, and 9 meters from Well 100D. Kasevich et al. [17] also proved that the RF producer used could efficiently focus its radiation pattern into the desired region by measuring return loss and electromagnetic radiation. In well T10, situated 3 meters from Well 100D, the medium temperature was increased from 293 K to 393 K over 20 hours of EM heating.

3.2.2. Avintaquin Canyon and Asphalt Ridge (Utah). In 1980, one of the most detailed studies on electromagnetic-heat-based oil recovery was done at the Illinois Institute of Technology Research Institute (IITRI) by Bridges et al. [11]. They carried out extensive research work on the use of the different types of electromagnetic heating for different types of deposits, oil shale, and tar sand.

Bridges et al. [11] tested their IITRI technique of RF electromagnetic heating with two field experiments in Avintaquin Canyon, Utah, USA. Shale that was 6 meters thick was found in outcrops convenient for relatively cheap horizontal experiments. So arrays of holes were drilled and electrodes were inserted to a depth of 1 meter. These tests allowed the researchers to gain experience and to prove it was possible to achieve in situ pyrolysis of oil shale, thus increasing its thermal maturity. The power applied to the formation ranged from 5 kW to 20 kW, with a frequency of 13.56 MHz. As a result of EM heating, temperatures rose to 673 K and 20–30 % of the oil content was collected. However, it should be noted that the amount of produced oil was badly affected by the presence of cracks which allowed light hydrocarbons to escape (evaporate).

In 1981, Bridges et al. [11] conducted field tests on tar sand at Asphalt Ridge, Utah, USA. The first experiment tested the gravity drive bitumen recovery process, designed to prove EM heating concepts and improve equipment design. This experiment used vertical electrode placement and a mined collection chamber and tunnel. It was equipped with a 200 kW radio transmitter and heated 25 m^3 of tar sand. In the first experiment, the roof of the mined chamber was not supported well enough, which resulted in early termination of the experiment. Therefore, the situation had to be fixed by constructing a concrete arch for the second pilot test.

Table 4: Characteristics of the reservoir in the Wildmere field, Alberta, Canada.

Parameter	Value
Net thickness, m	1
Depth, m	600
Density, kg/m^3	987
Oil viscosity at 20°C, Pa·s	20

Spencer [18].

The heating power used on tar sand varied from 40 kW to 75 kW with a frequency of 13.56 MHz.

The second test quantified the results of heating over a longer period and at higher temperatures. In this experiment, temperatures exceeded 473 K, and 30 to 35% recovery was achieved in just 20 days. This was encouraging because continuation of heating could have resulted in even higher recovery. The power loss was minimal in all the experiments, which proved the efficiency of heating.

4. Canada

4.1. The Wildmere Field, Alberta, Canada. According to Spencer [18] commercial EM heating was first introduced in the field at Wildmere, Alberta, Canada (Table 4). The first well was drilled in January 1986 and began producing oil in March of the same year. Before EM heating began in May, the well was producing about 0.95 tonnes/day. After EM heating commenced, production rates increased and soon settled at the level of 3.18 tonnes/day until November 1986, when the well was closed due to technical reasons. Another well in this field increased in production from 1.59 tonnes/day to an average of 4.77 tonnes/day, with the maximum flow rate reaching 9.54 tonnes/day.

4.2. The Lloydminster Heavy Oil Area, Saskatchewan, Canada. In 1988-1989, two electromagnetic stimulation projects were conducted by Davidson [22] in the Lloydminster heavy oil area in Saskatchewan, Canada. Unfortunately the economic potential of the process could not be evaluated from either pilot test, because long-term heating could not be achieved due to equipment failure (casing insulation) and special reservoir conditions. However, the technical results looked promising.

The first pilot well was in Northminster (Saskatchewan, Canada) and produced 11.4 API oil from the Sparky formation. The power was applied to the well in a pulsating manner with a baseline of 20 kW with four-hour spikes of 30 kW (2 daily), in order to reduce the risk of significant damage to the insulation. Later, the power was increased to a 25 kW baseline with 35 kW pulses and finally to a 30 kW baseline with 50 kW peaks. At that point the insulation failed, and the power rates came down to 28 kW. Power rates were later leveled at 47 kW and stayed that way until terminated.

As can be seen from Table 5, water cut and production both reacted positively to electromagnetic stimulation. However, it should be noted that some portion of the increased

Table 5: Oil field performance for the Northminster pilot: primary and achieved by EM heating.

Parameter	Primary	EM heating
Production rate, m^3/day	10–12	20
Water cut, %	15–20	10–12
Productivity index, bbl/psi	0.33	0.42
Stimulation ratio		1.27

Davidson [22].

production is related to the increase in pump speed. Water cut drop can be related directly to the EM effect and improvement in oil mobility. Once the heating has been terminated, technical parameters return rapidly to their initial states.

The second pilot well was situated in Lashburn (Saskatchewan, Canada) and produced very viscous 11.4 API crude oil from the Sparky formation. During the reservoir heating phase, the power ranged from 13 to 18 kW. This well has high sand cuts and before electrical power was applied its production was not stable and regularly had to be stimulated by flushing the wellbore. The peak production reached 5.0 m^3/day and had begun to decline before electromagnetic heating was applied. Electromagnetic heating reduced the water cut, but the well was still prone to high water cuts after shut-in periods. Oil production also increased when electromagnetic forces were applied, until it reached 9 m^3/day. During the initial heating phase of the reservoir, the temperature in the bottom hole increased steadily from 295 K to 309 K, but temperatures began dropping immediately after the power was turned off or when power delivery systems failed.

4.3. ESEIEH (Alberta, Canada). Enhanced Solvent Extraction Incorporating Electromagnetic Heating technology (ESEIEH) has been patented and is currently undergoing tests, as reported by Rassenfoss [20]. The ESEIEH consortium is relying on three oil company partners to help with this testing: Laricina Energy, Nexen, and Suncor Energy. The pilot project is planned to take three years and currently is in the first stage. Field application is expected to start later in 2013. The ESEIEH method combines the familiar horizontal well pair design commonly used in the Canadian oil sands, coupled with heating using RF-EM waves and solvents, such as butane or propane. The company aims to heat the reservoir by running an antenna underground that emits enough energy to raise the temperature to 50°C (120°F).

5. Conclusion

A review of electromagnetic heating for enhanced oil recovery was presented in this paper. A number of studies show that electromagnetic heating is a promising method of enhanced oil recovery. However, the studies to date are limited, and only a few field trials have been reported. Most of the current research is based on laboratory experiments or numerical models. It should be noted that this paper did not cover the computer simulations carried out to research the effectiveness of EM heating.

Better understanding of the in situ electromagnetic process is essential and can be achieved by combining laboratory, numerical, and field-scale tests. At the moment it is not possible to assess the efficiency of EM heating or the opportunities for economic applications of it alone or in combination with traditional methods; therefore, more global studies should be conducted.

Even though sustainability of this technology has not yet been completely evaluated, the method definitely should not be overlooked by the industry because of its enormous potential. Attempts should be made to develop viable screening criteria for possible production of heavy oil, oil shale, and tar sand deposits.

References

[1] S. Chistyakov, F. Sayakhov, and G. Balabyan, "Experimental study of formations dielectric properties under the influence of high-frequency electromagnetic fields," in *University Investigations: Geology and Exploration*, pp. 153–156, 1971.

[2] F. Sayakhov, "Particular properties of filtration and fluid flow under the influence of high-frequency electromagnetic field," in *Joint University Scientific Book*, pp. 108–120, 1980.

[3] B. Savinikh, V. Dyakonov, and A. Usmanov, "The influence of alternating electric currents on the thermal conductivity of dielectric fluids," *Journal of Engineering Physics and Thermophysics*, no. 2, pp. 269–276, 1981 (Russian).

[4] A. Davletbaev and L. Kovaleva, "Combined RF EM/solvent treatment technique: heavy/extra-heavy oil production model case study," in *Proceedings of the 10th Annual International Conference Petroleum Phase Behavior and Fouling*, Rio de Janeiro, Brazil, 2009.

[5] M. A. Fatikhov, "Experimental study of bitumen initial pressure gradient in the electromagnetic field," *University Investigations: Oil and Gas*, no. 5, pp. 93–94, 1990 (Russian).

[6] L. Kovaleva, A. Davletbaev, T. Babadagli, and Z. Stepanova, "Effects of electrical and radio-frequency electromagnetic heating on the mass-transfer process during miscible injection for heavy-oil recovery," *Energy and Fuels*, vol. 25, no. 2, pp. 482–486, 2011.

[7] G. Malofeev, O. Mirsaetov, and I. Cholovskaya, "Injection of hot fluids for enhanced oil recovery and well stimulation," in *Regular and Chaotic Dynamics*, Institute of Computer Science, RussiaIgevsk, Russia, 2008.

[8] F. Sayakhov, R. Bulgakov, V. Dyblenko, B. Deshura, and M. Bykov, "About HF heating of bitumen reservoirs," *Petroleum Engineering*, no. 1, pp. 5–8, 1980 (Russian).

[9] F. L. Sayakhov, L. A. Kovaleva, M. A. Fatikhov, and G. A. Khalikov, "Method of thermal effect on oil-bearing formation," SU Patent 1723314, 1992.

[10] F. Sayakhov, I. Habibullin, M. Yagudin, and M. Fatyhov, "Technique and technology of thermal well stimulation on the basis electro-thermo-chemical and electromagnetic effects," *University Investigations: Oil and Gas*, no. 2, pp. 33–42, 1992 (Russian).

[11] J. E. Bridges, J. J. Krstansky, A. Taflove, and G. C. Sresty, "The IITRI in situ RF fuel recovery process," *Journal of Microwave Power*, vol. 18, no. 1, pp. 3–14, 1983.

[12] J. Bridges, "Method for in-situ heat processing of hydrocarbonaceous formation," US Patent 4140180, 1979.

[13] A. D. Haagensen, "Oil well microwave tools," Patent USA 3170119, 1965.

[14] H. W. Ritchey, "Radiation Heating System, US Patent," Tech. Rep. 2757738, 1956.

[15] G. C. Sresty, R. H. Snow, and J. E. Bridges, "Recovery of liquid hydrocarbons from oil shale by electromagnetic heating in-situ," US Patent 4485869, 1984.

[16] R. Wilson, "Well production method using microwave heating," US Patent 4485868, 1987.

[17] R. S. Kasevich, S. L. Price, D. L. Faust, and M. F. Fontaine, "Pilot testing of a radio frequency heating system for enhanced oil recovery from diatomaceous earth," in *Proceedings of the SPE Annual Technical Conference & Exhibition*, pp. 105–113, New Orleans, La, USA, September 1994.

[18] H. L. Spencer, "Electromagnetic Oil Recovery, Ltd," Calgary, Canada, 1987.

[19] F. E. Vermeulen and F. S. Chute, "Electromagnetic techniques in the in-situ recovery of heavy oils," *Journal of Microwave Power*, vol. 18, no. 1, pp. 15–29, 1983.

[20] S. Rassenfoss, "Seeking more oil, fewer emissions," *Journal of Petroleum Technology*, vol. 64, no. 9, pp. 34–38, 2012.

[21] B. C. W. Mcgee and F. E. Vermeulen, "The mechanisms of electrical heating for the recovery of bitumen from oil sands," *Journal of Canadian Petroleum Technology*, vol. 46, no. 1, pp. 28–34, 2007.

[22] R. J. Davidson, "Electromagnetic stimulation of Lloydminster heavy oil reservoirs: field test results," *Journal of Canadian Petroleum Technology*, vol. 34, no. 4, pp. 15–24, 1995.

[23] A. Chakma and K. N. Jha, "Heavy-oil recovery from thin pay zones by electromagnetic heating, paper SPE 24817," in *Proceedings of the Annual Technical Conference and Exhibition*, Society of Petroleum Engineers, Washington, DC, USA, October 1992.

[24] B. Hascakir, C. Acar, Schlumberger, B. Demiral, and S. Akin, "Microwave assisted gravity drainage of heavy oils," in *Proceedings of the International Petroleum Technology Conference (IPTC '08)*, pp. 1908–1916, Kuala Lumpur, Malaysia, December 2008.

[25] B. Hascakir, T. Babadagli, and S. Akin, "Experimental and numerical modeling of heavy-oil recovery by electrical heating, paper SPE 117669," in *Proceedings of the International Thermal Operations and Heavy Oil Symposium (ITOHOS '08)*, p. 14, Society of Petroleum Engineers, Alberta, Canada, October 2008.

[26] M. Koolman, N. Huber, D. Diehl, and B. Wacker, "Electromagnetic heating method to improve steam assisted gravity drainage, paper 1177481," in *Proceedings of the International Thermal Operations and Heavy Oil Symposium (ITOHOS '08)*, pp. 327–338, Society of Petroleum Engineers, Alberta, Canada, October 2008.

[27] K. A. Jha, N. Joshi, and A. Singh, "Applicability and assessment of micro-wave assisted gravity drainage (MWAGD) applications in Mehsana heavy oil field, paper SPE 14591," in *Proceedings of the SPE Heavy Oil Conference and Exhibition*, Society of Petroleum Engineers, Kuwait City, Kuwait, December 2011.

[28] J. R. Kershaw, G. Barrass, and D. Gray, "Chemical nature of coal hydrogenation oils part I. The effect of catalysts," *Fuel Processing Technology*, vol. 3, no. 2, pp. 115–129, 1980.

[29] S. Odenbach, "Ferrofluids—magnetically controlled suspensions," *Colloids and Surfaces A*, vol. 217, no. 1-3, pp. 171–178, 2003.

[30] C. Ovalles, A. Fonseca, A. Lara et al., "Opportunities of downhole dielectric heating in Venezuela: three case studies involving medium, heavy and extra-heavy crude oil reservoirs, paper SPE 78980," in *Proceedings of the International Thermal Operations and Heavy Oil Symposium and International Horizontal Well Technology Conference*, Alberta, Canada, November 2002.

[31] M. A. Ayrapetyan, "About oil fields development prospects by high-frequency currents electrical fields," in *Materials of KSSR Institute of Oil*, pp. 38–52, 1958.

[32] M. A. Ayrapetyan, V. S. Velikanov, and E. Ya. Magnikov, "Reservoir high-frequency heating investigations," in *Materials of KSSR Institute of Oil*, pp. 113–124, 1959.

[33] M. A. Carrizales, L. W. Lake, and R. T. Johns, "Production improvement of heavy-oil recovery by using electromagnetic heating, paper SPE 115723," in *Proceedings of the SPE Annual Technical Conference and Exhibition (ATCE '08)*, Denver, Colo, USA, September 2008.

[34] A. D. Hiebert, F. E. Vermeulen, F. S. Chute, and C. E. Capjack, "Numerical simulation results for the electrical heating of Athabasca oil-sand formations," *SPE Reservoir Engineering*, vol. 1, no. 1, pp. 76–84, 1986.

[35] J. Burge, P. Surio, and M. Combarnu, *Thermal Methods of Enhanced Oil Recovery*, Nedra Publishing, Moscow, Russia, 1988.

11

Simulation of Formation Damage after Long-Term Water Flooding

Liu He,[1] Gao Yang,[1] Li Guoxin,[2] and Li Yiliang[1]

[1] *RIPED, PetroChina, Beijing 100083, China*
[2] *Exploration & Production Company, PetroChina, Beijing 100083, China*

Correspondence should be addressed to Liu He; liuhe@petrochina.com.cn

Academic Editor: Alireza Bahadori

Water flooding is a commonly used technology for enhancing oil recovery. Its main mechanism is to maintain higher pressure to sweep oil towards production wells. However, the strong water flooding will cause higher compression pressure around the injection wellbore. This high pressure in the reservoir causes stress redistribution and higher stress near the wellbore which induces material damage and permeability change. We developed a fluid-solid coupling finite element model to simulate and quantitatively analyze the pressure evolution in the reservoir as well as damage and permeability change in the formation during long-term water flooding process. The obtained results offer theoretical understanding of the benefits (pore pressure increase in the simulation domain), rock damage, permeability change of long-term water flooding, and the insights of how to detect and prevent wellbore failure and collapse due to water flooding.

1. Introduction

Water flooding is a worldwide used technology for enhancing oil recovery. Its main mechanism is to maintain higher pressure to sweep oil towards production wells. The efficiency and key considerations about the technology have been studied [1, 2]. However the injected water will cause the pressure around the injection wellbore to be much higher than that farther away from the well [3]. This high pressure in the reservoir causes stress redistribution and higher stress near the wellbore which induces material damage and permeability change. Experimental results show that dramatic increase of permeability occurs after rock damage [4–6]. Different mechanical models of describing the rock damage and permeability change have been established. Rutqvist et al. [7] proposed that the permeability is an exponential function of the equivalent strain based on the field measurement data of tunnel excavation problem. According to the uniaxial compression tests of rock material, Tang et al. [5] found that permeability jumps up steeply accompanying rock damage. Other damage models can be consulted in the literature such as Loland damage model, Mazars damage model, and Sidoroff damage model [8].

In the present paper, an axial symmetrical fluid-solid coupling finite element model is established to simulate the water flooding process. The pressure evolution in the reservoir as well as damage and permeability change in the formation during long-term water flooding procedure is studied. The numerical results demonstrate that after 5-year water flooding, the maximum pore pressure is increased remarkably. The formation damages gradually near the water injection wellbore and in the damaged area, the reservoir permeability jumps up. The area of damage is enlarged during the water flooding process. The efficiency of water injection is significant. Rock damage and permeability change near the wellbore should be considered when planning a water injection well due to the potential risks.

Fully coupled fluid-solid finite element simulation of water-injection is rarely seen in the literature to our knowledge. The present model can be used to predict the formation pressure evolution and the formation property variation caused by water injection. It is valuable for engineering design.

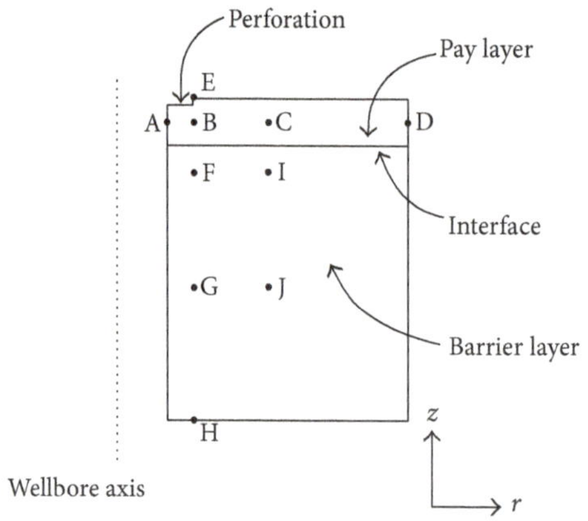

FIGURE 1: The computational model (a meridian plane).

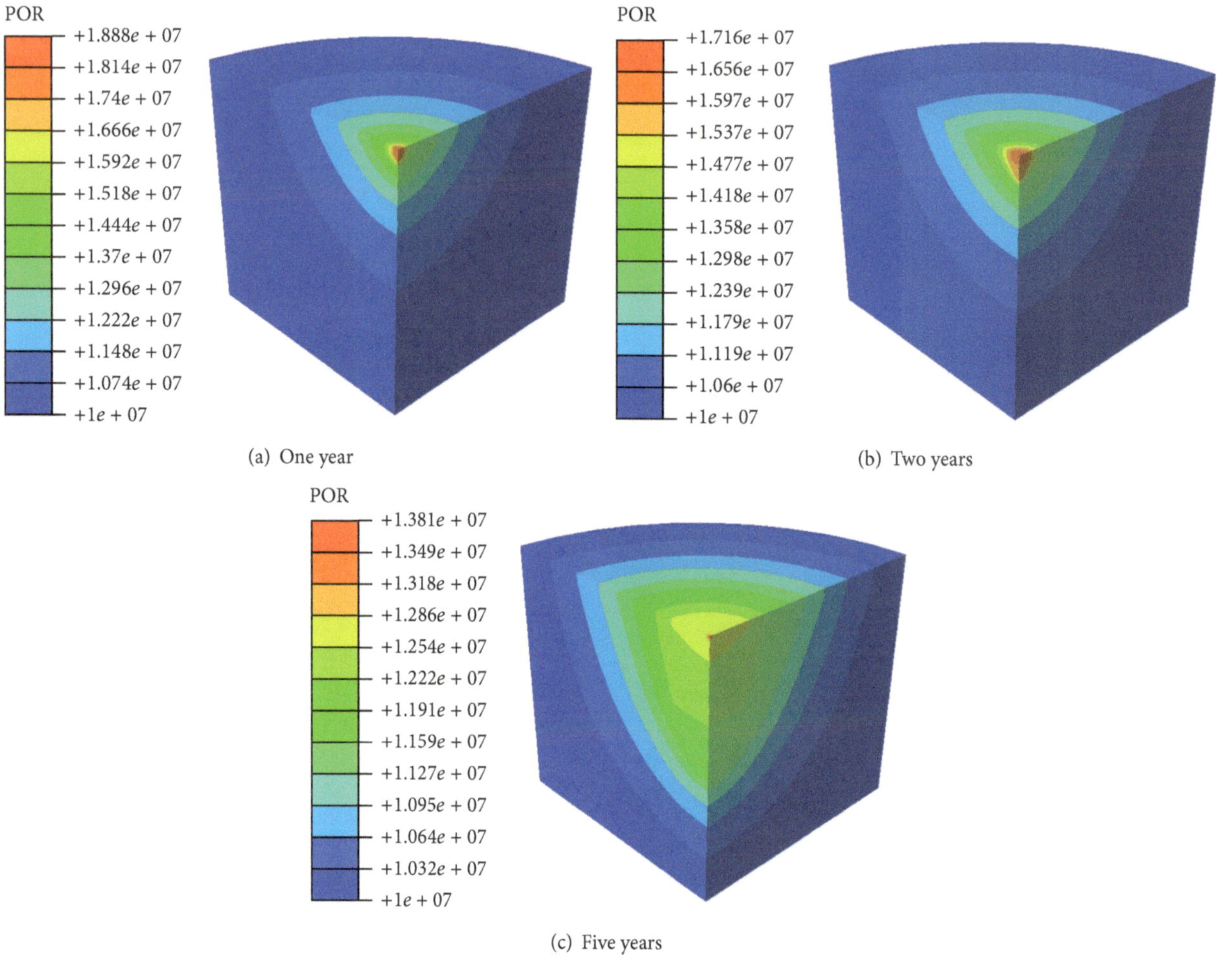

FIGURE 2: Pore pressure distributions at different injection years (The model is rotated by 90°).

2. Mathematical Model

(1) *Governing Equations of Stress-Seepage Flow Coupling Field in Formation.* In the current configuration, the equilibrium equation of porous formation can be written as [9]

$$\nabla \cdot \boldsymbol{\sigma} + \mathbf{f} = \mathbf{0}, \tag{1}$$

where $\boldsymbol{\sigma}$ is the total stress matrix in the formation and $\mathbf{f}$ is the body force vector.

The effective stress couples solid deformation with fluid flow by the following formulation [10]:

$$\overline{\boldsymbol{\sigma}} = \boldsymbol{\sigma} - p_w \mathbf{I}, \tag{2}$$

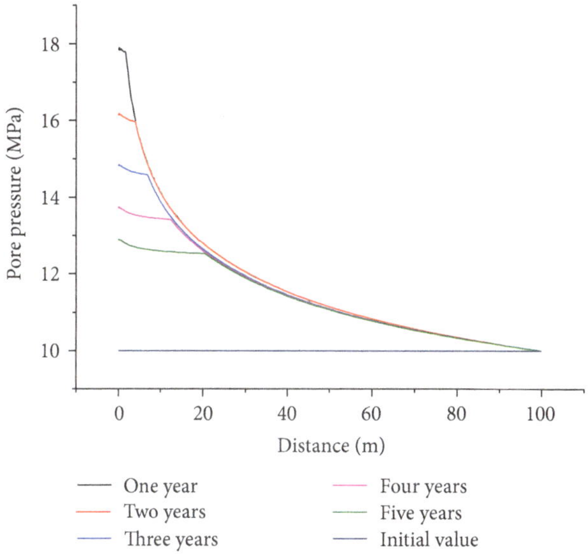

FIGURE 3: Pore pressure evolution along AD line.

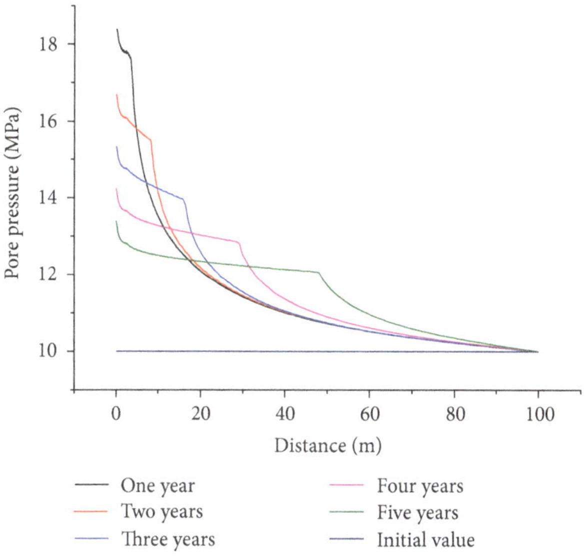

FIGURE 4: Pore pressure evolution along EH line.

where $\overline{\boldsymbol{\sigma}}$, p_w, and $\mathbf{I}$ are the effective stress matrix, the pore pressure, and the identity matrix, respectively.

Continuity equation of seepage flow considering large deformation of porous formation is [9]

$$\frac{\partial}{\partial t}(\rho_w n_w) + \nabla \cdot (\rho_w n_w \mathbf{v}_w) = 0, \quad (3)$$

where ρ_w, n_w, and $\mathbf{v}_w$ are the porous fluid density, the rock porosity, and the seepage flow velocity vector, respectively.

The kinetics equation of seepage flow is the Darcy's law in the following form [11]:

$$\mathbf{v}_w = -\frac{1}{n_w \mu}\mathbf{k} \cdot (\nabla p_w - \rho_w \mathbf{g}), \quad (4)$$

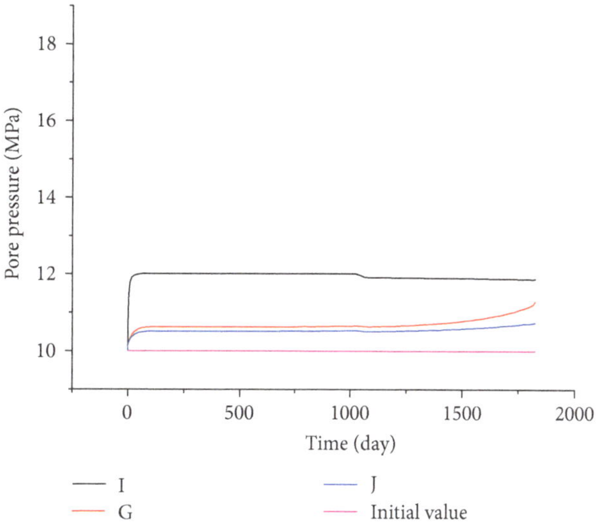

FIGURE 5: Pore pressure evolutions of points I, J, and G.

where μ, $\mathbf{k}$, and $\mathbf{g}$ represent the fluid viscosity, the permeability matrix, and the gravity acceleration vector, respectively.

(2) *Damage Model and Permeability Change Description.* When the deformation of rock material reaches a criterion value, the elastic modulus may degrade gradually as damage progresses. Based on elastic damage mechanics, the damaged elastic modulus E could be related to the original elastic modulus E_0 with the following formula [12]

$$E = (1 - D) E_0. \quad (5)$$

D in the above formula is the damage factor of the material and it can be written as [5]

$$D = \begin{cases} 0 & \varepsilon \le \varepsilon_0 \\ 1 - \dfrac{\varepsilon_0}{\varepsilon} & \varepsilon_0 \le \varepsilon, \end{cases} \quad (6)$$

where ε and ε_0 are the minimum principle strain and the threshold strain value of damage initiation.

Experimental results show that rock permeability may change when suffering external loading. Once the rock is damaged, the permeability may jump up steeply. These phenomena can be described as [5]

$$k = \begin{cases} k_0 \exp(-\beta \overline{\sigma}_{ii}) & D = 0 \\ \xi k_0 \exp(-\beta \overline{\sigma}_{ii}) & D > 0, \end{cases} \quad (7)$$

where k_0 and $\overline{\sigma}_{ii}$ are the original permeability and the first invariant of effective stress matrix. β and ξ are fitting coefficients.

The deformation, strain, stress, and the fluid seepage in the formation during water injection are described by (1)–(7) which couple each other and are highly nonlinear. The 4 node quadrilateral elements are used for describing deformation and seepage in formation. There are 3 nodal unknowns at each node, that is, incremental displacements in 2 directions of cylindrical coordinate system and incremental

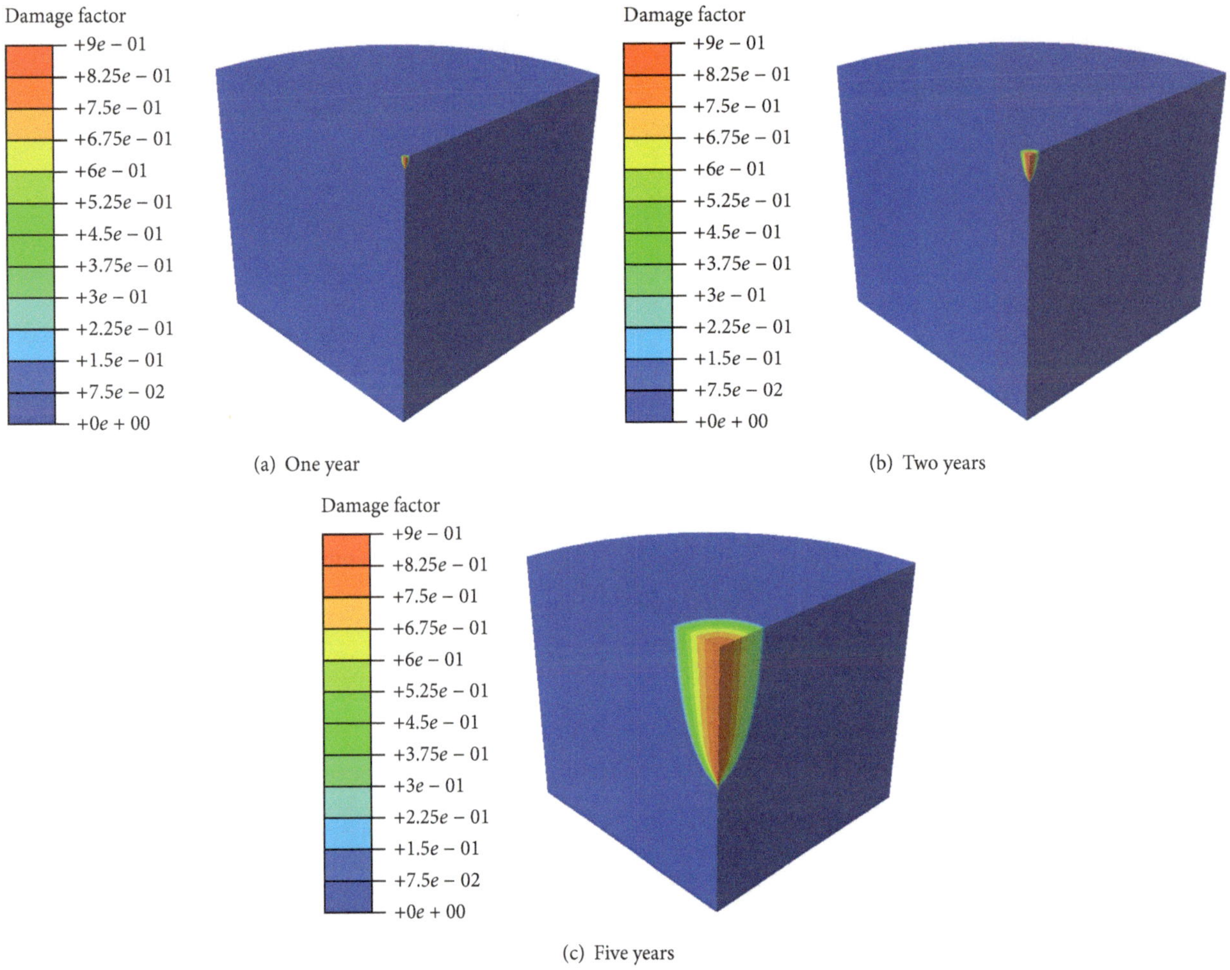

FIGURE 6: Damage factor distributions (The model is rotated by 90°).

pore pressure. A corresponding incremental finite element formula was derived in detail [13], and the formula should be solved with the well-known Newton-Raphson iteration scheme. In order to keep the convergence of the simulation, the upper value of damage factor is set as 0.9.

3. Computational Model Description

Most of the wells (including water injection and production wells) are perforated in the pay layer to weaken the casing before normal operation starts. The perforations are helically distributed, and usually its density is rather high (e.g. 8–16 perforations/meter). Each perforation in a set of helically distributed perforations may develop a damage (or fracture) zone during injecting. The configurations of the damaged zones are very different. The damaged zones connect together, and finally a roughly circular dish-like distributed damaged area forms [14]. Therefore in the paper article we proposed an axial symmetrical fluid-solid coupling finite element model to simulate the long-term water flooding process. Our self-developed user subroutines are incorporated into ABAQUS code to solve (1)–(7) simultaneously. This model can essentially represent the mechanism of the process, but the computation burden is greatly reduced compared to 3D model.

The rock type of the reservoir is sandstone. The simulation domain is constructed of three layers: one pay layer sandwiched between two barrier layers. The perforation is in the middle of pay layer, and its diameter and length are 8.8 mm and 0.5 m, respectively. The thickness of pay layer is 5 m. Since the simulation domain is symmetric, the computational model only takes the low half of the whole model as shown in Figure 1.

The radius and height of the computational model are both 100 m. Fluids can flow throughout the whole simulated domain. The diameter of wellbore is 124 mm. Other basic model parameters are initial elastic modulus = 20 GPa (for pay layer) and 25 GPa (for barrier layers), Poisson's ratio = 0.2 (for pay layer) and 0.25 (for barrier layers), initial permeability = 1 mD (for pay layer) and 0.01 mD (for barrier layers), void ratio = 0.1 (for pay layer) and 0.01 (for barrier layers), initial pore pressure = 10 MPa, the water injection rate = 6.23 m^3/day, water injection duration = 5 years, threshold strain value of damage initiation = 0.0001, and fitting coefficient $\xi = 20$.

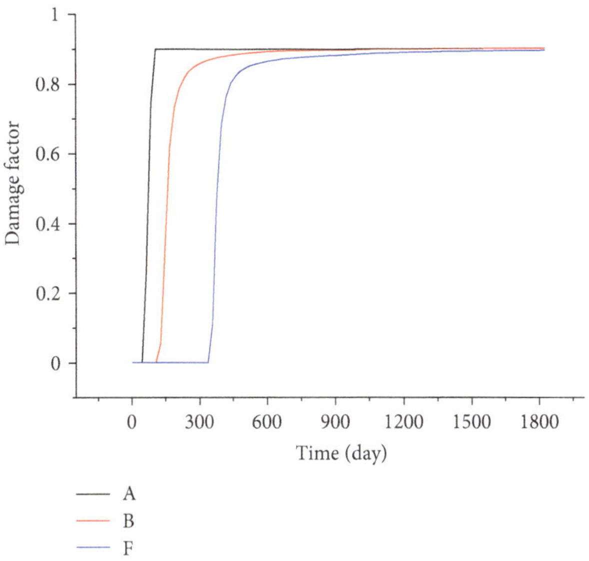

FIGURE 7: Damage factor evolutions of points A, B, and F.

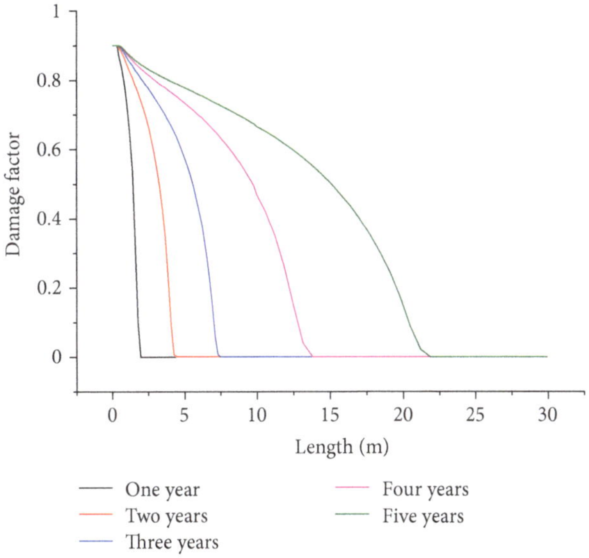

FIGURE 8: Damage factor evolution along AC line.

Ten points A, B, C, D, E, F, G, H, I, and J are designated for the following discussion. A is located at the left bound of the model, and the distance from A to interface is 1.25 m. B, C, and D are in the same horizontal line with A. The distances from B and C to A are 0.5 m and 30 m, respectively. D is located at the right bound of the model. E, F, G, and H are in the same vertical line with B. The distances from F and G to interface are 1 m and 52 m, respectively. E and H are located at the top and bottom bounds of the model, respectively. I, J, and C are in the same vertical line. I and F are in the same horizontal line. J and G are in the same horizontal line.

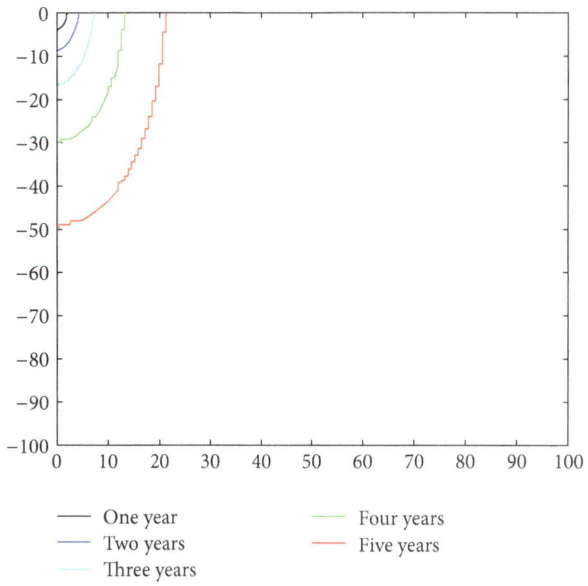

FIGURE 9: Intersections of damaged area and undamaged area at different time.

4. Results and Discussion

Figure 2 shows the pressure distributions in the model after different injection years. Figures 3 and 4 depict the pore pressure evolution along AD line and EH line, respectively. The following observation can be conducted from the figures. Pore pressure increases (compared to the initial value) due to water injection. The maximum values of pore pressure decrease, and the influenced area enlarges during water injecting. The inflection points in Figures 3 and 4 are the intersection points of damaged and undamaged areas. Figure 5 depicts the pore pressure evolution of points I, J, and G. After 5-year water injection, the maximum pore pressure is 13.8 MPa, which is 38% higher than the initial value. In the area that is 30 m far away from wellbore (such as points I and G), the pressure is higher than the initial value by over 10%. Higher pressure in reservoir is maintained, and the oil could be swept from the reservoir and pushed towards production wells. This is the main purpose of water flooding.

Figure 6 shows the damage factor distributions after different injection years. Figure 7 depicts the damage factor evolutions of points A, B, and F. Figure 8 depicts the damage factor distribution along AC line (the horizontal axis represents the distance to point A). Figure 9 depicts the intersection of damaged area and undamaged area. Obviously, the damaged zone is enlarging as water flooding proceeds. Point A is located on the wall of wellbore, where the compression stress is a lot higher, leading to the damage factor that firstly reaches the maximum value. In the region far away from wellbore, the damage factor is comparatively smaller.

Figure 10 presents the first invariant of effective stress distribution after different injection years. It should be pointed out that the values in the figure are the increased parts caused by water injection. With careful observations of the figures

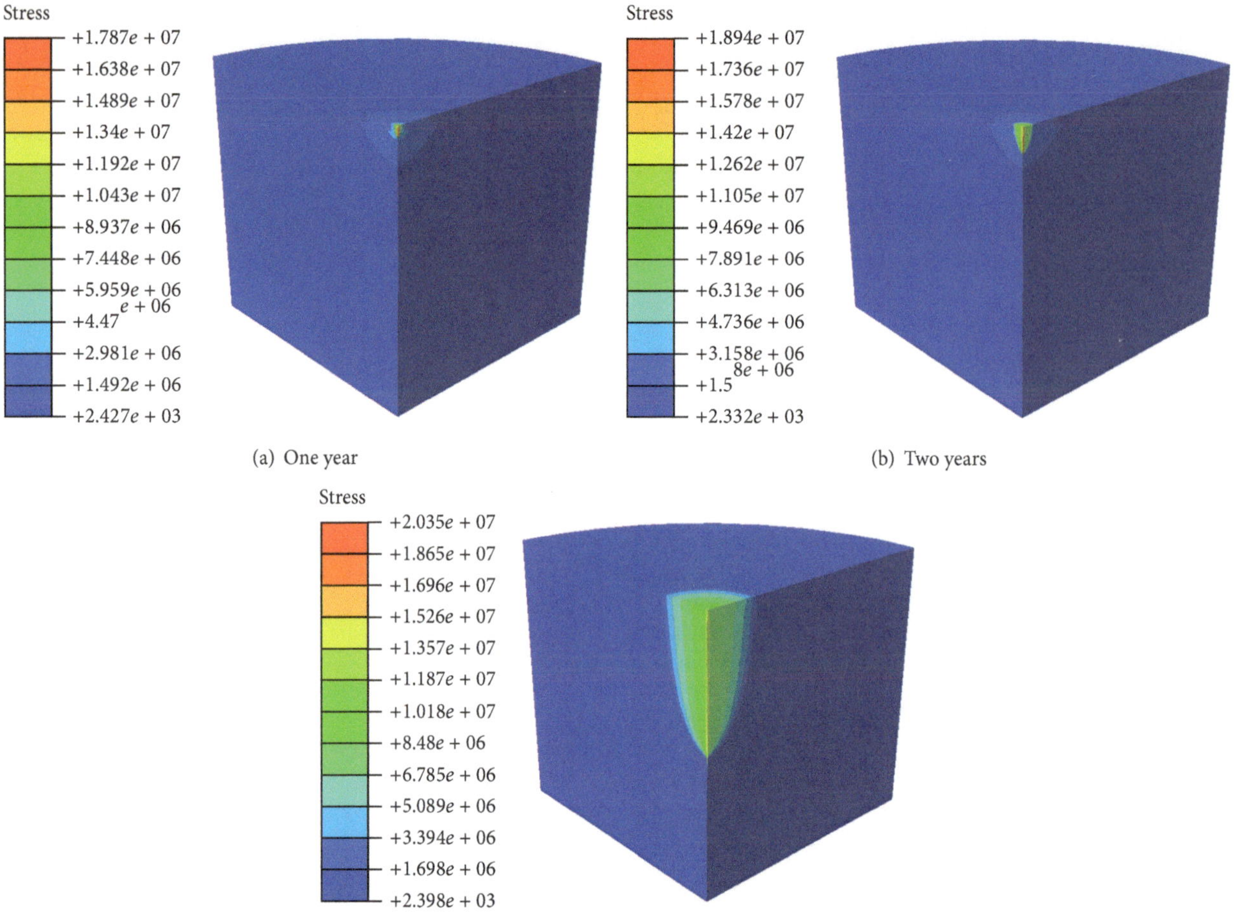

(a) One year

(b) Two years

(c) Five years

FIGURE 10: The first invariant of effective stress distribution induced by water injection (The model is rotated by 90°).

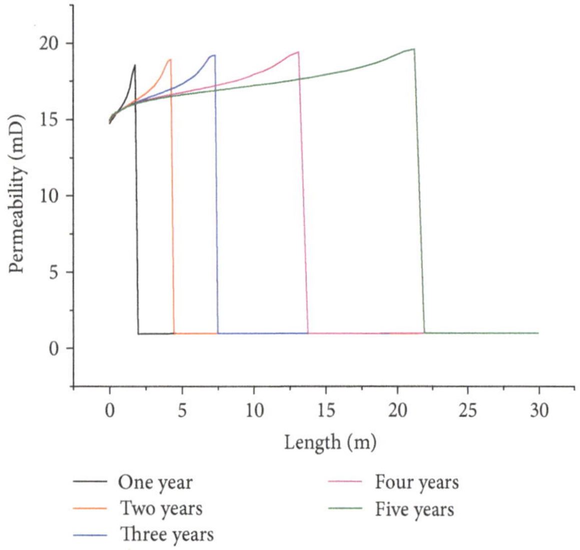

FIGURE 11: The permeability distribution along AC line.

we can find that the maximum values of the first invariant are 17.9 MPa, 18.9 MPa, and 20.4 MPa at the injection point after one, two, and five years of injection, respectively. This high stress concentration results in material damage and may cause casing failure.

Figures 11 and 12 show the permeability along AC and FG lines, respectively. It can be seen that as the distance increases, the permeability firstly increases and then drops. In the zone near wellbore, the stress is higher and accompanying material damage; as a result, the permeability jumps up. But in the undamaged zone, the permeability drops and the value is close to the initial value.

5. Conclusions

An axial symmetrical fluid-solid coupling finite element model is established to simulate water flooding process. Our developed user subroutines are incorporated into ABAQUS code to solve the mathematical equations.

Compared to the initial value, water flooding will cause the pore to pressure increase, and the influence area enlarges gradually. After 5-year water injection, the maximum pore pressure may be 38% higher than the initial value. In the area that is 30 m far away from wellbore, the pressure is higher than the initial value by over 10%. The purpose of water injection is achieved.

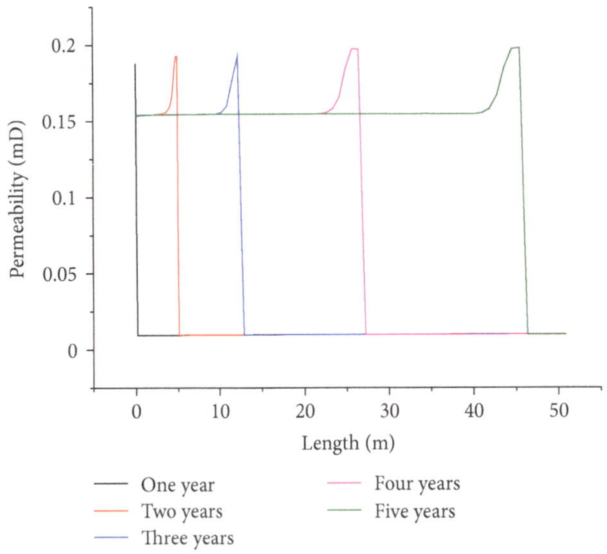

FIGURE 12: The permeability distribution along FG line.

Water injection may cause the formation property change. As water injection proceeds, the compression stress will redistribute and enlarge. As a result the rock material may damage, and the damaged area enlarges during water injecting. In this area, the formation permeability jumps up, leading to the pore pressure decrease. High stress concentration near injection point exists. Attention should be paid on the risk related to casing failure and/or formation fracture in design and performance of water injection.

Nomenclature

D: Damage factor, dimensionless
E: Damaged elastic modulus of the rock material, m/Lt^2, Pa
E_0: Original elastic modulus of the rock material, m/Lt^2, Pa
$\mathbf{f}$: Body force vector, m/L^{2}t^2, Pa/m
$\mathbf{g}$: Gravity acceleration vector, L/t^2, m/s^2
$\mathbf{I}$: Identity matrix, dimensionless
$\mathbf{k}$: Permeability matrix, L^2, m^2
k_0: Initial permeability, L^2, m^2
n_w: Porosity, dimensionless
p_w: Porous pressure, m/Lt^2, pa
$\mathbf{v}_\mathbf{w}$: Seepage flow velocity vector, L/t, m/s
β : Fitting coefficient, dimensionless
ε : Minimum principle strain and the threshold value of damage initiation, dimensionless
ε_0: Threshold value of ε, dimensionless
μ : Fluid viscosity, m/Lt^3, Pa.s
ρ_w: Density of porous fluid, m/L^3, kg/m^3
$\boldsymbol{\sigma}$: Total stress matrix, m/Lt^2, Pa
$\overline{\boldsymbol{\sigma}}$: Effective stress matrix, m/Lt^2, Pa
$\overline{\sigma}_{ii}$: First invariant of effective stress tensor, m/Lt^2, Pa
ξ : Fitting coefficient, dimensionless.

Acknowledgments

The authors highly appreciate the financial support of State 863 Projects and PetroChina Innovation Foundation.

References

[1] M. Latil, *Enhanced Oil Recovery*, Editions Technip, Paris, France, 1980.

[2] B. A. Stenger, A. B. Al-Katheeri, H. H. Hafez, and S. H. Al-Kendi, "Short-term and long-term aspects of a water injection strategy," in *Proceedings of the 13th Abu Dhabi International Petroleum Exhibition and Conference (ADIPEC '08)*, vol. 116989, pp. 74–86, Abu Dhabi, United Arab Emirates, November 2008.

[3] E. E. Kristian, C. Jerey, I. Orkhan, N. Julian, and A. A. K. Vegard, "Improved oil recovery with water injection," EiT-Gullfaks Village, Norwegian University of Science and Technology in cooperation with Statoil ASA, 2011.

[4] J. A. Wang and H. D. Park, "Fluid permeability of sedimentary rocks in a complete stress-strain process," *Engineering Geology*, vol. 63, no. 3-4, pp. 291–300, 2002.

[5] C. A. Tang, L. G. Tham, P. K. K. Lee, T. H. Yang, and L. C. Li, "Coupled analysis of flow, stress and damage (FSD) in rock failure," *International Journal of Rock Mechanics and Mining Sciences*, vol. 39, no. 4, pp. 477–489, 2002.

[6] A. S. Abou-Sayed, K. S. Zaki, G. G. Wang, and M. D. Sarfare, "A mechanistic model for formation damage and fracture propagation during water injection," in *Proceedings of the SPE 6th European Formation Damage Conference: Delivering Sustained Production (EFDC '05)*, vol. 94606, pp. 223–230, Scheveningen, The Netherlands, May 2005.

[7] J. Rutqvist, L. Börgesson, M. Chijimatsu et al., "Modeling of damage, permeability changes and pressure responses during excavation of the TSX tunnel in granitic rock at URL, Canada," *Environmental Geology*, vol. 57, no. 6, pp. 1263–1274, 2009.

[8] X. R. Ge, J. X. Ren, Y. B. Pu, W. Ma, and H. Sun, *Macroscopic and Meso Laboratory Research of Rock and Soil Damage Mechanics*, Scientific Press, Beijing, China, 2004.

[9] L. E. Malvern, *Introduction to the Mechanics of a Continuous Medium*, Prentice-Hall, Upper Saddle River, NJ, USA, 1969.

[10] M. J. Economides and K. G. Nolte, *Reservoir Stimulation*, John Wiley & Sons, New York, NY, USA, 3rd edition, 2000.

[11] M. A. Marino and J. N. Luthin, *Seepage and Groundwater*, Elsevier, New York, NY, USA, 1982.

[12] A. Turon, P. P. Camanho, J. Costa, and C. G. Dávila, "A damage model for the simulation of delamination in advanced composites under variable-mode loading," *Mechanics of Materials*, vol. 38, no. 11, pp. 1072–1089, 2006.

[13] G. M. Zhang, H. Liu, J. Zhang, H. A. Wu, and X. X. Wang, "Mathematical model and nonlinear finite element equation for reservoir fluid-solid coupling," *Rock and Soil Mechanics*, vol. 31, no. 5, pp. 1657–1662, 2010 (Chinese).

[14] F. J. Biao, *A study of numerical simulation on hydraulic fracture propagation of horizontal fracture*, University of Science and Technology of China, 2010.

12

Rethinking Petroleum Products Certification

Thiago Feital,[1] Públio Lima,[2] José Carlos Pinto,[1] Maurício B. de Souza Jr.,[3] Gilberto Xavier,[4,5] Mario Jorge Lima,[4] and Marcel Joly[4,6]

[1] *Programa de Engenharia Química/COPPE, Universidade Federal do Rio de Janeiro, 68502 Rio de Janeiro, RJ, Brazil*
[2] *Escola Politécnica, Universidade Federal do Rio de Janeiro, 68529 Rio de Janeiro, RJ, Brazil*
[3] *Departamento de Engenharia Química/EQ, Universidade Federal do Rio de Janeiro, 68542 Rio de Janeiro, RJ, Brazil*
[4] *PETROBRAS Petróleo Brasileiro S.A., Rio de Janeiro, RJ, Brazil*
[5] *Faculdade de Tecnologia de Resende, Universidade do Estado do Rio de Janeiro, Rodovia Presidente Dutra Km 298, Resende, RJ, Brazil*
[6] *Departamento de Engenharia Química, Universidade de São Paulo, São Paulo, SP, Brazil*

Correspondence should be addressed to Thiago Feital; tfeital@peq.coppe.ufrj.br

Academic Editor: Alireza Bahadori

Facing various challenges in the everchanging refining landscape, it is essential that refiners raise their operations to new levels of performance. Advances in in-line blending (ILB) technology accuracy and reliability have encouraged refiners to take a step forward. Having ILB as a precursor, a new methodology is in concern: the so-called in-line certification (ILC) procedure. Blending processes make use of in-line measurements which, at least in principle, can be used to certificate the product, if the precision and accuracy of available in-line measurements are comparable to measurements provided by standard off-line tests. Such procedure may allow for significant reduction in refinery's tank farming and product inventory, increase of process flexibility, and reliability with benefits to company image. The main limitations for real-world ILC applications in the oil industry remain at the legal and technological levels. This paper proposes novel concepts and foundations of a basic in-line certification model for petroleum products regarding current interdisciplinary challenges and promising solutions.

1. Introduction

Petroleum refining is one of the most important industries, comprising many different and complicated processes with various possible configurations. Globally, it processes more materials than any other industry [1]. Due to the scale and significance of this industry, it becomes more crucial to address the considerable challenges that the industry faces today and in the future.

Production planning and scheduling optimization are essential tasks to maximize refinery's profit margins and to remain in the competitive market. In this sense, several opportunities strictly related to refinery production optimization could be identified. At the strategic level, the supply chain optimization may be posed as a master problem, since there are numerous trade-offs between decisions made at the various nodes of its superstructure in which the refinery planning is embedded. At the tactical level, two major challenges should be addressed by refiners [2].

The first one is that refining industry needs to effectively evaluate process performances to identify options for producing desirable products and meeting increasing constrained environment regulations. The second challenge refers to major transformation from an industry of mainly producing fuels for transportation to one that makes a wide set of value-added products, including chemicals, speciality products, electricity, and hydrogen.

In a scenario in which the world economic growth slowed down and demand for petroleum products flattened or even decreased, it becomes clear that *flexibility* is the keyword for everybody involved in the oil sector [3]. In this sense, online remote characterization and real-time optimization of petroleum products emerge as instrumental technologies for competitive refineries [4]. Such technologies may have a special impact on refining profitability with environmental benefits when considered to provide chemical information of great importance for real-time adjustment of short-term scheduling decisions concerned with product formulation and refinery dispatch logistics.

Blending consolidates as the refinery's last chance to impact profitability at low investment levels and approximately 50% of about 700 refineries worldwide have implemented such technology [5]. Besides positively impacting the overall refinery production scheduling, in-line blending (ILB) operations at the end of the refinery allow for substantial benefits including end product giveaway minimization, lower inventory levels, optimized logistics and, therefore, reduced utility consummations and atmospheric emissions, economic savings in new hardware investment and maintenance, and operational safety.

Apart from typical economical benefits ranging from US$ 0.15 to 0.30 per crude oil processed [6], ILB represents the precursor technology for allowing an additional efficiency jump in refinery scheduling optimization: the so-called in-line certification (ILC) (Figure 1).

As the fundamental premise, an ILC-based approach should consider product dispatch to a third party before sending the official (i.e., legal) product quality certificate to it. ILC procedures are primarily intended to minimize refinery scheduling bottlenecks. However, it is not completely clear yet how large the benefits provided by ILC procedures can be. Since products can be delivered directly from processing to clients, some potential advantages that can probably be related to the implementation of ILC schemes are as follows.

(i) Suppression of conventional operations, as tank homogenization, sampling, and laboratory testing, and this way saves manpower and shortening the refining production cycle time (Figure 2).

(ii) Dead inventory minimization.

(iii) Tankage minimization and associated maintenance costs optimization.

(iv) Improvement of the process reliability and optimization (as process control techniques must necessarily be implemented in-line and in real time).

(v) Enhanced flexibility to tightly specify different product grades according to real-time customer needs.

(vi) Shipping demurrage minimization.

(vii) Enhanced energetic and environmental performances of the refinery storage area provided by an optimized scheduling.

(viii) Staff safety and productivity increase provided by automation replacing manual operations.

Such benefits must compensate the investment on technology and process upgrade required by ILC procedures, if actual implementation at plant site is sought. Apart from legal issues, the refinery's logistic environment will determine how expressive the opportunity for implementing an ILC-based operational philosophy is.

This paper is structured as follows. Section 2 proposes basic concepts and discusses an ILC-based business model.

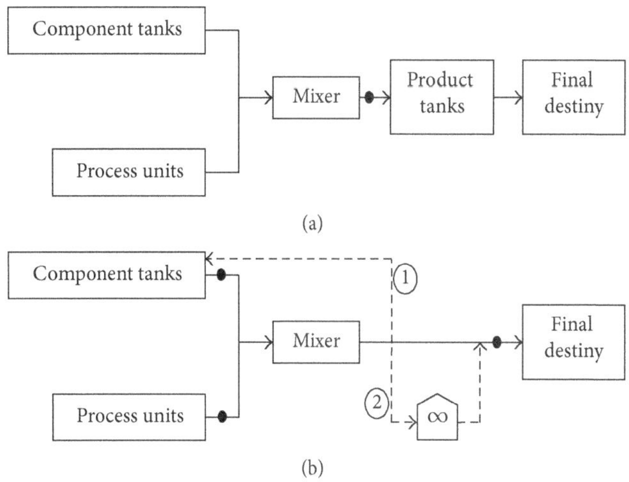

FIGURE 1: (a) Typical configuration of ILB units; (b) a virtual basic scheme for an ILC unit using two possible routes for off-spec fuels: (1) reblending or (2) infinite dilution. In both schemes, dashed circles represent on-line analyzers.

Some practical ILC implementation issues, with identification of current technical limitations, promising solutions, and required investments, are also discussed here. Section 3 discusses how the ILC scheme should work out in practice in order to be accepted by regulatory agencies and customers in general whereas Section 4 outlines some identified prerequisites to a generic ILC implementation. Section 5 presents the conclusions of the paper.

2. An ILC-Based Custody Transfer Model

Custody transfer in the oil industry refers to the transactions involving transporting physical substance from one operator to another. Since presupposing product *releasing before testing* to a customer, an ILC-based approach implies new technical-commercial paradigms.

2.1. Model Definition. The first step required to support a new proposal or procedure is the definition of important related terms and the development of the basic ideas behind it. *In-line certification* is defined here as a computer-controlled procedure based on real-time data from on-line analyzers able to automatically measure stream qualities and perform actions over manipulated process variables to ensure product specification and optimization this way enabling its prompt delivery from the refinery to third parts prior to the official issuing of the conventional, laboratory-based, *certificate of analysis* (CoA). Here, the term *quality* encompasses any physicochemical property related to the legal specification of a given finished product. Similarly, the *on-line analyzers* terminology encompasses American Society for Testing and Materials (ASTM) on-line analyzers and near-infrared (NIR) systems.

If the in-line certification is thought to simultaneously be performed by an in-line blending unit, one can refer to it as an *in-line blending certification* (ILBC) unit (Figure 3). However,

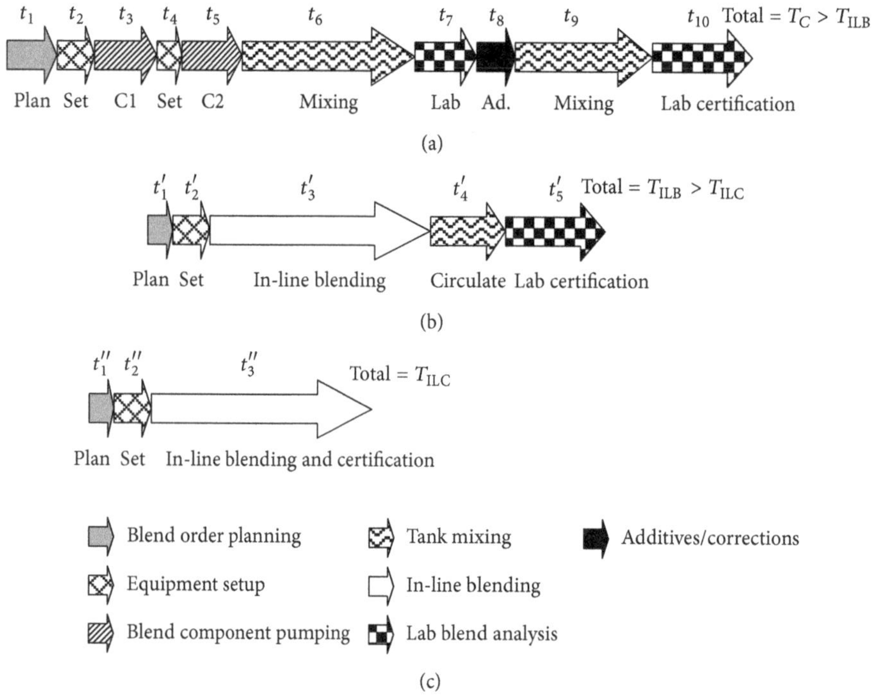

FIGURE 2: Time saving with ILC [7]. (a) Conventional time (T_C) operation based on manual tank blending and certification; (b) in-line blending time (T_{ILB}) and conventional tank certification; (c) in-line certification time (T_{ILC}) of the finished blend product.

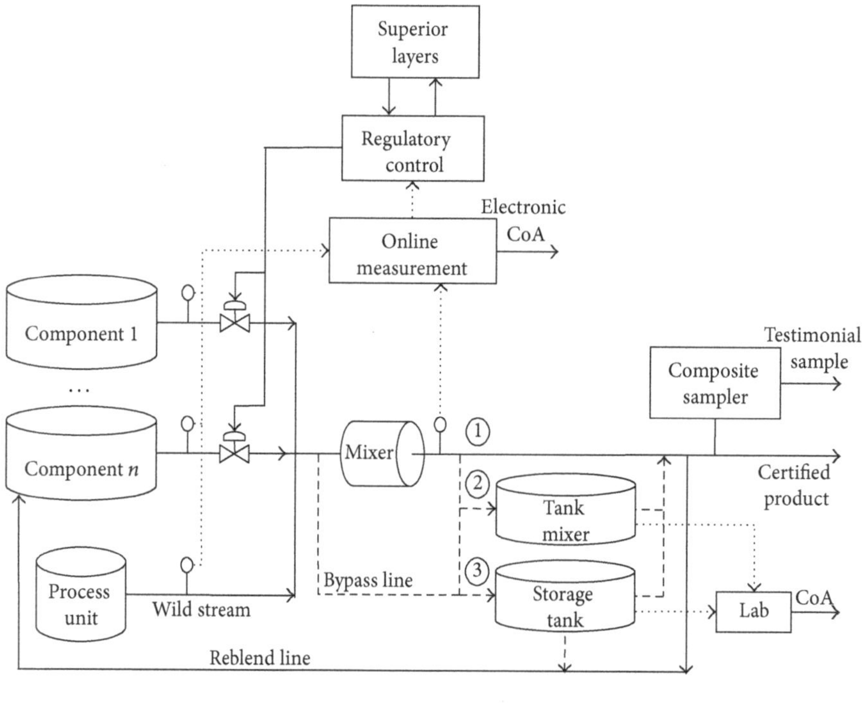

FIGURE 3: An ILBC unit model using three routes: (1) normal operation condition; (2) blend through precertified tank during periods of abnormal operation condition; and (3) blend to product tank during maintenance periods.

for the sake of simplicity, ILC and ILBC units are generically referred to as ILBC thereafter in this paper. Such integrated engineering approach should be understood as a real-time optimization technology.

Minimal requirements for an industrial ILBC application may also include on-line analyzer validity checking, automatic composite sampler for spot and final checks, and automatic generation of product info reports and their archiving [5].

Once homogenized in a static mixer, the finished blend product is analyzed and then sampled by an auto composite sampler, which retains a *representative sample* (RS) of the overall product delivered to a third part. According to a typical hierarchical planning superstructure, the material quantity that cross the refinery frontier should be defined *a priori* at scheduling level (Figure 4). It defines the *fiscal batch*, whose quality must satisfy the required specification. Right after the fiscal batch production is concluded, an electronic certificate of analysis (ECoA, based on measurements from on-line analyzers) is available and may be sent to the third part, prior to the release of the conventional laboratory-based CoA.

As the basic ILBC procedure premise accounts for automatic product custody transfer right after production, further extrapolations can be suggested, as *segmented fiscal batch*. More specifically, once the transferred volume of product is known and already certificated, there is no reason to wait for completion of the product transferring for product certification. This idea follows a divide-and-conquer strategy, where the whole amount of commercialized product is divided into small fractions that can be certified independently of the remaining ones.

Ideally, the full set of specification qualities should be evaluated in-line at real time by on-line analyzers. However, this may not always be technically feasible since some product specifications are based on experimental assays that are inherently time demanding ones (e.g., gum content for gasoline) or require specific experimental apparatus (e.g., jet fuel smoke point) or even are not quantitative measurements (e.g., cooper corrosion for liquefied petroleum gas (LPG)). In the general case, an ILBC approach should consider a subset of critical (constrained) commercial specifications [8] and indirect measures for well-known behavior or not critical remaining quality variables. In this sense, research on physicochemical inferences development and process statistic control techniques are prolific fields for investigation and applied developments, including novel in-line testing methodologies. In addition to immediate benefits, such new methodologies could serve as truly precursors for promoting interdisciplinary discussions oriented to upgrade legal requirements concerning oil product quality evaluation. In this sense, the US Environmental Protection Agency (EPA) has indeed taken actions to allow refineries and laboratories to use more current and improved fuel testing procedures for ASTM analytical test methods [9]. On the other hand, emerging techniques, as molecular characterization of oil products [10], will stress the need for revising obsolete testing methodologies replacing them by more reliable and accurate ones, potentially interesting for real-time applications [11].

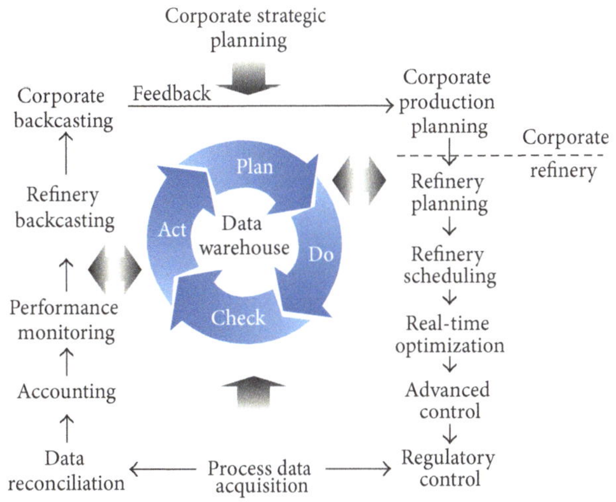

FIGURE 4: A typical hierarchical decision-making framework based on Moro [3].

After the fiscal batch custody transfer is concluded, a CoA is then produced and sent to the third part. A CoA must present the complete physicochemical characterization of the delivered product, as required by regulatory agencies and usually performed by standard tests, such as those recommended by the American Society for Testing and Materials (ASTM).

In principle, the CoA may be based on either auto composite sampling (Figure 3—configuration 1) or manual tank (Figure 3—configurations 2 and 3) and laboratory testing. Conversely, on-line analyzer measurements have also been considered to compose part of the CoA data. In this case, on-line analyzers should use approved certification test methods, as those defined by EPA 40CFR80.46 (mostly ASTM tests), or correlated to them [5]. If real-time data is considered to final certification, a hybrid or *combined* CoA results (Figure 3). This may be attractive in the sense of optimizing laboratory resources by limiting laboratory testing only for qualities that can not be in-line evaluated. Anyway, each emitted CoA must count on testimonial sample for eventual counterproofing. Testimonials must be representative and consider proper storage against possible degradation reactions.

For pure substances, as the case of petrochemicals, the arrangement of Figure 3 may be simplified if a single component exists and no mixing are required. In this case, the computer-controlled system does not actuate over blending optimization variables but essentially (and directly) over process manipulated variables. This may be the case of propylene plants, a typical first generation petrochemical. Nevertheless, it should be pointed out that even such single systems may consider blending opportunities with off-spec product.

While producing a fiscal batch, the ILBC unit on-line analyzer measurements are integrated by the computer-controlled system. Anytime, if mean product quality does not comply with the specification requirements and there is no expectation of correcting it before batch production conclusion, then the product release should be interrupted. Hence, ILC procedures must also consider fall-back strategies to circumvent such undesired situations. A reblend or reprocess line after on-line analyzer may be provided for ready (automatic) production deviation.

In-line certification should not be confounded with *in-line delivery*, which is exclusively based on the conventional certification of the products after delivery, usually performed to save time and storage at the plant. In-line certification, on the other hand, seeks for a definitive certification solution, as finished products are transferred directly to a third part as a fiscal batch and available in-line instrumentation and techniques are supposed to provide enough information about the product quality in reasonable time (real time) and with high confidence.

Currently, when not legally prohibited, both procedures are typically conditioned by regulatory agencies and are periodically subject to ad-hoc independent audit programs concerned with the computer-controlled system operation [5]. The goal of the audit program is to insure that fiscal batches data and information reported to the governmental entity are accurate and valid. Despite the lack of explicit conditions, regulatory agencies usually reserve their right to make inspections to assure compliance with the requirements (EPA 40CFR80.4).

In legal terms, the US legislation is virtually the most comprehensive one. Technical and legal procedure steps are described by US EPA in USA Code of Federal Regulation, 40CFR80, part 65, which allows refiners legally to blend directly to a third party before finishing product testing and CoA delivery. In Europe, shipping before testing practices have also been mentioned [8] but seem to be mediated by bilateral agreement between refiners and third parties. However, in both US and European cases, the number of refineries operating ILBC-like custody transfer is very limited and the third party is usually, if not always, a transporter company (pipeline or vessel), not a final client (distributor company). Conversely, from the aforementioned cases, in Brazil, the Petroleum National Agency, a governmental entity that regulates the Brazilian oil sector, currently does not allow for finished product shipping before the CoA emission.

2.2. The ILBC Hardware. An ILBC implementation presupposes process hardware properly designed with excellent instrumentation. Blend headers, static mixers, lines, pumps, and valves design should consider operational feasibility and safety accessories during transitory, steady-state, and fall-back conditions. Coriolis flowmeters have been extensively recommended for in-line blending systems [5, 8] and may help to persuade regulatory agencies to approve novel philosophies for custody transfer procedures, as ILBC. In order to reduce stream quality variability from (running) tank components, a tank mixer should be provided with the objective of allowing for inventory homogenization (during) before the fiscal batch production (route 2 in Figure 3). Whenever applicable, filters and coalescer devices for solid and free water removal should also be considered as basic blending system accessory.

At the design level, analytical redundancy should be a project premise. Since real data conciliation may be a valuable tool to the system reliability and early warning of problems, on-line analyzers at each component should be provided in order to produce redundant analytical information whenever possible. Temperature-controlled sampling systems should consider sampling points not susceptible to contaminants and located at critical process points, as components lines, blended fuel line as well as at the refinery boundary if the in-line certified product is then stocked in a tank or passes through it before shipping. Certified reference materials and prototype fuels (or *protofuels*), whose properties are known, require dedicated installations. Fast loops are required to allow for sample movement from process lines to the analyzer house (shelter) and vice versa (Figure 5).

2.3. The ILBC Software. As a real-time optimization business layer, the ILBC operation must agree with the refinery's objectives that are defined at the strategic and tactical levels. This is achievable through capturing the business knowledge in advanced decision-making tools, which should allow for increasing visibility into plant operations and communicating economic drivers to operations allowing them to identify and eliminate constraints by better understanding the planning economic objectives [12].

A schematic representation for ILBC software applications is proposed in Figure 6. In this figure, blending scheduling refers to an off-line optimization layer in which the refinery tactical planning is detailed into short-term operational decisions. Here, fiscal batches are defined and sequenced to execution throughout a time horizon aiming at satisfying the foreseen product requests by customers subject to refinery's operational, logistic, and economic constraints. The *Blend Property Control* (BPC) comprises at least two real-time applications: (a) the multivariable predictive control (MPC), which is responsible by the real-time blended product quality control, and (b) the on-line optimization process model, which should maximize or minimize some performance criteria subject to real-time constraints regarding the whole fiscal batch production. On the other hand, the *Regulatory Blend Control* (RBC) is responsible for manipulating component flows to match an optimal blend recipe defined at the BPC layer. For each predefined fiscal batch volume, the BPC application functions should include field equipment selection and lineup, pump sequencing control, appropriate ramps for blending start-up and conclusion, and additive injections. In addition, the RBC should also be responsible for implementing a trim fall-back strategy in case of system failure. *The Key Performance Indicator* (KPI) is a monitoring module from which the fiscal batch data is collected to compose the ECoA right after the batch conclusion.

It seems clear that the implementation of advanced control schemes must certainly be considered during development of ILBC procedures, as process disturbances are inherent to process operation and lead to some degree of unavoidable process variability. Therefore, process variability must be rigorously considered during the implementation of

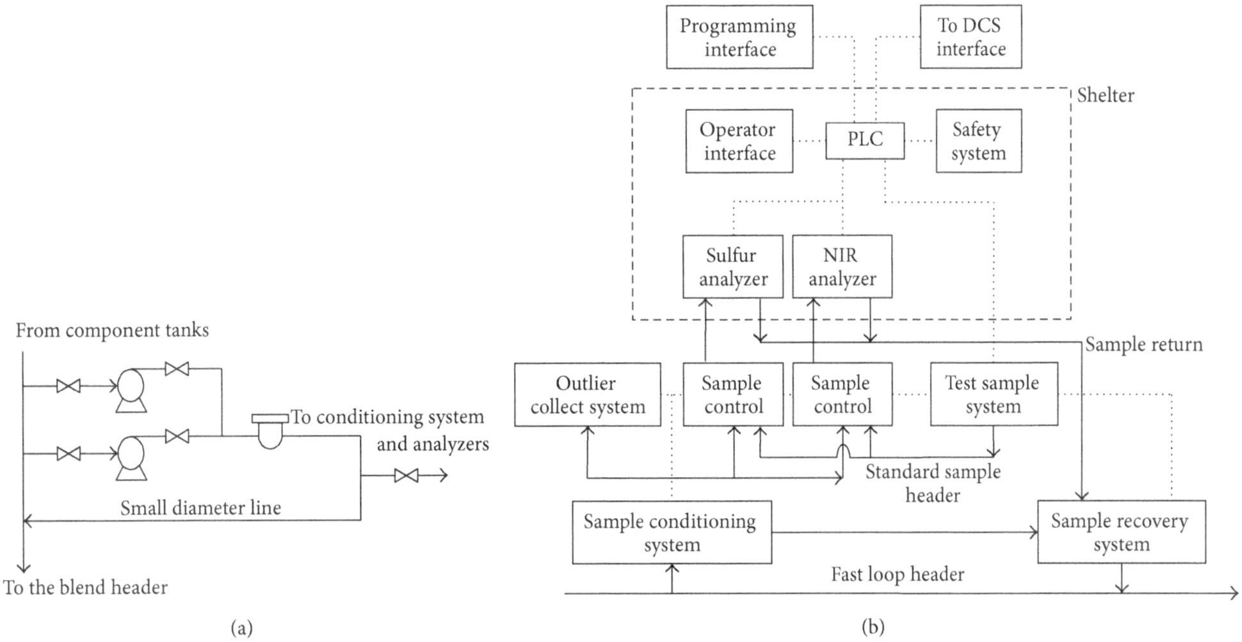

FIGURE 5: A typical fast loop scheme for refinery ILB. (a) Pumping and filtering; (b) the sample conditioning system.

the ILBC scheme, with the help of sound statistical analyses of measured data performed in-line and on real time.

Similarly, as instrument and operation faults cannot be completely avoided at plant site, the implementation of advanced tools for fault management (perhaps including the in-line control reconfiguration) must also be required during real ILBC implementations, as the process operation must be continued in spite of the faulty event, such as leaking, valve sticking, instrumentation failure, and feed contamination. As a consequence, protection actions and protection mechanisms must be employed to support the ILC decisions and the mechanisms used to maintain the product certification scheme.

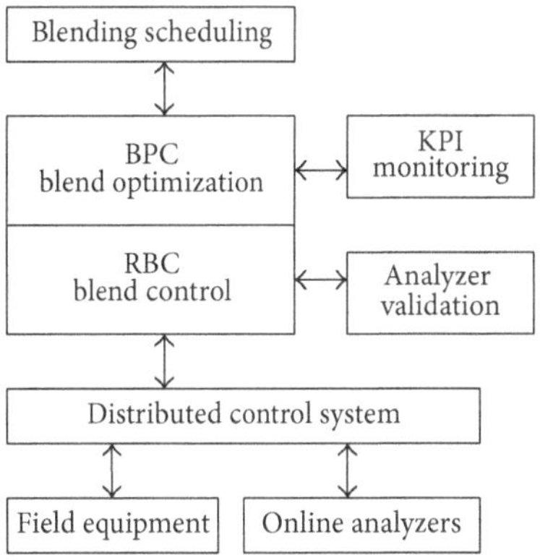

FIGURE 6: The ILBC software applications.

3. The ILBC Unit Operation

The ILC procedure implicitly breaks the paradigm of the batch operation and, by doing so, suggests a new *modus operandi* for the process operation. It is expected that stream properties in process lines present larger variability, when compared to bulk properties of stored products. Therefore, for the sake of certification, this new vigorous dynamic operation status certainly requires tighter control. However, the proper in-line measurement of the whole suit of fuel properties, necessary to certification, is one of the two major technical challenges. The other one is the analytical accuracy and reliability.

Efforts have been made to address the control problem, but the in-line instrumentation performance has seldom been compared to the off-line standard test ones for the sake of product certification (using repeatability and reprodutibility); correlation measures are provided instead [4, 13, 14]. In this sense, data validation is the key to assure the measurement precision, enhancing the signals of the sensors.

Since instruments must operate in-line for long periods of time, the long-term behavior of process instrumentation must be properly evaluated and characterized. In addition, part of the required properties may be not available in-line due to technological or economic issues. The remaining of this section is dedicated to examine expected problems related to the ILBC low-level operational layers of the superstructure schematized in Figure 4.

3.1. Measurement. Before reviewing the main techniques and instrumentation used to monitor fuel properties, two major practical issues must be discussed: sample representativeness and sensibility to measured properties.

Sample representativeness is especially critical when analyzers require long times to perform the analyses. More specifically, most property analyzers do not operate on a real-time basis since the analyses are complex and time demanding. The product flow rate and the sample size must be defined in such a mode that the on-line analysis shows volumetric compliance without incuring into any constraint on the system production.

Addressing in-line measurement, ASTM D4177 standard focuses on practices for automatic sampling of petroleum and petroleum products whereas the ASTM D6624 standard is concerned with practices for determination of average property values of product streams (FPAPV). Lastly, the ASTM D7453 standard accounts for practices for sampling petroleum products by process stream analyzers.

Another important question refers to the analyzer sensitivity to a particular property. Such a question has been extensively studied in the literature, for example, using NIR technology [13], and different commercial solutions addressing this issue have been released in the market, for example, the FUELex analyzer of the Bruker Optics company. In fact, the search for surrogated variables and correlation development are encouraged. Agoston et al. [15, 16] introduced alternative sensors which enable in-line oil viscosity and corrosiveness monitoring, respectively. Such sensors surely agree with the ILC-based operational philosophy. The ASTM D6708 standard defines some statistical procedures used for assessment and improvement of the expected agreement between two test methods that are assumed to measure the same property of a material.

Concerning the technology used for in-line measurement, in-line specialized analyzers constitute the first reasonable option, at least for simple properties, such as density and conductivity. These analyzers are usually cheap, reliable, and fast, making them appropriate for composing an ILBC scheme. For properties hard to measure in line, inferences based on combinations of electromagnetic spectroscopy and inference models have become very popular both in academic and industrial applications [13]. The combination of NIR spectrometers and multivariate calibration techniques certainly constitutes the most popular alternative for inference development due to fast response times and high precision. In general, NIR systems present better precision (repeatability) than ASTM analyzers and same accuracy (reproducibility) as the calibration method [17] and can produce one complete set of quality determinations in 30 seconds [5, 8]. Real-world NIR in-line applications have become very attractive, since 10 or more product qualities can efficiently be evaluated with NIR technology (diesel or gasoline) herein its ability to predict blend indices which are extensively used in off-line applications, as refinery production planning and scheduling. Practices for infrared multivariate and quantitative analyses are addressed by ASTM E1655. On the other hand, while equipments based on nuclear magnetic resonance (NMR) have also been investigated with comparable performance [18], less applications have been verified.

As pointed out by some authors [19], NIR-based analyzers require careful building of the inferential property model. However, suppliers of NIR-based analyzers offer poor modeling support and do not provide guarantees for measurement performance [20]. Acceptable performance assumes that the defined ASTM precision is reached at least 90% of the time during normal operation with no more than two model updates every year [20]. Other common complaints include filters clogging and product film deposition on optical parts. Proper instrument installation standards are described in ASTM D6122, which recommends sample conditioning systems design including automatic probe wash and regular model validation (daily or twice a day) [20].

Alternative manners to cope with the aforementioned problems are based on identification and analyses of the actual sources of property variability: the sample composition [21]. It is well known that the composition of a mixture can drive its chemical and physical behavior [14]. Recently, a promising branch of analytical chemistry based on advanced mass spectrometry, usually named FT-ICR-MS or simply FTMS [21], has been applied to determine the fuel properties based on a detailed composition of the sample. This has led to development of a new family of analytical techniques called *petroleomics*.

Although these techniques are still incipient and practically restricted to laboratories, typical analyses last from 5 to 15 minutes, making feasible in-line implementations. More mature techniques that can be used to provide the detailed composition of oil products include bidimensional (2d) chromatography, especially GC×GC and GC×MS [22]. However, the main disadvantage in this case is the trade-off between data resolution and the time required for analyses, around 80 minutes. Specifically, chromatographers can utilize very different columns, with different lengths and diameters, which influence resolution, analysis time, maintenance (clogging), and so forth.

Since the abovementioned equipments measure compositional variables, it is necessary to implement particular knowledge to convert such data into specified properties. In this subject, quantitative structure-property relationships have shown very promising results [23, 24], despite the sparse data of pure compounds. Table 1 shows a summary of the measurement technologies discussed here.

A plausible model for producing a combined CoA may be based on Figure 7 schematics, where the NIR predictions are first generated during in-line blending and the QSPR predictions are then generated, using the compositional analyses carried out by chromatographers or spectrometers.

3.2. Validation. At the second level of the operation hierarchy, validation of obtained measurements must be performed in order to assure product certification. Such control layer enhances the measurement quality by identifying outliers, noise, or data trends and, if necessary, performing data reconciliation. Moreover, in case of identification of off-spec materials, protection mechanisms must detect and execute the necessary actions to ensure process safety and product quality.

Data validation is usually carried out by characterizing the measurement error, frequently measuring protofuels. Alternatively, results obtained by on-line analyzers can be compared to standard laboratory results according to statistical techniques aiming at detecting and quantifying

Table 1: Technologies to integrate an ILBC-based operation model.

Technologies	References	Pros	Cons
Spectroscopy	Gilbert et al., 2004 [18] Morris et al., 2009 [13]	Fast analysis; high precision; low prices	Subject to local and periodic calibrations; insensitive sulfur compounds
(2d) chromatography	Yang et al., 2002 [25] Wang et al., 2005 [22]	Compositional data; established technology; varied configurations	Trade-off between analysis time and data resolution for real-time applications
Mass spectrometry	Marshall and Rodgers, 2004 [21]	Fast analysis; ultrahigh resolution; detailed characterization;	Qualitative data, due to ionization deficiencies; high prices
Quantitative structure-property relationships	Creton et al., 2010 [23] Saldana et al., 2011 [24]	Fundamental knowledge; universal applications	Few pure compound properties known; small data bases

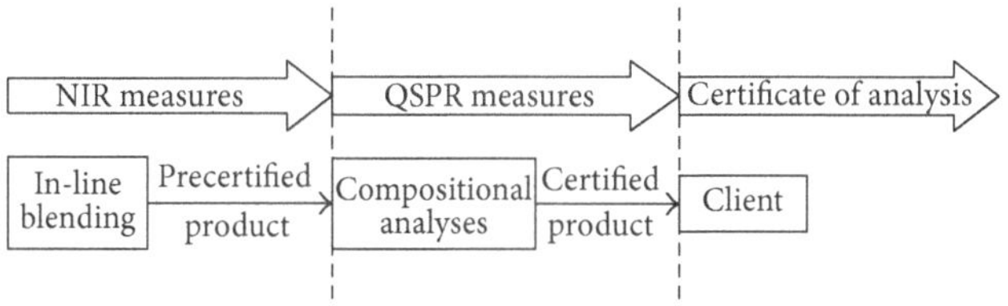

Figure 7: Basic ILBC analytical steps.

discrepancies. The main standard practices available for validating on-line analyzers are ASTM D3764, for performance validation of process stream analyzers, ASTM D6299, for statistical quality assurance and control charting techniques of measurement systems, and ASTM D6122, for validation of multivariate on-line, at-line, and laboratory NIR-based analyzers.

Due to well-known limitations of conventional validation practices, which are usually performed with long time intervals, the measurement reliability is not ensured during the whole period of product transferring. Besides, precision and accuracy of nonconventional on-line analyzers are insufficient for product certification or have not been characterized. In this scenario, simple multivariate techniques, such as principal component analysis (PCA) and canonical variate analysis (CVA), can be of great interest to provide process data at short time intervals and reduce uncertainties, improving data repeatability and reproducibility [26].

Basically, multivariate techniques use correlations of properties measured by different on-line analyzers to increase data redundancy and to enhance capabilities for process fault detection. Abnormal behaviors can also be detected by comparing current data with successful historical runs. For example, NIR and FTMS analyzers can provide electromagnetic and mass spectra that can be used directly to provide operation states and monitor the process behavior [27]. In case of sensor malfunction or maintenance, multivariate techniques can be used to estimate the missing properties based on correlations with measurements provided by other sensors and/or historical models, allowing for acceptable ILC robustness [26]. In case of process fault, automatic protection actions must be triggered, such as segregation of the off-spec components or product, in order to guarantee the ILBC premises.

Figure 8 illustrates the worst validation scenario of the proposed ILBC scheme, in which all components behave abnormally, yielding an off-spec product that must be isolated from the jetty line. After problem detection with the help of electromagnetic spectra and petroleomic tools, the product is then sent back to an off-spec product storage tank through a recycle line, for posterior reblending or reprocessing.

3.3. The Human Organization. Changing attitudes, revising previous assumptions, and rethinking best practices for refinery operation are particular challenges of all in performing optimization.

Actually, rather than an interdisciplinary activity in nature, ILBC operation must be understood as a very specialized optimization activity inside refinery. A major challenge to successfully implement an ILBC technology may indeed be educating the refinery's personnel to work according to new paradigms and work process.

Since at the refinery technical-commercial boundary, the ILBC operation is not only strongly dependent on the refinery production scheduling but also provides a vital feedback for its daily updating. Such collaborative scheduling formulation may be especially important for refineries embedded in a very competitive environment, in which scheduling is usually market oriented. In this sense, rather than any computational systems, the quality of the staff may be posed as the prime success factor, along with their ability to work as a single team, despite focusing on distinct problems and tasks [12].

Best work processes in turn enable the personnel to consistently achieve production plan objectives. Therefore, work process should be independent of personnel changes. This is accomplished via structuring a refinery team responsible for the ILBC unit performance. Some critical characteristics of a successful team are identified. As pointed out by Barsamian

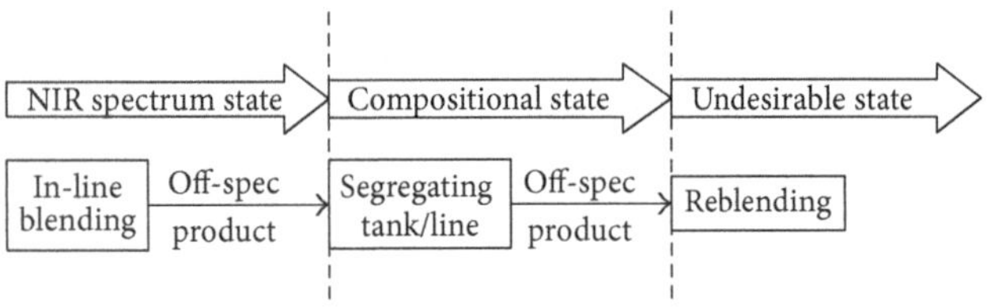

FIGURE 8: Segregating route of ILC procedure.

[5], thinking and working oriented to "do things right the first time" should drive local raining people's attitudes and mindsets aiming at supporting the ILBC unit. Rapid communications, faster response times (analyzer repairs with high priority), key people on call, and management motivation of ILBC team members are mandatory practices that motivate a matrix type organization in which the individuals remain physically in their respective areas, but their daily work responsibility is to the team (Figure 9).

The ILBC team is responsible for the daily operation of the ILBC unit. It comprises tight monitoring of performance indicators including critical analysis of the nature and frequency of operational issues and incidents, definition of priorities related to system maintenance and upgrading, resource forecasting to ensure high factors of system availability and reliability, and managerial follow-up.

The most prominent change in the refinery's personnel profile is certainly experimented by the laboratory staff. Not only a new work philosophy must be learned, but also a set of more added-value activities should be performed. Since now focusing on the on-line analyzers performance, the laboratory's staff must be qualified to locally provide process streams and fuel additives spectral modeling, protofuel management, research on physicochemical inferences development, and equipment monitoring and maintenance. As product certification process is expected to be less demanding for laboratory ASTM analysis of manual operational routines, as periodic line and tank sampling, mechanical work should be replaced by challenging activities performed by high-profile professionals. On the other hand, it is uncertain to say that ILBC automation will cause the reduction of laboratory activities. Anyway, it is worthwhile to note that those refiners that did not reduce their technical staff when automation was introduced have experienced gains in operational safety [28].

Since such technologies are a major concern as they impact the entire plant schedule and real-time optimization process applications, a highly qualified and integrated blending team becomes instrumental to the modern refining industry.

3.4. Superior Control Layers. The remaining control layers of the proposed ILBC operation hierarchy are concerned with the blending process operations. However, it is important to note that (a) product delivery underscored by ILBC technology is faster than the conventional one (Figure 2) and (b) on-line properties measurements present more vigorous dynamics and this way requires tighter control. Once the property mixture rules and indexes are known and the

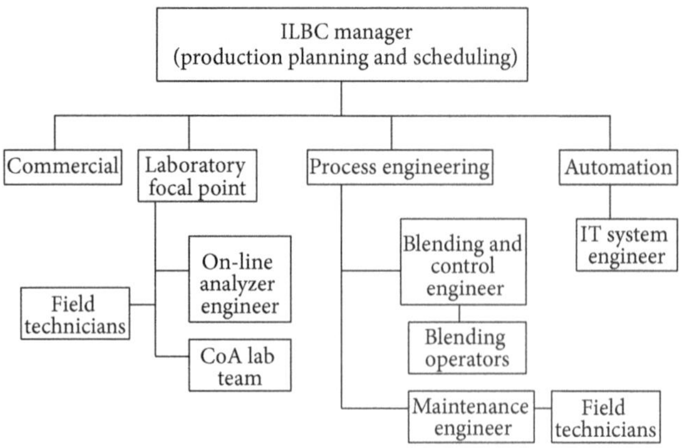

FIGURE 9: A ILBC team model [5].

inferential models are implemented, NIR-based analyzers can quickly measure the feed and product properties [19]. Hence, it allows for a more sensitive control using both feedback and feed forward control structures. An important feature of on-line measurements of oil properties is the higher frequency of analyses, which enables improved process operation and more frequent optimization of process setpoints. For instance, both electromagnetic and mass spectra can offer essential information for definition of the best recipe for a particular production run [27].

4. ILBC Prerequisites

Custody transfer involves industry and legal metrology standards, contractual agreements between the parties, and governmental regulation and taxation. Therefore, any concrete investment decision on ILBC implementation should be preceded by an interdisciplinary analysis of the refinery and the environment in which it is inserted. Major issues are addressed in this section, as follows.

4.1. Technical Issues. At the technical viewpoint, an ILBC unit may be understood as a forward step of a reliable ILB unit. The success of an ILBC implementation is conditioned to the maturity stage of the refinery ILB automation technology as well as its previous experience in dealing with ILB operations. Table 2 summarizes major qualitative characteristics expected to further raise a successful ILBC application.

4.2. Legal Issues. The first question that must be considered prior to legal implementation of an ILBC procedure is the following: is it legally permitted?

Under the preceding US EPA proposition for shipping before testing based only on on-line analyzer control of properties of interest to EPA (USA Code of Federal Regulation, 40CFR80, part 65), governmental regulatory legislation worldwide should be expected to be revised in the light of the fast development of ILBC technology as well as due to environmental pressures, since it may optimize volatile organic compound emissions at the end of refinery. As a result, the formulation of new *modus operandi* and mechanisms to

TABLE 2: Some critical technical requirements for an ILBC-based operation model.

Discipline	Item	Prerequisite
Analytical	ASTM analyzers	Reliability
	On-line analyzers	Robustness Sample condition system Redundancy with on-line ASTM analyzers
	Sampling	Sampling conditioning systems (on-line measurements) Automated composite sampling
	Laboratory	Laboratory NIR spectrometer Reference materials/protofuels Spectral database (protofuels, streams, and additives)
Information technology	Real-time database	Process streams data storage (extensive/intensive properties) Discrete and continuous plant data storage (e.g., on/off flows, tank heels)
	Off-line database	Process configuration (e.g., equipment operational limits)
Process engineering	System design	Allowing for redundant measurements
	Process modeling and control	Reliable inferences to noncritical product qualities MPC regarding feed forward on components blend indices Blending RTO technology Ratio control including fall-back strategies
	Oil movement system	Integration to refinery production scheduling Automated plant operation management (i) Tanks: swings, heel qualities, and stratification detection (ii) Line ups: flushing, valve position, and pipe quality tracking
	Data validation	Real-time data reconciliation on flows On-line analyzer data validation on ASTM analyzers

ensure consistent ILBC implementation and customer rights may be propelled by prospective studies and pilot projects by the oil industry.

4.3. Commercial Issues. The ILBC approach may be particularly attractive for vicinal companies operating local continuous-flow product custody transfer among them. This may be the case of integrated industrial pools involving refinery and petrochemical plants.

Once the ILBC operation mode is identified to potentially bring substantial benefits to both the buyer and the seller, a number of issues should be addressed. Since representing the ultimate manufacturing step, severe consequences may arise in case of system failure including economics and reputation damages. In this sense, a comprehensive analysis of hazardous events should be done and translated to contractual agreements if not properly addressed by the governmental legislation.

Rather than spot commercial contacts related to demand forecasting and CoA emission, such real-time integrated operation implies a high level of multidisciplinary human interactions between the personnel of both parts, including collaborative technological efforts to ensure reliability of the continuous operation and monitoring, including common fall-back strategies.

5. Conclusions

Facing various challenges in the everchanging refining landscape, it is essential that refiners raise their operations to new levels of performance. To help refineries improve its competitive edge and smoothly change to future refining operation, this work has introduced a new paradigm concerned with the high performance refining: in-line certification.

This subject promises to revolutionize the whole picture of refineries production by upgrading the blending process operations to new ends and legal threshold. The use of old-fashioned and unreliable analyzers, which perform long-run analyses, may be in close connection with the historical background of the petroleum business but should not necessarily remain the same in the near future, given the fast technological development of new instruments and novel analytical techniques. Compound identification can be programmed more easily and cause the reduction of uncertainties and subjectivity, suggesting that future ILBC procedures may have close connection with petroleomics.

This revolution can bring a series of advantages that can allow for implementation of an ILBC approach in the near future. Technical-economic advantages as well as interdisciplinary challenges to be addressed were outlined and discussed. In-line measurement was identified to be the key issue for a successful certification model. The inherent measurement complexity of some specified properties demands very dedicated analyzers, which generate uncertain and/or subjective data after long analysis time. Potential surrogates for these analyzers were discussed, including the well-known NIR-based and new petroleomics-based techniques. Data validation and reconciliation were pointed out to have an important role in the ILBC approach, ensuring the required reliability for a real-world application. Complementary discussion on promising process control strategies and human organization for supporting ILBC operation were also presented.

At the managerial level, clear refinery business objectives including minimum inventory policies, flexible, nevertheless optimized, scheduling to efficiently capture commercial spot opportunities and low demurrage aiming at minimizing jetties occupancy should be praised as well as the commitment of the refinery staff with a realistic scheduling and its economic performance. Therefore, new work processes based on an interdisciplinary dedicated team whose core skills include (a) fast communication according to an efficient organization workflow in which reliable information (data and models) is available on time, (b) know how autonomy to promptly solve maintenance problems locally at the refinery, and (c) new procedures regarding chemometric model maintenance, reference samples management, analyzers validation, on-line sampling, fall-back strategies implementation, and customer claims.

Acknowledgments

The authors would like to thank Geraldo Márcio Diniz Santos (Petrobras Headquarters), Magno Schiavolin and Wallace Nascimento (Petrobras REPLAN Refinery), and Antonio Katata and Aerenton Bueno (Petrobras REVAP Refinery) for valuable collaboration with this work.

References

[1] J. R. Katzer, M. P. Ramage, and A. V. Sapre, "Petroleum refining: poised for profound changes," *Chemical Engineering Progress*, vol. 96, no. 7, pp. 41–51, 2000.

[2] S. Hu and X. X. Zhu, "A general framework for incorporating molecular modeling into overall refinery optimization," *Applied Thermal Engineering*, vol. 21, no. 13-14, pp. 1331–1348, 2001.

[3] L. F. L. Moro, "Optimization in the petroleum refining industry—the virtual refinery," in *10th International Symposium on Process Systems Engineering—PSE 2009*, R. M. D. B. Alves, C. A. O. do Nascimento, and E. C. Biscaia Jr., Eds., Elsevier, Salvador, Brazil, 2009.

[4] P. A. Pantoja, J. López-Gejo, G. A. C. Le Roux, F. H. Quina, and C. A. O. Nascimento, "Prediction of crude oil properties and chemical composition by means of steady-state and time-resolved fluorescence," *Energy & Fuels*, vol. 25, no. 8, pp. 3598–3604, 2011.

[5] A. Barsamian, "Gasoline blending: technology, operations, economics," Technical Seminar, Refinery Automation Institute LLC, Morristown, NJ, USA, 2007.

[6] J. -P. Favennec, *Petroleum Refining: Refinery Operation and Management*, vol. 5, Editions Technip, Paris, France, 2001.

[7] M. Joly, "Refinery planning and scheduling: the refining core business," *Brazilian Journal of Chemical Engineering*, vol. 29, no. 2, pp. 371–384, 2012.

[8] Technip, "In Line certification of petroleum products," Technical Seminar, Technip Advanced Systems Engineering, Paris, France, 2007.

[9] EPA, "40CFRPart80 regulation of fuel and fuel additives: gasoline and diesel fuel test methods," EPA Report OAR2005-0048, Environmental Protection Agency, Washington, DC, USA, 2006.

[10] Y. Wu and N. Zhang, "Molecular characterization of gasoline and diesel streams," *Industrial and Engineering Chemistry Research*, vol. 49, no. 24, pp. 12773–12782, 2010.

[11] M. M. S. Aye and N. Zhang, "A novel methodology in transforming bulk properties of refining streams into molecular information," *Chemical Engineering Science*, vol. 60, no. 23, pp. 6702–6717, 2005.

[12] N. Zhang and M. Valleur, "Refinery planning and scheduling," in *ASTM Handbook of Petroleum Refining and Natural Gas Processing*, chapter 18, ASTM International, Conshohocken, Pa, USA, 2009.

[13] R. E. Morris, M. H. Hammond, J. A. Cramer et al., "Rapid fuel quality surveillance through chemometric modeling of near-infrared spectra," *Energy & Fuels*, vol. 23, no. 3, pp. 1610–1618, 2009.

[14] D. J. Cookson, C. P. Lloyd, and B. E. Smith, "Investigation of the chemical basis of diesel fuel properties," *Energy & Fuels*, vol. 2, no. 6, pp. 854–860, 1988.

[15] A. Agoston, C. Ötsch, and B. Jakoby, "Viscosity sensors for engine oil condition monitoring—application and interpretation of results," *Sensors and Actuators A*, vol. 121, no. 2, pp. 327–332, 2005.

[16] A. Agoston, N. Dörr, and B. Jakoby, "Corrosion sensors for engine oils—laboratory evaluation and field tests," *Sensors and Actuators B*, vol. 127, no. 1, pp. 15–21, 2007.

[17] A. Barsamian, "Get the most out of your NIR analyzers," *Hydrocarbon Processing*, vol. 80, no. 1, pp. 69–72, 2001.

[18] W. R. Gilbert, F. S. G. de Lima, and A. F. Bueno, "Comparison of NIR and NMR spectra chemometrics for FCC feed online characterization," *Studies in Surface Science and Catalysis*, vol. 149, pp. 203–215, 2004.

[19] A. Barsamian, "Consider near infrared methods for inline blending," *Hydrocarbon Processing*, vol. 84, no. 6, pp. 97–100, 2005.

[20] A. Barsamian, "Optimize fuels blending with advanced online analyzers," *Hydrocarbon Processing*, vol. 87, no. 9, pp. 121–122, 2008.

[21] A. G. Marshall and R. P. Rodgers, "Petroleomics: the next grand challenge for chemical analysis," *Accounts of Chemical Research*, vol. 37, no. 1, pp. 53–59, 2004.

[22] F. C.-Y. Wang, K. Qian, and L. A. Green, "GC×MS of diesel: a two-dimensional separation approach," *Analytical Chemistry*, vol. 77, no. 9, pp. 2777–2785, 2005.

[23] B. Creton, C. Dartiguelongue, T. De Bruin, and H. Toulhoat, "Prediction of the cetane number of diesel compounds using the quantitative structure property relationship," *Energy & Fuels*, vol. 24, no. 10, pp. 5396–5403, 2010.

[24] D. A. Saldana, L. Starck, P. Mougin et al., "Flash point and cetane number predictions for fuel compounds using quantitative structure property relationship (QSPR) methods," *Energy & Fuels*, vol. 25, no. 9, pp. 3900–3908, 2011.

[25] H. Yang, Z. Ring, Y. Briker, N. McLean, W. Friesen, and C. Fairbridge, "Neural network prediction of cetane number and density of diesel fuel from its chemical composition determined by LC and GC-MS," *Fuel*, vol. 81, no. 1, pp. 65–74, 2002.

[26] T. N. Pranatyasto and S. J. Qin, "Sensor validation and process fault diagnosis for FCC units under MPC feedback," *Control Engineering Practice*, vol. 9, no. 8, pp. 877–888, 2001.

[27] M. Hur, I. Yeo, E. Park et al., "Combination of statistical methods and Fourier transform ion cyclotron resonance mass spectrometry for more comprehensive, molecular-level interpretations of petroleum samples," *Analytical Chemistry*, vol. 82, no. 1, pp. 211–218, 2010.

[28] J. P. Favennec, *Le Raffinage du Petrole: Exploitation et Gestion de la Raffinerie*, vol. 5, Tome 5, Technip Editions, Institut Francais du Petrole, Paris, France, 1998.

Modeling of Partially Hydrolyzed Polyacrylamide-Hexamine-Hydroquinone Gel System Used for Profile Modification Jobs in the Oil Field

Upendra Singh Yadav and Vikas Mahto

Department of Petroleum Engineering, Indian School of Mines, Dhanbad 826004, India

Correspondence should be addressed to Upendra Singh Yadav, yadavupend@gmail.com

Academic Editor: Serhat Akin

The cross-linked polymer gel systems are being used increasingly to redirect or modify reservoir fluid movement in the vicinity of injection wells for the purpose of permeability/profile modification job in the oil field due to their high temperature stability and capability to provide rigid gel having high mechanical strength. In this study, a partially hydrolyzed polyacrylamide-hexamine-hydroquinonegel is used for the development of polymer gel system. The experimental investigation demonstrates that the gelation time varies with polymer and crosslinker concentration and the temperature. The mathematical model is developed with the help of gelation kinetics of polymer gel and using Arrhenius equation, which relates the gelation time with polymer, crosslinker concentrations, and temperature. The developed model is solved with the help of multivariate regression method. It is observed in this study that the theoretical values of gelation time have good agreement with the experimental values.

1. Introduction

The reservoir heterogeneity, permeability variations, and presence of fractures and fracture network in the reservoir are the main hurdles of the water flooding operations used to enhance the oil recovery from matured oil fields. Normally during water flooding, water sweeps through more permeable sections of the reservoir leaving back oil in low permeability channels leading to low oil recovery and early water breakthrough. This excessive water production from the producers leads to rise in handling and disposal costs and reduces the economic life of the well. The polymer gels block or reduce the permeability in high permeability channels and divert the injected water through the low permeability sections, which were not flooded or swept earlier, leading to improvement in oil recovery. This technique is known as profile modification or permeability modification technique and the proper application of this technique is essential for the success of water flooding projects in the oil fields [1–3].

Basically two types of polymers have been used for profile modification jobs. These are polyacrylamides with different degrees of hydrolysis and polysaccharide such as xanthan biopolymer. These are cross linked with inorganic and organic crosslinkers to produce a three-dimensional polymer structure of the gel [4–7]. The inorganic gel system based on the crosslinking of the carboxylate groups on the partially hydrolyzed polyacrylamide chain (PHPA) with a trivalent cation like Cr (III). This crosslinking is believed to rely on coordination covalent bonding. It should be mentioned that Cr (III) based polymer gel system were reported to be stable at temperatures up to 300°F [8–10]. The organically crosslinked gels developed using phenol and formaldehyde crosslinkers are thermally stable under harsh environmental conditions. But, these crosslinkers are not environment friendly. To overcome the toxicity issues associated with formaldehyde and phenol, the formaldehyde can be replaced with less toxic derivatives like hexamethylenetetramine, glyoxal, paraformaldehyde, acetaldehyde, propionaldehyde, butyraldehyde and so forth and phenol can be replaced with hydroquinone, resorcinol, pyrogallol, phenyl salicylate and so forth [11–13]. Other system is based on PEI crosslinker and a copolymer of acrylamide and t-butyl acrylate (PA-t-BA). PA-t-BA is a relatively low molecular-weight polymer is expected

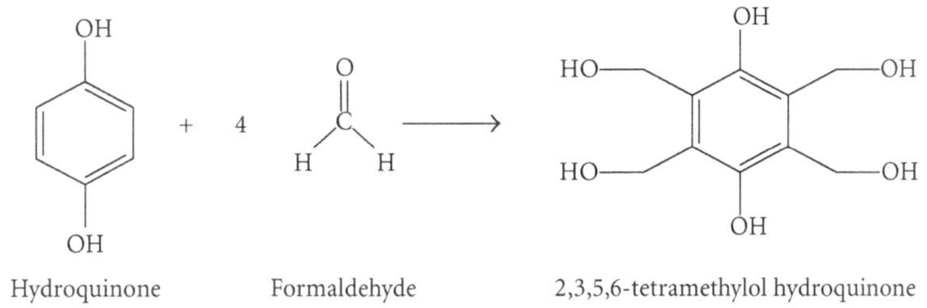

Scheme 1

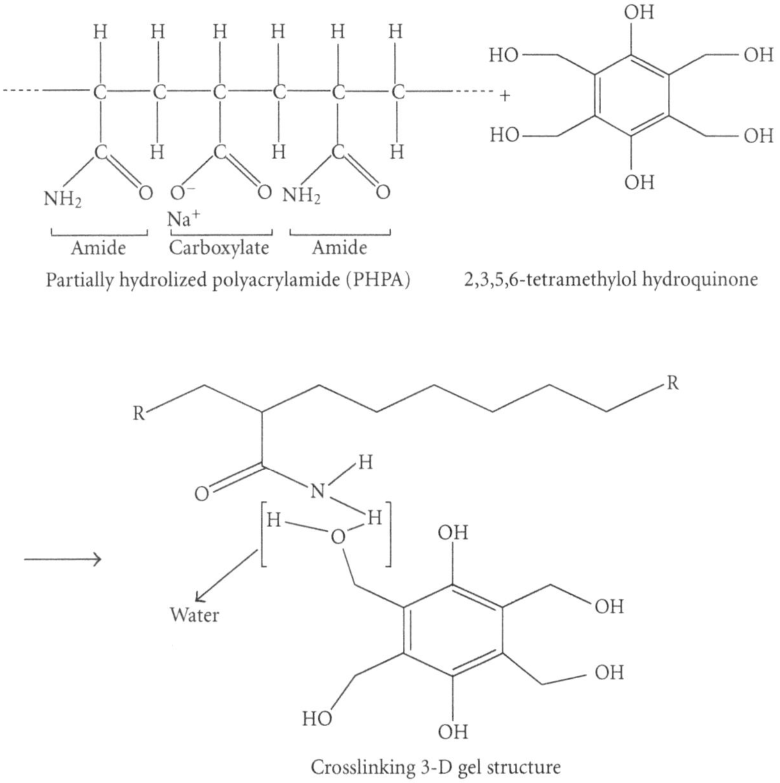

Scheme 2

to provide rigid ringing gels. Polyethyleneimine (PEI) crosslinking a copolymer of acrylamide and tert-butylacrylate (PAtBA) as water shutoff gels has been widely used in recent years [14].

The gelation mechanism of organic crosslinkers is done by covalent bonding, which is much more stable than the ionic bonds. Organic crosslinkers were introduced to obtained gels that can remain stable over a wide temperature range. For high temperature (>90°C) reservoirs, acrylamide-based copolymers with organic crosslinking agents, such as phenol-formaldehyde and its derivatives, can be used to form a gel with thermal stability, and the gelation time is adjustable [15, 16].

The different methods are presented in the literature for the determination of the physical and chemical properties of polymer gel system such as bottle testing method, sealed tube method, dynamic shear method (rheometer), and static shear method (viscometer). Out of these methods, the bottle testing method is most suitable, faster, and inexpensive method to study the gelation kinetics [17].

Several models have been reported in the literature to relate the gelation time with temperature. Jordan found that the gelation time of a specified system decreases as the temperature is increased. Plots of the logarithm of gelation time versus the reciprocal of the absolute reaction temperature were found to be linear for the systems studied. This correlation showed that the system apparently followed the Hurd and Letteron model which assumed rate dependence on only one species. Hurd and Letteron studied the effect of temperature on the formation of silicic acid gels and found empirically that plot of the logarithm of the gelation time versus the reciprocal of the absolute temperature linear. They developed an equation, similar to the Arrhenius equation, which relates gelation time and the temperature and support the linear relationship [15]. But very few or rare models relate the gelation time with polymer and crosslinker concentration

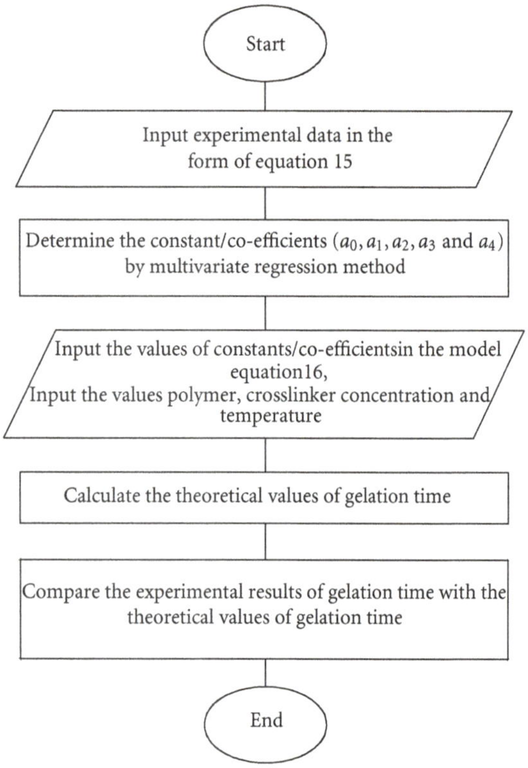

FIGURE 1: Flow chart for the solution of developed model.

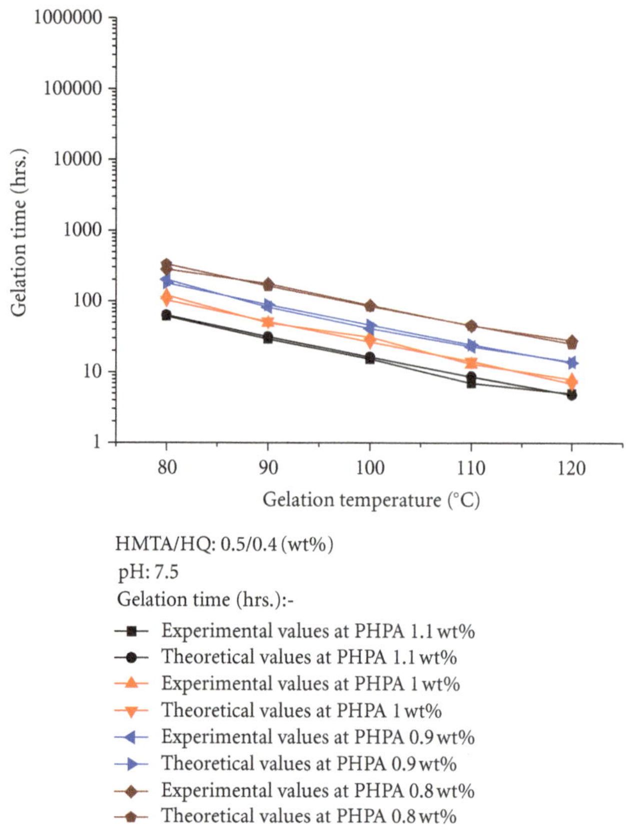

FIGURE 2: Plot of gelation temperature versus experimental and theoretical values of gelation time at different polymer concentration, constant crosslinker concentration (HMTA/HQ 0.5/0.4 wt%), and pH 7.5.

and temperature in case of polymer gel system used in the oil fields. For the successful design of the profile modification job the mathematical model equation consists of polymer and crosslinkers concentration as well as temperature is essential.

The present paper involves bulk gelation studies of the polymer gelant prepared from partially hydrolyzed polyacrylamide (PHPA) and hexamine (HMTA)/hydroquinone (HQ). The gelation time determined from the bulk gelation studies at different temperatures is useful because its knowledge at reservoir temperature is required to find out the depth up to which the gel can be placed in the petroleum reservoir rock. The gelation time and gel strength mainly depend on polymer and crosslinker concentrations and reservoir temperature were also reported in this paper. On the basis of the kinetics of polymer gel system, the mathematical model for the gelation behavior is also proposed which relates the gelation time with polymer-crosslinkers concentrations and temperature. The theoretical values of gelation time obtained from the proposed model matches with the experimental data.

2. Gelation Mechanism

The hexamine on hydrolysis yields formaldehyde which then combines with hydroquinone and form 2,3,5,6-tetramethylol hydroquinone. Further, partially hydrolyzed polyacrylamide reacts with 2,3,5,6-tetramethylol hydroquinone and forms three dimensional networks of polymer gel, the different steps of which are as follows.

Step I. In the initial step, hexamine hydrolyses to yield formaldehyde

$$\underset{\text{HMTA}}{(CH_2)_6N_4} + \underset{\text{Water}}{H_2O} \longrightarrow \underset{\text{Methane diol}}{6OH\text{–}CH_2\text{–}OH} + 4NH_3 \quad (1)$$

$$OH\text{–}CH_2\text{–}OH \longrightarrow \underset{\text{Formaldehyde}}{H\text{–}CH{=}O} + \underset{\text{water}}{H_2O} \quad (2)$$

Step II. Formaldehyde produced in Step I then react with hydroquinone to form a condensed structure known as 2,3,5,6-tetramethylolhydroquinone (see Scheme 1).

Step III. The condensed molecule formed in Step II then reacts with PHPA to form the 3-dimensional gel structure which helps in profile modification jobs (see Scheme 2).

3. Development of Mathematical Model

The rate of gelation process for partially hydrolyzed polyacrylamide-hexamine-hydroquinone system may be present as follows.

$$\alpha A + \beta B + \Upsilon C \rightarrow \text{3D gel structure}$$

$$r_A = -\frac{dC_A}{dt} = kC_A{}^a C_B{}^b C_C{}^c, \quad (3)$$

where k is the reaction rate constant, C_A, C_B, and C_C denotes the concentration of partially hydrolyzed polyacrylamide,

TABLE 1: Experimental values of gelation time at different polymer concentration, temperature, and pH 7.5 (constant crosslinker concentration HMTA/HQ 0.5/0.4 wt%).

Sl. No.	Polymer PHPA (wt%)	Crosslinkers HMTA (wt%)	Crosslinkers HQ (wt%)	Experimental values of gelation time (hrs.)	Theoretical values of gelation time (hrs.)	Temperature (°C)
1	0.8	0.5	0.4	280	328.8475	80
2	0.9	0.5	0.4	200	178.3855	
3	1.0	0.5	0.4	120	103.3145	
4	1.1	0.5	0.4	61	62.9199	
5	0.8	0.5	0.4	174.3	163.9435	90
6	0.9	0.5	0.4	82.3	88.9322	
7	1.0	0.5	0.4	49	51.4565	
8	1.1	0.5	0.4	29.3	31.3681	
9	0.8	0.5	0.4	87	84.8392	100
10	0.9	0.5	0.4	41	46.0215	
11	1.0	0.5	0.4	31	26.6282	
12	1.1	0.5	0.4	15.3	16.2326	
13	0.8	0.5	0.4	45.3	45.4394	110
14	0.9	0.5	0.4	23	24.6489	
15	1.0	0.5	0.4	13	14.2619	
16	1.1	0.5	0.4	7	8.6941	
17	0.8	0.5	0.4	28	25.1225	120
18	0.9	0.5	0.4	14	13.6278	
19	1.0	0.5	0.4	8	6.8917	
20	1.1	0.5	0.4	5	4.80681	

hexamine and hydroquinone and a, b, and c are the order with respect to A, B, and C, respectively. If the reactants are present in their stoichiometric ratios, they will remain at that ratio throughout the reaction [19]. Thus for reactants A, B, and C at any time,

$$\frac{C_B}{C_A} = \frac{\beta}{\alpha} \Longrightarrow C_B = \frac{\beta}{\alpha} C_A, \qquad \frac{C_C}{C_A} = \frac{\gamma}{\alpha} \Longrightarrow C_c = \frac{\gamma}{\alpha} C_A, \tag{4}$$

where α, β, and γ = reactant of species A, B, and C present in the reaction, respectively

$$\begin{aligned} r_A = -\frac{dC_A}{dt} &= kC_A{}^a \cdot \left(\frac{\beta}{\alpha} C_A\right)^b \cdot \left(\frac{\gamma}{\alpha} C_A\right)^c \\ &= k\left(\frac{\beta}{\alpha}\right)^b \left(\frac{\gamma}{\alpha}\right)^c \cdot C_A{}^{a+b+c}. \end{aligned} \tag{5}$$

The above equation can be also written as

$$r_A = -\frac{dC_A}{dt} = k' C_A{}^n. \tag{6}$$

Taking integration both side w.r.t. time, we have

$$t = \frac{1}{(n-1) k'} \left\{ \frac{1}{C_A{}^{n-1}} - \frac{1}{C_{Ao}{}^{n-1}} \right\}. \tag{7}$$

Rearranging the above equation and putting the value of k' and n,

$$t = \frac{1}{k\{(a+b+c)-1\}} \left\{ \frac{C_A}{C_A{}^a \cdot C_A{}^b (\beta/\alpha)^b \cdot C_A{}^c (\gamma/\alpha)^c} - \frac{C_{Ao}}{C_{Ao}{}^a \cdot C_{Ao}{}^b (\beta/\alpha)^b \cdot C_{Ao}{}^c (\gamma/\alpha)^c} \right\}. \tag{8}$$

Rearranging the above equation, we get

$$t = \frac{1}{k\{(a+b+c)-1\}} \left\{ \frac{C_A}{C_A{}^a C_B{}^b C_C{}^c} - \frac{C_{Ao}}{C_{Ao}{}^a C_{Bo}{}^b C_{Co}{}^c} \right\}, \tag{9}$$

where C_{Ao}, C_{Bo}, and C_{Co} = initial concentrations of polymer, hexamine & hydroquinone.

By generalizing the above equation, it can be written as

$$t \propto \frac{1}{C_A{}^a C_B{}^b C_C{}^c}. \tag{10}$$

The effect of temperature (T in degree kelvin) can also be incorporated in (8) and Arrhenius model [19] can be utilized to connect k with T which is as follows:

$$k = A \exp\left(\frac{-E_a}{RT}\right), \tag{11}$$

TABLE 2: Experimental values of gelation time at different polymer concentration, temperature, and pH 7.5 (constant crosslinker concentration HMTA/HQ 0.5/0.3 wt%).

Sl. No.	Polymer PHPA (wt%)	Crosslinkers HMTA (wt%)	Crosslinkers HQ (wt%)	Experimental values of gelation time (hrs.)	Theoretical values of gelation time (hrs.)	Temperature (°C)
1	0.8	0.5	0.3	312	376.7173	80
2	0.9	0.5	0.3	230	204.3528	
3	1.0	0.5	0.3	134	118.2393	
4	1.1	0.5	0.3	68	72.0791	
5	0.8	0.5	0.3	207	187.8085	90
6	0.9	0.5	0.3	91	101.8779	
7	1.0	0.5	0.3	55.3	58.9469	
8	1.1	0.5	0.3	33.3	35.9343	
9	0.8	0.5	0.3	101	97.1891	100
10	0.9	0.5	0.3	55.3	52.7208	
11	1.0	0.5	0.3	34	30.5045	
12	1.1	0.5	0.3	19	18.5956	
13	0.8	0.5	0.3	50	52.0539	110
14	0.9	0.5	0.3	25.3	28.237	
15	1.0	0.5	0.3	14.3	16.338	
16	1.1	0.5	0.3	8	9.9597	
17	0.8	0.5	0.3	32	28.7795	120
18	0.9	0.5	0.3	14.3	15.6116	
19	1.0	0.5	0.3	8.3	9.0329	
20	1.1	0.5	0.3	6.3	5.5065	

where E_a is an apparent activation energy and R is the gas constant. Therefore, the combination of (10) and (11), then we get

$$t \propto \frac{1}{C_A{}^a C_B{}^b C_C{}^c} \times \exp\left(\frac{E_a}{RT}\right). \quad (12)$$

The above equation can be written as follows, k'' is the proportionality constant:

$$t = k'' \frac{1}{C_A{}^a C_B{}^b C_C{}^c} \times \exp\left(\frac{E_a}{RT}\right). \quad (13)$$

Taking natural logarithm and introducing the coefficients a, b, c then we get,

$$\ln t = \ln\left\{k'' \frac{1}{C_A{}^a C_B{}^b C_C{}^c} \times \exp\left(\frac{E_a}{RT}\right)\right\}, \quad (14)$$

$$\ln t = a_0 + a_1 \ln C_A + a_2 \ln C_B + a_3 \ln C_C + \frac{a_4}{T}. \quad (15)$$

Taking antilog both sides and assume t then we get

$$t = C_A{}^{a_1} \times C_B{}^{a_2} \times C_C{}^{a_3} \times \exp\left(a_0 + \frac{a_4}{T}\right). \quad (16)$$

Equation (16) is the general equation for the gelation behavior of partially hydrolyzed polyacrylamide-hexamine-hydroquinone gel system. The value of constant depends upon the polymer and crosslinkers composition and temperature.

4. Solution Procedure of the Developed Model

The mathematical model for gelation time consists of four variable parameters (C_A, C_B, C_C, and T) and five constants (a_0, a_1, a_2, a_3, and a_4). For the determination of these constants, five simultaneous equations are generated by multiplying the variables in both side of the model equation and utilizing the experimental data, the flow diagram for solution of proposed model is shown in Figure 1. Further, these equations are solved by multivariate regression method and values of the constants a_0, a_1, a_2, a_3, and a_4 are determined. After putting these values in the model equation, the gelation time are calculated by varying the different parameters (C_A, C_B, C_C, and T). Finally, these calculated values are compared with the actual values obtained from the laboratory work.

5. Experimental Work

5.1. Material Used. The materials used for this work are partially hydrolyzed polyacrylamide, hexamine, hydroquinone, sodium chloride, hydrochloric acid, and sodium hydroxide. Partially hydrolyzed polyacrylamide is procured from Oil and Natural Gas Corporation Limited, Mumbai, India. Hexamine is purchased from Otto Kemi Mumbai, India and hydroquinone is procured from Ranbaxy Fine Chemicals Ltd., New Delhi, India. Hydrochloric acid is purchased from Central Drug house (P) Ltd. New Delhi, India. Sodium chloride is purchased from Nice Chemical Pvt. Ltd. Cochin, India and

TABLE 3: Experimental values of gelation time at different polymer concentration, temperature, and pH 7.5 (constant crosslinker concentration HMTA/HQ 0.4/0.4 wt%).

Sl. No.	Polymer PHPA (wt%)	Crosslinkers HMTA (wt%)	HQ (wt%)	Experimental values of gelation time (hrs.)	Theoretical values of gelation time (hrs.)	Temperature (°C)
1	0.8	0.4	0.4	331	411.5734	80
2	0.9	0.4	0.4	243	223.2606	
3	1.0	0.4	0.4	140	129.1795	
4	1.1	0.4	0.4	77	78.7483	
5	0.8	0.4	0.4	226	205.1856	90
6	0.9	0.4	0.4	105	111.3043	
7	1.0	0.4	0.4	71	64.4011	
8	1.1	0.4	0.4	37.3	39.2591	
9	0.8	0.4	0.4	115	106.1816	100
10	0.9	0.4	0.4	61	57.5989	
11	1.0	0.4	0.4	37	33.3269	
12	1.1	0.4	0.4	23	20.3162	
13	0.8	0.4	0.4	56	56.8703	110
14	0.9	0.4	0.4	30	30.8496	
15	1.0	0.4	0.4	17	17.8497	
16	1.1	0.4	0.4	10	10.8812	
17	0.8	0.4	0.4	36	31.4424	120
18	0.9	0.4	0.4	17	17.0561	
19	1.0	0.4	0.4	9	9.8687	
20	1.1	0.4	0.4	7	6.016	

sodium hydroxide is purchased from S. D. Fine-Chem Ltd. Mumbai, India.

5.2. Experimental Procedure. Initially stock solution of partially hydrolyzed polyacrylamide was prepared in brine solution and kept for aging at ambient temperature for 24 hrs and further fresh solution of hexamine and hydroquinone were also prepared in brine. The appropriate solution of partially hydrolyzed polyacrylamide, hexamine, hydroquinone, and brine were thoroughly mixed at room temperature by magnetic stirrer. The pH of the gelant solution was measured by Century CP-901 digital pH meter and the pH of the gelant solution was adjusted by using 1 N sodium hydroxide and 1 N hydrochloric acid. Finally, the gelant solution was transferred into small glass bottle and kept at the desire temperature in the hot air oven (temperature ranges from 80°C to 120°C). Due to the increased temperature and the crosslinking reaction, gel formation takes place. The quality of gel was visually inspected at regular intervals and gelation time was noted. Here, the time for the formation of stiff/rigid gel was only considered.

6. Results and Discussions

The gelation behavior of the polymer gel system largely depends on the polymer-crosslinker concentrations and the temperature. The effects of these parameters on gelation time are described below.

6.1. Effect of Temperature on Gelation Time. Different gelling solutions were prepared with different concentrations of polymer and crosslinking agent at pH 7.5 and kept for gelation at different temperature. The reaction rate between amide group and methylol group was accelerated by increasing temperature and the gelation time decreased. Increasing the gelation temperature results in a decrease in gelation time since at higher temperature gels are formed in lesser time due to rapid crosslinking as is depicted in Figures 2 and 3 and Tables 1, 2, 3, 4, and 5. A possible explanation for rapid cross-linking is either due to an increase in molecular mobility or formation of new cross-linking sites as a result of gelation reaction. It is a known fact that degree of hydrolysis of the polymer increases at elevated temperatures which in turn increases the number of cross-linking sites that is found increasing the reaction rate and decreasing the gelation time.

6.2. Effect of Polymer Concentration on Gelation Time. The polymer concentration has a significant effect on the physical properties of gel. The polymer concentration increases and gelation time decreases were shown in Figure 4. The polymer concentration ranges from 0.8 to 1.1 wt% and constant concentration of crosslinker (0.5 to 0.3 wt% HMTA and 0.4 to 0.3 wt% HQ) at pH 7.5 are depicted in Tables 1 to 5. As the

TABLE 4: Experimental values of gelation time at different polymer concentration, temperature, and pH 7.5 (constant crosslinker concentration HMTA/HQ 0.4/0.3 wt%).

Sl. No.	Polymer PHPA (wt%)	Crosslinkers HMTA (wt%)	HQ (wt%)	Experimental values of gelation time (hrs.)	Theoretical values of gelation time (hrs.)	Temperature (°C)
1	0.8	0.4	0.3	370	471.4855	80
2	0.9	0.4	0.3	286	255.7604	
3	1.0	0.4	0.3	168	147.9839	
4	1.1	0.4	0.3	84	90.2116	
5	0.8	0.4	0.3	266	235.0542	90
6	0.9	0.4	0.3	135.3	127.5067	
7	1.0	0.4	0.3	78	73.7758	
8	1.1	0.4	0.3	42.3	44.974	
9	0.8	0.4	0.3	133	121.6383	100
10	0.9	0.4	0.3	71	65.9835	
11	1.0	0.4	0.3	42	38.1783	
12	1.1	0.4	0.3	27	23.2736	
13	0.8	0.4	0.3	60	65.1488	110
14	0.9	0.4	0.3	35	35.3404	
15	1.0	0.4	0.3	20.3	20.4481	
16	1.1	0.4	0.3	12	12.4652	
17	0.8	0.4	0.3	39	36.0194	120
18	0.9	0.4	0.3	19	19.539	
19	1.0	0.4	0.3	10	11.3053	
20	1.1	0.4	0.3	8	6.8917	

polymer concentration increases, it means more crosslinking sites are available for the fast crosslinking reaction takes place. Thus, the gel formation reaction increases, which leads to the decreases of gelation time. This trend is expected to be the same at all gelation temperatures under study. Thus, the required time to obtain a non-flowing polymer gel with a tolerable strength decreased when the polymer concentration was increased.

6.3. Effect of Cross-Linker Concentration on Gelation Time. Hexamine and hydroquinone crosslinkers are a multifunctional group which can build a complex network with amide groups of partially hydrolyzed polyacrylamide and form a 3-dimentional gel network structure. Crosslinker concentration has a significant effect on the gel strength. The crosslinker concentration increases and gelation time decreases were shown in Figure 5. The several samples were prepared to investigate the effect of crosslinker concentration on the network strength. Bottle testing results shown in Tables 1 to 5 indicate that when the concentration of both crosslinker was decreased the gelation rate and gel quality is also decreased. In other words, when crosslinking agent concentration was increased, the stage of polymer gel changed from a state of flowing gel to one of deformable non-flowing gel, because crosslinking sites are increased for the formation of gel in lesser time intervals.

7. Model Discussion

Equation (12) shows the relation of gelation time with polymer concentration, cross linker concentrations and temperature. Here, C_A, C_B, and C_C are polymers, HMTA and hydroquinone concentrations, respectively, in gm/litre and temperature, are taken in degree kelvin. This equation indicates that the gelation time is inversely proportional to the concentrations of polymer and crosslinking agent, which reflects that the increasing the concentration of polymer and crosslinking agent decreases the gelation time. The proposed model for the study of the gelation behavior of partially hydrolyzed polyacrylamide-hexamine-hydroquinone polymer gel system is shown in (15)-(16). The values of the constants/coefficients a_0, a_1, a_2, a_3, and a_4 are determined by multivariate regression method.

Equation (15) can be rewritten as follows:

$$y = a_0 + a_1 \cdot x_1 + a_2 \cdot x_2 + a_3 \cdot x_3 + a_4 \cdot x_4, \qquad (17)$$

where

$$y = \ln t; \quad x_1 = \ln C_A; \quad x_2 = \ln C_B, \quad x_3 = \ln C_C, \quad x_4 = \frac{1}{T}. \qquad (18)$$

The above equation solved by multivariate regression method and the values of constants a_0, a_1, a_2, a_3, and a_4 are calculated.

TABLE 5: Experimental values of gelation time at different polymer concentration, temperature, and pH 7.5 (constant crosslinker concentration HMTA/HQ 0.3/0.3 wt%).

Sl. No.	Polymer PHPA (wt%)	Crosslinkers HMTA (wt%)	HQ (wt%)	Experimental values of gelation time (hrs.)	Theoretical values of gelation time (hrs.)	Temperature (°C)
1	0.8	0.3	0.3	451	629.6609	80
2	0.9	0.3	0.3	362	341.5637	
3	1.0	0.3	0.3	216	197.63	
4	1.1	0.3	0.3	111	120.476	
5	0.8	0.3	0.3	311	313.9109	90
6	0.9	0.3	0.3	158	170.283	
7	1.0	0.3	0.3	102	98.5264	
8	1.1	0.3	0.3	55.3	60.0621	
9	0.8	0.3	0.3	170	162.4459	100
10	0.9	0.3	0.3	82	88.1198	
11	1.0	0.3	0.3	58	50.9864	
12	1.1	0.3	0.3	30	31.0815	
13	0.8	0.3	0.3	86	87.0051	110
14	0.9	0.3	0.3	66	57.1604	
15	1.0	0.3	0.3	24.3	27.308	
16	1.1	0.3	0.3	15	16.6471	
17	0.8	0.3	0.3	45	48.1034	120
18	0.9	0.3	0.3	23	26.094	
19	1.0	0.3	0.3	14	15.098	
20	1.1	0.3	0.3	10	9.2038	

TABLE 6: Comparison of gelation time of developed model and Arrhenius type equation.

Temperature (°C)	Experimental gelation time (hrs.)	Theoretical gelation time (hrs.) by Arrhenius type equation {Civan et al. [18]} $\ln(t) = 8288.47281/T\ (^\circ\text{K}) - 17.63044$ (0.8 wt% PHPA, 0.5 wt% HMTA and 0.3 wt% hydroquinone)	Theoretical gelation time (hrs.) by Developed Model (0.8 wt% PHPA, 0.5 wt% HMTA and 0.3 wt% hydroquinone)
80	312	343.66928	376.66426
90	207	180.07746	187.78207
100	101	97.68365	97.17542
110	50	54.70794	52.04664
120	32	31.55635	28.77554

The values of the constants a_0, a_1, a_2, a_3, and a_4 are found to be 1.77831, −5.19301, −1.00564, −0.47246 and 8926.87612, respectively and finally the gelation time is calculated from the following equation:

$$t = C_A^{-5.19301} \times C_B^{-1.00564} \times C_C^{-0.47246} \times e^{(1.77831+(8926.87612/T))}. \quad (19)$$

The value of R^2 is 0.99122, which shows the good agreement between experimental and theoretical values of gelation time. This study reveals that the developed model may be used for the study of gelation behavior of partially hydrolyzed polyacrylamide-hexamine-hydroquinone gel system for its application in profile modification jobs. The theoretical values calculated from above equation are shown in Tables 1–5 at different polymer-crosslinker concentration and temperature.

7.1. Comparison between Developed Model and Civan et al. [18]. The developed model in the present study relates stoichiometric relation between reactants involved in the gelation reaction as well as Arrhenius equation. In this case, gelation time is dependent on polymer and crosslinker concentration as well as temperature. However, model proposed by Civan et al. is simply Arrhenius type equation which is dependent on temperature only. Figure 3 shows the theoretical values of gelation time for different concentration of polymers. It was found that there are several variations

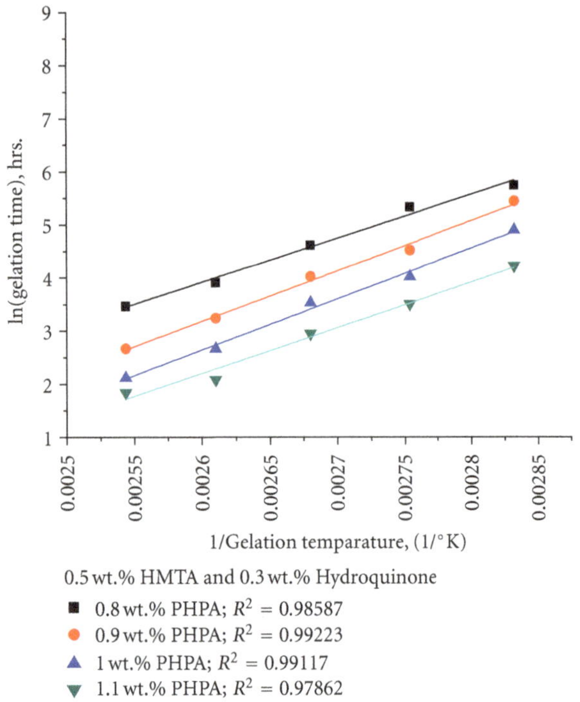

FIGURE 3: Plot for the temperature dependence of gelation time according to Arrhenius-type equation.

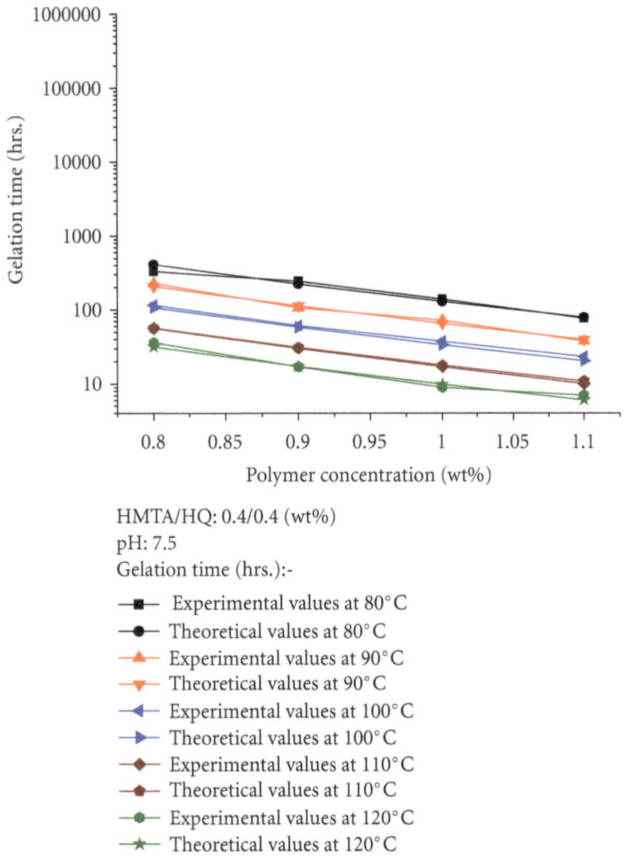

FIGURE 4: Plot of polymer concentration versus experimental and theoretical values of gelation time at different temperature, constant crosslinker concentration (HMTA/HQ 0.4/0.4 wt%) and pH 7.5.

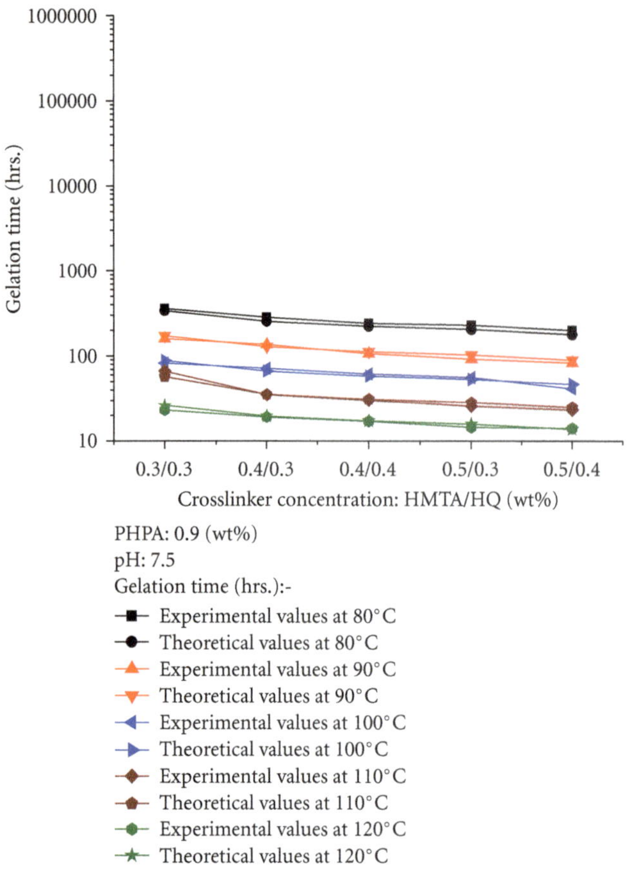

FIGURE 5: Plot of crosslinker concentration versus experimental and theoretical values of gelation time at different temperature, constant polymer concentration (PHPA 0.9 wt%), and pH 7.5.

in the values of slopes and intercepts as well as R^2 and may have different activation energy for different concentrations of polymers. The comparison of theoretical gelation time calculated using Civan et al. model and our developed model at the concentration (0.8% polymer) are approximately the same which is shown in Table 6 and have good agreement with that particular experimental data. But, the developed model presents the same values of coefficients and activation energy for whole experimental study and all data are fitted in that mathematical equation. Hence, it may be more realistic in nature.

8. Conclusion

The following conclusions are drawn from the present study.

(1) The gelation time of partially hydrolyzed polyacrylamide-hexamine-hydroquinone gel system decreases with increasing the cross linker concentration and temperature.

(2) On the basis of kinetics of gelation behavior the mathematical model for the partially hydrolyzed

polyacrylamide-hexamine-hydroquinone gel system is the following:

$$t = C_A^{a_1} \times C_B^{a_2} \times C_C^{a_3} \times e^{(a_0+(a_4/T))}. \quad (20)$$

(3) The values of constants a_0, a_1, a_2, a_3, and a_4 for this study are found to be 1.77831, −5.19301, −1.00564, −0.47246, and 8926.87612, respectively.

(4) The final mathematical model for the present study is as follows:

$$t = C_A^{-5.19301} \times C_B^{-1.00564} \times C_C^{-0.47246} \times e^{(1.77831+(8926.87612/T))}. \quad (21)$$

(5) A values show good agreement between experimental and theoretical values of gelation time. This study reveals that the developed model may be used for the study of gelation behavior of partially hydrolyzed polyacrylamide-hexamine-hydroquinone gel system for its application in profile modification jobs.

Nomenclature

C_A:	Concentration of partially hydrolyzed polyacrylamide (PHPA) (gm/liter)
C_B:	Concentration of hexamine (HMTA) (gm/liter)
C_C:	Concentration of hydroquinone (gm/liter)
t:	Gelation time (second)
T:	Temperature (°K)
k, k', k'':	Reaction rate/proportionality constants
C_{Ao}:	Initial concentrations of polymer
C_{Bo}:	Initial concentrations of hexamine
C_{Co}:	Initial concentrations of hydroquinone
E_a:	Activation energy
R:	Gas constant
a_0, a_1, a_2, a_3 and a_4:	Constants
a, b and c:	Reaction order with respect to A, B, and C
α, β and γ:	Reactant of species A, B and C present in the reaction respectively.

Acknowledgment

The authors are very thankful to Council of Scientific and Industrial Research (CSIR), New Delhi, India for financial assistance to carry out the research work.

References

[1] Y. Liu, B. Bai, and Y. Wang, "Applied technologies and prospects of conformance control treatments in China," *Oil and Gas Science and Technology*, vol. 65, no. 6, pp. 859–878, 2010.

[2] R. Jain, C. S. McCool, D. W. Green, G. P. Willhite, and M. J. Michnick, "Reaction kinetics of the uptake of chromium (III) acetate by polyacrylamide," *Society of Petroleum Engineers Journal*, vol. 10, no. 3, pp. 247–255, 2004.

[3] C. A. Grattoni, H. H. Al-Sharji, C. Yang, A. H. Muggeridge, and R. W. Zimmerman, "Rheology and permeability of crosslinked polyacrylamide gel," *Journal of Colloid and Interface Science*, vol. 240, no. 2, pp. 601–607, 2001.

[4] A. Moradi-Araghi, "A review of thermally stable gels for fluid diversion in petroleum production," *Journal of Petroleum Science and Engineering*, vol. 26, no. 1–4, pp. 1–10, 2000.

[5] A. Stavland and H. C. Jonsbraten, "New insight into aluminum citrate/polyacrylamide gels for fluid control," in *Proceedings of the SPE/DOE 10th Symposium on Improved Oil Recovery*, Paper SPE/DOE 35381, pp. 347–356, Tulsa, Okla, USA, April 1996.

[6] S. L. Bryant, G. P. Borghi, M. Bartosek, and T. P. Lockhart, "Experimental investigation on the injectivity of phenol-formaldehyde/polymer gelants," in *Proceedings of the SPE International Symposium on Oilfield Chemistry*, Paper SPE 37244, pp. 335–343, Houston, Tex, USA, February 1997.

[7] H. Jia, W. F. Pu, J. Z. Zhao, and R. Liao, "Experimental investigation of the novel phenol-formaldehyde cross-linking HPAM gel system: based on the secondary cross-linking method of organic cross-linkers and its gelation performance study after flowing through porous media," *Energy and Fuels*, vol. 25, no. 2, pp. 727–736, 2011.

[8] H. Jia, W. F. Pu, J. Z. Zhao, and F. Y. Jin, "Research on the gelation performance of low toxic PEI cross-linking PHPAM gel systems as water shutoff agents in low temperature reservoirs," *Industrial and Engineering Chemistry Research*, vol. 49, no. 20, pp. 9618–9624, 2010.

[9] H. T. Dovan, R. D. Hutchins, and B. B. Sandiford, "Delaying gelation of aqueous polymers at elevated temperatures using novel organic crosslinkers," in *Proceedings of the SPE International Symposium on Oilfield Chemistry*, Paper SPE 37246, pp. 361–371, Houston, Tex, USA, February 1997.

[10] R. D. Hutchins, H. T. Dovan, and B. B. Sandiford, "Field applications of high temperature organic gels for water control," in *Proceedings of the 10th Symposium of Improved Oil Recovery*, Paper SPE/DOE 35444, Tulsa, Okla, USA, April 1996.

[11] L. Eoff, D. Dalrymple, and D. Everett, "Global field results of a polymeric gel system in conformance applications," in *Proceedings of SPE Russian Oil and Gas Technical Conference and Exhibition*, Paper SPE 101822, pp. 280–285, Moscow, Russia, October 2006.

[12] J. Vasquez, E. D. Dalrymple, L. Eoff, B. R. Reddy, and F. Civan, "Development and evaluation of high-temperature conformance polymer systems," in *Proceedings of the SPE International Symposium on Oilfield Chemistry*, Paper SPE 93156, Houston, Tex, USA, February 2005.

[13] G. A. Al-Muntasheri, H. A. Nasr-El-Din, and I. A. Hussein, "A rheological investigation of a high temperature organic gel used for water shut-off treatments," *Journal of Petroleum Science and Engineering*, vol. 59, no. 1-2, pp. 73–83, 2007.

[14] G. A. Al-Muntasheri, "A study of polyacrylamide-based gels crosslinked with polyethyleneimine," in *Proceedings of the SPE International Symposium on Oilfield Chemistry*, Paper SPE 105925, Houston, Tex, USA, February 2007.

[15] S. D. Jordan, D. W. Green, E. R. Terry, and G. P. Willhite, "The effect of temperature on gelation time for polyacrylamide/chromium (III) systems," *Society of Petroleum Engineers Journal*, vol. 22, no. 4, pp. 463–471, 1982.

[16] G. A. Al-Muntasheri, H. A. Nasr-El-Din, and P. L. J. Zitha, "Gelation kinetic and performance evaluation of an organically

crosslinked gel at high temperature and pressure," in *Proceedings of the 1st International Oil Conference and Exhibition*, Paper SPE 104071, September 2006, Cancun, Mexico.

[17] M. Simjoo, M. Vafaie Sefti, A. Dadvand Koohi, R. Hasheminasab, and V. Sajadian, "Polyacrylamide gel polymer as water shut-off system: preparation and investigation of physical and chemical properties in one of the iranian oil reservoirs conditions," *Iranian Journal of Chemistry and Chemical Engineering*, vol. 26, no. 4, pp. 99–108, 2007.

[18] F. Civan, J. Vasquez, D. Dalrymple, L. Eoff, and B. R. Reddy, "Laboratory and theoretical evaluation of gelation time data for water-based polymer systems for water control," *Petroleum Science and Technology*, vol. 25, no. 3, pp. 353–371, 2007.

[19] O. Levenspiel, *Chemical Reaction Engineering*, John Wiley & Sons, New York, NY, USA, 3rd edition, 1999.

Knowledge Discovery for Classification of Three-Phase Vertical Flow Patterns of Heavy Oil from Pressure Drop and Flow Rate Data

Adriane B. S. Serapião[1] and Antonio C. Bannwart[2]

[1] *DEMAC, IGCE, UNESP, CP 178, 13506-900 Rio Claro, SP, Brazil*
[2] *DEP, FEM, UNICAMP, CP 6122, 13081-970 Campinas, SP, Brazil*

Correspondence should be addressed to Adriane B. S. Serapião; adriane@rc.unesp.br

Academic Editor: Guillaume Galliero

This paper focuses on the use of artificial intelligence (AI) techniques to identify flow patterns acquired and recorded from experimental data of vertical upward three-phase pipe flow of heavy oil, air, and water at several different combinations, in which water is injected to work as the continuous phase (water-assisted flow). We investigate the use of data mining algorithms with rule and tree methods for classifying real data generated by a laboratory scale apparatus. The data presented in this paper represent different heavy oil flow conditions in a real production pipe.

1. Introduction

The design of oil production pipelines involves evaluation of flow lines subject to multiphase flow of oil, water, and gas, where oscillations in pressure, temperature, and phase concentration typically occur. Furthermore, the phases usually flow on different geometrical distributions inside the pipe, named flow patterns. The identification of flow patterns is essential for the economic evaluation of the project, such as pressure drop and flow rate along the pipeline. These aspects are critical on offshore production conditions, where extensive distances and high costs are involved. Flow pattern identification is an important step to design separation equipments, slug catchers, gas lift operations, wellhead gathering systems, and production management and control. With the discovery of heavy oil reservoirs the lack of tools and methodologies for flow pattern identification deserves attention because the existing multiphase flow correlations are made for low API oils, where the oil-water mixture may be treated as a single liquid phase with average properties. However, for water-continuous flow of heavy oil below bubble-point, three distinct phases are present, that is, oil, water, and gas, thus making the traditional approach of flow pattern classification and pressure drop prediction in three-phase flow may have poor accuracy.

Basically, there are two types of models for flow pattern prediction: empirical and mechanistic. Empirical models are related to experimental data, where flow pattern maps are experimentally determined and analyzed with respect to mathematical relations representing the boundaries between the flow pattern regions. These relations depend on the amount of experimental data used and on the coordinate system in which the data are presented. Mechanistic models are based on balance equations [1]. Notwithstanding, these models are formulated to describe single or two-phase flows, and they cannot be highly extended for oil-gas-water mixtures, when the phases are considered heterogeneous, as in the "water-assisted flow" case, because of the complexity of phase interaction in the multi-phase flow. The same inadequate extension from two- to three-phase representation occurs for correlation models.

Pressure drop control for heavy oil production and transportation can be achieved by injecting water in the pipeline so as to create a water-continuous flow known as "water-assisted flow." In view of the complexity of three-phase flow, the development of an objective flow pattern identification scheme is of fundamental importance to obtain useful information about the flow nature and control purposes as well.

In a previous paper [2], the three-phase water-assisted flow of heavy crude oil with gas (air) in a vertical pipe was investigated at a low pressure laboratory set up. For each trio of flow rates, flow patterns were identified by means of direct visualization (with the help of movie recording, if necessary) and total pressure drop was measured for comparison with existing correlations.

The main difficulty in visual observation is that the picture is often confusing and difficult to interpret, especially when dealing with high velocity flows. So, in an automatic flow pattern identification tool is broadly interesting and useful for online applications in the oil industry, particularly when the flow visualization is not available.

In the present work, we investigate the automatic classification of the three-phase flow patterns described in [2] from the correspondent flow rates and pressure drop data. This topic is suitable for field applications when direct visualization or tomography (impedance, ultrasonic, optical, or other) of the flow is not possible or impracticable. So, the automatic classification can be used, for instance, as a first step of a decision making system for analyses of real heavy oil production pipelines. It might be regarded as an alternative to the traditional physical modeling of flow pattern transitions in rather complex flows such as three-phase, for which a unified predicting model is, to the best of our knowledge, unavailable.

As an alternative to overcome the prediction difficulties with empirical and mechanistic modeling, neural networks and other data mining techniques can be used at pattern recognition and trend prediction for processes that are nonlinear, poorly understood, and too complex for accurate mathematical modeling, as three-phase flow pattern classification/identification.

Classification is one of the important data mining tasks. Data mining is a progress to extract implicit, nontrivial, previously unknown and potentially useful information (such as knowledge rules, constraints, regularities) from data in database. Data mining is an interdisciplinary field. The obtained results must conform to three major requisites: accuracy, comprehensibility, and interest for the user [3].

The classification corresponding to the vertical upward flow patterns were elaborated by visual inspection by a petroleum engineer and some data mining methods were trained to learn this classification: PSO/ACO2 algorithm, neural networks, support vector machines (SVM), decision tree learning algorithms (J48, REPTree and LMT), rule induction algorithm (JRIP) [4], and rules built by an expert.

Many classification algorithms have been constructed and applied to discover knowledge from data in different applications, yet many suffer from poor performance in prediction accuracy in many practical domains. While it seems unlikely to have an algorithm to perform best in all the domains, it may well be possible to produce classifiers that perform better on a wide variety of real-world domains [5].

In this paper we have also two additional purposes. The first one is to study the contribution of recent natural computing algorithms to solve real-world applications in the petroleum engineering area. For this goal, the PSO/ACO2 algorithm receives special attention.

Rule and tree based methods are particularly recommended in applications where knowledge comprehensibility is very important, such as in engineering processes—where discovered knowledge should be carefully validated and interpreted by experts before they are actually used to diagnose a problem, suggest an intervention procedure or recognize a pattern, rather than blindly trust the result provided by an algorithm. In this context, the second purpose is to find the algorithm that leads to a best comprehensibility of the real problem, because it is very important topic whenever discovered knowledge will be used for supporting a decision made by a human user. When properly designed and trained, a classification tool for applications in multiphase flow systems can potentially improve on-line monitoring and diagnostics, playing the role of a reliable, objective, and quantitative indicator of flow regime.

2. Experiments in Three-Phase Water-Assisted Flow of Heavy Oil with Gas

In the following we summarize the test conditions and the main results obtained in the experiments described by Bannwart et al. [2].

The test section consisted of a 2.84 cm i.d., 2.5 m long vertical glass tubing for the three-phase flow. The oil flow rate was measured with a Coriolis mass flow meter, whereas the water and air flow rates were read in rotameters. Pressure data in the test section were measured with differential and absolute pressure transducers connected to a data acquisition system.

The oil utilized was a blend of crude dead oil with a viscosity of $\mu_o = 5,040$ mPa·s and a density of $\rho_o = 971\ \mathrm{kg/m^3}$ at 25°C. The oil phase was observed to be a w/o emulsion. The water used was tap water contained in the separator tank and the air was provided by an existing group of compressors.

The experiments consisted of simultaneously flowing water, crude oil and air at several flow rate combinations. For each set a video footage of the established flow pattern was taken with a high-speed camera (1000 frames/s) and pressure data were collected. The experimental superficial velocities (J) varied within the following ranges:

(i) oil: $0.02 < J_o < 1.2$ m/s,

(ii) air: $0.04 < J_g < 9$ m/s,

(iii) water: $0.04 < J_w < 0.5$ m/s.

The experiments took place at ambient temperature and near atmospheric pressure. In all runs, water was always injected first (in order to make sure that it would be the continuous phase), followed by oil and air. The glass pipe was never observed to be fouled (hydrophilic behavior). Steady state operation was achieved after the flow rates and average pressure drop readings were observed to be stable.

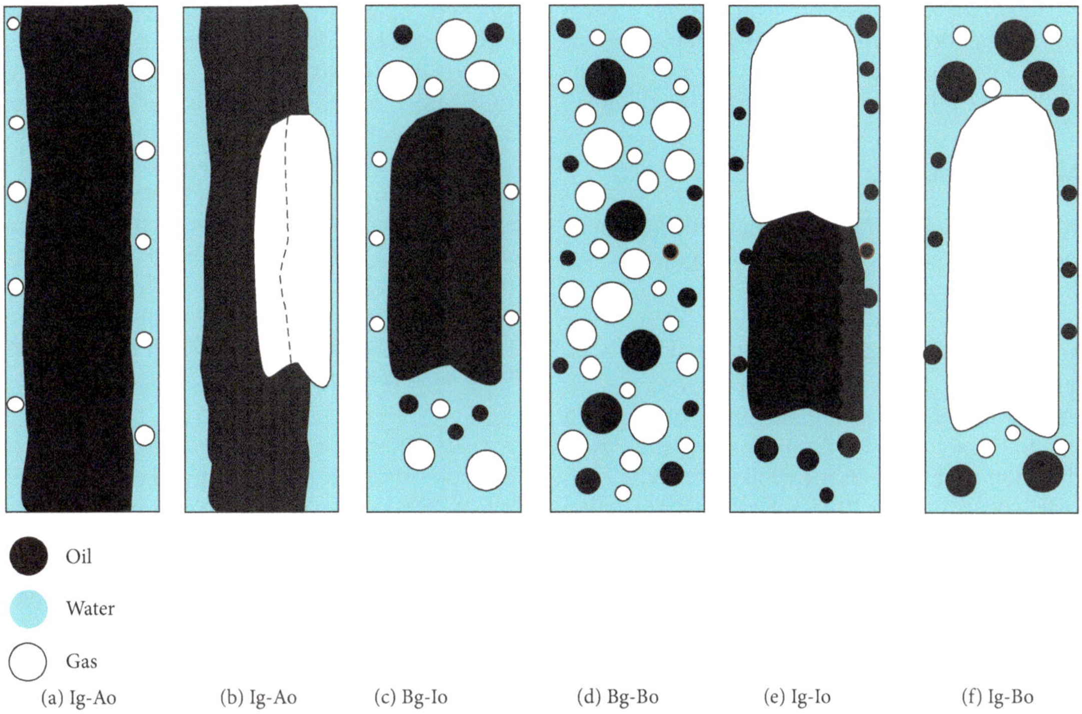

FIGURE 1: Three-phase flow patterns for vertical upward water-assisted flow of heavy oil in the presence of a free gas phase.

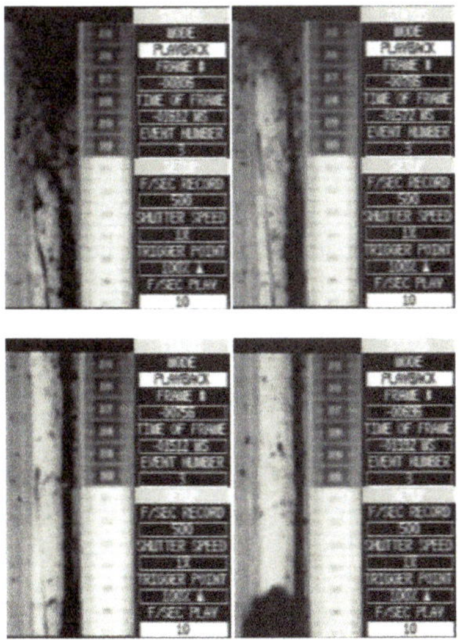

FIGURE 2: Illustration of the passage of large gas bubble in the Ig-Io flow pattern.

The recorded movies were played in slow motion in order to make possible the identification of the flow patterns. Figure 1 illustrates the six identified flow patterns, all of them water-continuous, which were named according to the gas and oil distributions within the water-continuous phase. Figure 2 provides details of one of the flow patterns observed in the test section.

Following a brief description of each flow pattern is presented.

(a) Bg-Ao: Bubbly Gas-Annular Oil. This pattern is similar to heavy oil-water core flow, except that here gas bubbles are seen in the water phase. The oil-water interface is typically sinuous. This pattern occurs for high oil and low gas superficial velocities.

(b) Ig-Ao: Intermittent Gas-Annular Oil. The gas phase forms large bubbles which partly surround a still continuous oil core. This pattern occurs for high oil and moderate gas superficial velocities.

(c) Bg-Io: Bubbly Gas-Intermittent Oil. The gas forms small bubbles and the oil forms large bubbles. This pattern occurs for moderate oil and low gas superficial velocities.

(d) Bg-Bo: Bubbly Gas-Bubbly Oil. This pattern was observed for low oil and gas superficial velocities, but only when the water superficial velocity was higher than about 0.3 m/s, which was enough to disperse the oil into bubbles.

(e) Ig-Io: Intermittent Gas-Intermittent Oil. The gas and the oil both form large bubbles which are very close to each other. Detailed observation shows that the oil bubble is sucked towards the low pressure wake behind the gas bubble. In this pattern, small oil and gas bubbles are spread in the water. An illustration of this flow pattern is shown in Figure 2.

This pattern occurs for high gas and oil superficial velocities, and also for moderate gas and oil superficial velocities.

(f) Ig-Bo: Intermittent Gas-Bubbly Oil. At high gas superficial velocities the gas forms large, high speed bubbles and the oil is dispersed into small bubbles. This pattern is typically pulsating, indicating a transition to annular gas-liquid flow.

3. Data Mining Methods

In this section, we present a brief description of the classification methods used in this work: PSO/ACO2 algorithm, J48, REPTree, LMT, JRIP, neural networks, and SVM.

3.1. PSO/ACO2 Algorithm. The algorithm PSO/ACO1, originally proposed in [6, 7], an algorithm proposed for the discovery of classification rules in data mining problems. It was designed to hierarchical classification, where the classes to be predicted are arranged in a tree-like hierarchy. This algorithm uses concepts of the Particle Swarm Optimization (PSO) algorithm, which is mainly inspired by social behavior patterns of organisms that live and interact within large groups [8], combined with Ant colony optimization (ACO), which takes inspiration from the foraging behavior of some real ant species [9].

The modified version of the PSO/ACO algorithm used in this paper, hereafter denoted PSO/ACO2 [10], uses a sequential covering approach to discover one-classification-rule-at-a-time. The modifications in this algorithm include changes in the pheromone update procedure, in the quality measure evaluation, and in the rule initialization method, as well as the splitting of the rule discovery process into two separated phases.

A rule consists of an antecedent (a set of attribute-values) and a consequent (class). The consequent of the rule is the class that is predicted by that rule. The antecedent consists of a set of terms. A term is defined by a triple ⟨*attribute, operator, value*⟩, where value is a value belonging to the domain of attribute. The operator used is "=" in the case of categorical/nominal attributes, or "≤" and ">" in the case of continuous attributes. In this context, each case is assigned to one predefined class according to the values of its attributes for the case. The discovered knowledge is expressed in the form of *IF-THEN* rules (1), as follows:

$$IF \ \langle attrib = value \rangle \ And \cdots And \ \langle attrib \geq value \rangle \ THEN \ \langle class \rangle . \quad (1)$$

3.2. J48. J48 is an implementation of the well-known Quinlan algorithm (C4.5) [11], which is an improvement of the basic ID3 algorithm. This classifier builds a decision tree for the given dataset, whose nodes represent discrimination rules acting on selective features by recursive partitioning of data, using depth-first strategy.

The algorithm uses the fact that each attribute of the data can be used to make a decision by splitting the data into smaller subsets. To make the decision, the algorithm considers all the possible tests that can split the data set and culls a test that gives the highest information gain. For each discrete attribute, one test with outcomes for each distinct value of the attribute is considered. For each continuous attribute, binary tests involving every distinct values of the attribute are considered. In order to gather the entropy gain of all these binary tests efficiently, the training data set belonging to the node in consideration is sorted for the values of the continuous attribute and the entropy gains of the binary cut based on each distinct values are calculated in one scan of the sorted data. Then a new feature is chosen and the splitting process is repeated for each attribute in a recursive manner until further splitting is not gainful. In the resulting tree structure, each inner node in the tree corresponds to an attribute, each branch represents a possible value or range of values of that attribute and each leaf represents the predicted value of target attribute.

3.3. REPTree. Reduced Error Pruning Tree (REPTree) is a simple procedure for learning and pruning decision trees. It can build a decision or regression tree using information gain as the splitting criterion and prunes trees using reduced-error pruning. It only sorts values for numeric attributes once.

The procedure covers the internal nodes of the tree from the bottom to the top and verifies each internal node. The goal is to state whether replacing it with the most frequent class does not reduce the trees accuracy. If accuracy is not reduced then the node is pruned. Pruning is used to find the best sub-tree of the initially grown tree with the minimum error for the test set. This process continues until any further pruning would decrease the accuracy. The procedure stops with the smallest accurate (lowest classification error) sub-tree with respect to a given pruning set.

3.4. LMT. Logistic Model Tree (LMT) combines a standard tree structure with logistic regression functions at the leaves using posterior class probabilities to produce a single decision tree [12]. LMT consists of a tree structure that is made up of a set of inner nodes and a set of leaves or terminal nodes in an instance space. A test on one of the attributes is associated with every inner node. For numeric attributes, the leaf node has two child nodes which is branched right and left depending on the threshold. If the value of the numeric attribute is smaller than the threshold it is sorted to left branch and value of attribute greater than the threshold it is sorted to right branch otherwise. The threshold is usually fixed by Logit Boost method. For nominal attributes with n values, the branch has n child nodes, so that the instances are sorted down one of the n branches, according to their value of the attribute.

An algorithm for building logistic model trees has the following three steps: growing the tree, building the logistic models, and pruning. Briefly, the tree induction process identifies subdivisions by recursively splitting the instance space in a divide-and-conquer fashion until further subdivisions are not gainful.

3.5. JRIP. JRIP implements *repeated incremental pruning to produce error reduction* (RIPPER) in Java, a prepositional rule learner, as proposed by Cohen [13]. The RIPPER rule learning algorithm is an extended version of learning algorithm IREP (*incremental reduced error pruning*). Rules are created for every class in the training set and are then pruned. In this algorithm, the discovered knowledge is represented in the form of *IF-THEN* prediction rules, which have the advantage of being a high-level and symbolic knowledge representation contributing towards the comprehensibility of the discovered knowledge [14].

The method is based on the construction of a rule set in which all positive examples are covered. Initially, the current set of training examples are partitioned into two subsets, a growing set and a pruning set. The rule is constructed from examples in the growing set. The rule set initiates with an empty rule set and rules are added incrementally to the rule set until no negative examples are covered. After that, JRIP replaces or revises individual rules by using reduced error pruning in order to increase the accuracy of rules. It replaces or revises individual rules by using reduced error pruning. To prune a rule the algorithm takes in account only a final sequence of conditions from the rule and sorts the deletion that maximizes the function.

3.6. Neural Networks. Artificial neural networks (ANNs) are adaptive systems with the power of a universal computer. They are massive parallel processors comprised of single interconnected artificial neurons and are able to realize an arbitrary mapping (association) of one vector space (inputs) to the other vector space (outputs) finding patterns in data, using a computational model for information processing. They are being used with increasing frequency for high dimensional problems, either to approximate a posteriori probabilities for classification or regression [4]. Neural networks are analytical techniques, which are capable of acquiring the knowledge of the complex process of an environment from the observed data, storing the knowledge of underlying process, and making it available to apply to new observations of same data types for pattern recognition purposes. Two commons neural networks' models used for these tasks are Multi-Layer Perceptron (MLP) and Radial Basis Function (RBF).

In the most common form, neural networks are composed by hidden layers of multiple artificial neurons connected to the inputs and outputs neurons with different weights, which correspond to synapses in the biological neuron. Neurons are processing units that that apply nonlinear activation functions to approximate complex functions in the data. The weights are iteratively adjusted during the training procedure using any given approximation technique, such as gradient descendent method, by comparing desired inputs with observed inputs, until a stop criterion is reached. Training or learning procedure is the process of finding the best set of weights for the neural network, mapping the relationships between predictor and target variables.

One key criticism of ANNs is that are a "black box." The nature of the relationship between independent (inputs) and dependent variables (outputs) is usually not revealed and the importance of each variable is not made explicit. Any functional form or map of the relationships is not supplied by ANNs, because of the complexity of the functions used in the neural network approximations.

3.7. Support Vector Machine. Support Vector Machine (SVM) is primarily a classifier method that performs grouping tasks by constructing hyperplanes in a multidimensional space that separates cases of different class labels. In order to construct an optimal hyperplane, SVM employees an iterative training algorithm, that is used to minimize an error function. The foundations of SVM have been developed by Vapnik [15] and are gaining popularity due to many attractive features, and promising empirical performance.

SVM finds a linear separating hyperplane with the maximal margin in this higher dimensional space. The kernel function allows the algorithm to fit the maximum-margin hyperplane in the transformed feature space. There are four basic kernels: linear, polynomial, radial basic function (RBF) and sigmoid. SVMs are based on the *structural risk minimization* (SRM) principle from statistical learning theory.

This classifier method is also considered a "black-box" and does not offer typically any interpretation about the relationship between independent and dependent variables.

4. Results and Discussion

The classifiers tested in this work were modeled using the four inputs, representing the independent variables, which in this case are the superficial velocities of oil, water and gas (respectively, J_o, J_w, J_g) and pressure gradient ($\Delta P/L$). The output is one of the six target class previously classified by the expert (Bg-Ao, Bg-Bo, Bg-Io, Ig-Ao, Ig-Bo, and Ig-Io) representing each flow pattern according to Figure 1.

Experimental records of three-phase flow of heavy oil with gas and water in a vertical pipe data set consisting of 119 samples were used for the training and evaluation of the implemented classifier. For the classifiers' training, the whole data set was randomly separated into two subsets: 75% as training subsets (89 samples) and 25% as testing subsets (30 samples) after training. The training set contains 9, 5, 6, 21, 38 and 10 samples, respectively, for Bg-Ao, Bg-Bo, Bg-Io, Ig-Ao, Ig-Bo, and Ig-Io classes. The correspondent distribution for the test dataset is 2, 1, 3, 5, 13, and 6 samples, respectively.

In the classification task, after the discovered rules from a set of training data, those rules must be applied to a set of test data (unseen during training), and hopefully predict the correct class in the test set. Thus, the accuracy rate for unseen samples is for us the most important index to evaluate the classifier's efficiency, because it proofs its generalization ability.

The performance of the PSO/ACO2 classifier employed to identify the flow patterns was assessed through comparisons between original and estimated outputs taken from the data subsets used both in training and in testing procedure samples for four different types of machine learning algorithms, one knowledge rule model developed by a petroleum engineer expert, two neural networks model and three SVM kernels. Experiments for classification with swarm intelligence techniques were conducted using PSO/ACO2 software developed by Nicholas Holden and provided in the SourceForge (available at http://sourceforge.net/projects/psoaco2/)

project site. Upon experimentation, the suitable numbers of particles and iterations were found to be both 100 and the number of uncovered examples per class (*MaxUncovExampPerClass*) to 2. For PSO algorithm the constriction factor is $\chi = 0.72984$, and cognitive and social learning coefficients are $c_1 = c_2 = 2.05$.

For the Machine Learning experiments with rule and tree based algorithms (JRip, J48, RepTree and LMT), we used well-known tree and rule based classification algorithms available in the Waikato Environment for Knowledge Analysis (WEKA) software package [14].

The neural networks (NN) models explored for this experiment are: Multi-Layer Perceptron (MLP) and Radial Basis Function (RBF). On MLP neural network, four neurons were used in input layer, representing the independent variables like PSO/ACO2, ten neurons were employed in hidden layer and six neurons were used in output layer, representing each flow pattern. The amount of neurons in the hidden layer was found by experimentation, after assessing which configuration would yield the least global training error. The activation function used at each level was sigmoid tangent for hidden and output layer neurons. The supervised training algorithm used was the Backpropagation with weights and biases updating according to Levenberg-Marquardt optimization algorithm. For the RBF network, 35 hidden layer neurons were employed using Gaussian function with coefficient of spread (σ) equal to 0.1, with a linear activation function had been used on the six output neurons. The software Matlab MathWorks Inc. v.7.4 was used to implement the neural networks models employed in this work.

We have also used multi-class SVM Type 1 for the classification task, considering the "one-against-one" method for decomposing multi-class problem in binary subproblems. Linear, polynomial, and RBF functions were tested as the mapping function (kernel) for the classification system. Matlab was used to implement the SVM model mentioned in this work. The SVM classifier was modeled using the four inputs, representing the independent variables and the output is one of the six target class, as defined before. The complete description of this procedure can be consulted in [15].

In this work, the data were also interpreted according to the set of preestablished rules by an expert. The human knowledge was represented by a set of *IF-THEN* rules. In order to build these rules, human knowledge regarding characteristics of each flow pattern and the relation of each pattern with the four parameters measured with the experiment was determined by inspection of the dataset. For sake of comparing with the other data mining methods, the data were separated in the same training and test sets for evaluation with the expert rules.

All the experiments were obtained on a Centrino Duo PC (CPU 1.83 GHZ, RAM 2 GB) for the same datasets.

Table 1 indicates the classification accuracy of PSO/ACO2 algorithm for training and test sets, in the classification task of vertical flow patterns, using the settings parameters mentioned earlier and the discovered rules, compared with others data mining methods. The PSO/ACO2 algorithm was able to achieve a success rate of 68.56% in the training set, the worst performance among the other methods, and 70% in the test set, one of the best performances among the other classifiers. The hybrid particle swarm optimization/ant colony optimization algorithm produced in the training phase 13 classification rules used to identify the vertical flow patterns as indicated in Algorithm 1. All patterns were covered with this rule set.

In contrast with JRIP, which is another classifier based on discovering rules, PSO/ACO2 presented a best performance for unseen data (test set). JRIP produced only six simple rules for classifying the flow patterns, as detailed in Algorithm 1. The methods based on decision trees, J48, RepTree, and LMT demonstrated better results to learn the patterns during the training, notwithstanding the scores for testing do not outperformed PSO/ACO2, except RepTree, which had the same efficiency (70%).

Figure 3 displays the knowledge tree provided by RepTree, containing 17 decision leaves. The variable J_w is referred only to distingue the Ig-Ao and Bg-Ao classes. In the tree we can observe that no rules for Bg-Bo and Ig-Io were created. Thus, the recognition of these patterns is completely wrong using this solution.

Expert rules have found 80% of success in identifying the test samples, with a lower rate for the training set (73.03%), which was the worst among all the other techniques, apart from PSO/ACO2 algorithm. The expert has used nine rules to represent his knowledge about de flow patterns. According to those rules, the variable J_w is used only to aid detecting the Bg-Bo class. A comprehensive observation of the rules appoint that a pressure drop (ΔP) superior to 9300 Pa/m causes almost always a bubble regime in the gas phase, where the oil regime is determined by the superficial velocities. Otherhand, a pressure drop lower than 8700 Pa/m originates certainly an intermittent regime in the gas phase, so that the oil regime is determined by the others variables. In the intermediate range of drop pressure, between 8700 and 9300 Pa/m, depending on the values of the four measured variables, the possible flow patterns are only Bg-Bo, Ig-Io and Ig-Ao.

Despite their ineptitude to generate descriptive information about the classification process, which could useful to better understand the physical phenomenon under study,

TABLE 1: Comparing accuracy of flow classifiers.

Classifier	Training set	Test set	Full set
PSO/ACO2	68.56%	70.0%	68.92%
JRIP	78.65%	56.67%	73.11%
J48	86.52%	56.67%	78.99%
RepTree	77.53%	70.0%	75.63%
LMT	79.78%	63.33%	75.63%
MLP-BP NN	89.89%	60.0%	82.35%
RBF NN	91.01%	66.67%	84.87%
Linear SVM	85.39%	70.0%	81.51%
Polynomial SVM	100%	73.33%	93.28%
RBF SVM	100%	70.0%	92.44%
Expert rules	73.03%	80.0%	74.79%

PSO/ACO2 rules:
R1: If ($J_o \le 0.639$) and ($J_g \ge 2.858$) and ($\Delta P \ge 4033.14$) then *class* = Ig-Bo
R2: If ($J_o \ge 0.148$) and ($J_w \le 0.292$) and ($J_g \le 5.351$) and ($6461.45 \le \Delta P \le 9831.71$) then *class* = Ig-Ao
R3: If ($J_o \ge 0.673$) and ($J_w \ge 0.108$) and ($J_g \le 4.108$) and ($\Delta P \le 10003$) then *class* = Ig-Io
R4: If ($J_o \ge 0.251$) and ($J_g \le 0.293$) and ($\Delta P \ge 9714.49$) then *class* = Bg-Ao
R5: If ($J_o \le 0.069$) and ($J_w \le 0.071$) and ($J_g \le 4.645$) and ($\Delta P \le 8329.8$) then *class* = Ig-Ao
R6: If ($J_o \le 0.075$) and ($J_w \le 0.32$) and ($\Delta P \ge 9364.72$) then *class* = Bg-Io
R7: If ($J_o \le 0.241$) and ($J_w \ge 0.323$) and ($8719.27 \le \Delta P \le 9964.14$) then *class* = Bg-Bo
R8: If ($0.12 \le J_w \le 0.675$) and ($6329.42 \le \Delta P \le 9332.34$) then *class* = Ig-Ao
R9: If ($J_o \le 0.144$) and ($J_w \ge 0.094$) and ($\Delta P \le 8801.66$) then *class* = Ig-Bo
R10: If ($0.045 \le J_o \le 0.1489$) and ($J_w \le 0.084$) and ($\Delta P \ge 9224.91$) then *class* = Bg-Ao
R11: If ($J_o \le 0.04$) and ($0.1276 \le J_w \le 0.322$) and ($J_g \le 0.626$) and ($\Delta P \le 8619.74$) then *class* = Ig-Io
R12: If ($J_o \ge 0.055$) and ($J_w \le 0.253$) and ($\Delta P \ge 6282.8$) then *class* = Ig-Ao
R13: Otherwise, *class* = Ig-Bo

JRIP rules:
R1: If ($\Delta P \ge 8786.5$) and ($J_o \le 0.069$) and ($J_w \ge 0.327$) then *class* = Bg-Bo
R2: If ($J_g \le 0.06$) and ($J_o \le 0.069$) then *class* = Bg-Io
R3: If ($J_g \le 0.051$) and ($\Delta P \ge 9758$) then *class* = Bg-Ao
R4: If ($J_o \ge 0.999$) and ($J_w \ge 0.316$) then *class* = Ig-Io
R5: If ($J_g \le 3.592$) then *class* = Ig-Ao
R6: Otherwise, *class* = Ig-Bo

Expert rules:
R1: If ($J_o > 0.2$) and ($3.4 < J_g < 8.6$) and ($\Delta P < 7900$) then *class* = Ig-Bo
R2: If ($J_o < 0.1$) and ($0.3 < J_g < 0.6$) and ($6500 < \Delta P < 8500$) then *class* = Ig-Bo
R3: If ($J_o < 0.1$) and ($J_g > 3.5$) and ($\Delta P < 6500$) then *class* = Ig-Bo
R4: If ($0.33 < J_g < 0.38$) and ($6300 < \Delta P < 9800$) then *class* = Ig-Io
R5: If ($J_o > 0.5$) and ($3.1 < J_g < 4.1$) and ($5700 < \Delta P < 9000$) then *class* = Ig-Io
R6: If ($J_g < 0.055$) and ($\Delta P > 9300$) then *class* = Bg-Ao
R7: If ($J_o < 0.25$) and ($J_g < 0.07$) and ($9300 < \Delta P < 10000$) then *class* = Bg-Io
R8: If ($J_o < 0.3$) and ($J_w > 0.3$) and ($J_g < 0.4$) and ($8700 < \Delta P < 9860$) then *class* = Bg-Bo
R9: Otherwise, *class* = Ig-Ao

ALGORITHM 1: Generated rules from PSO/ACO2 and JRIP algorithms.

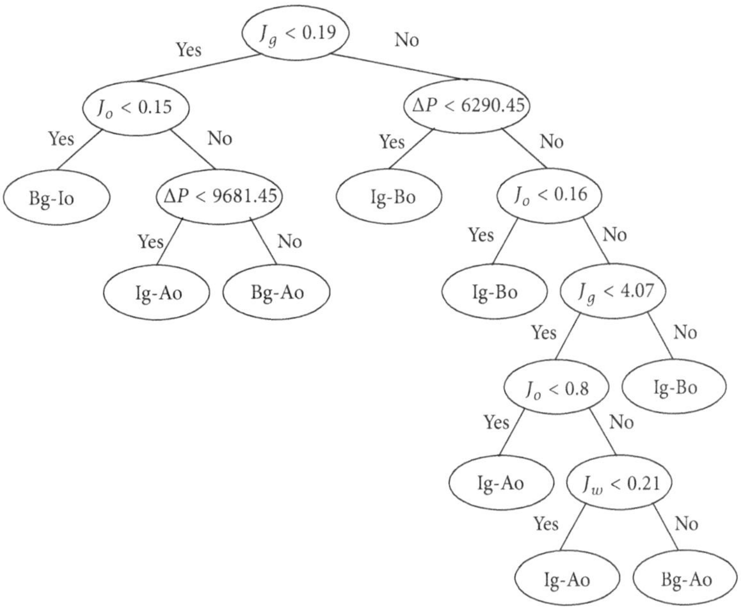

FIGURE 3: Knowledge tree generated by RepTree method.

neural networks and SVMs have presented superior prediction accuracy for both training and test sets than the meta-heuristic methods and expert rules, as observed in Table 1. Polynomial SVM presented the best recognition score among all methods, classifying correctly all samples of the training set and obtaining 73.33% of recognition score for the test set.

With regard to the moderate accuracy rate for the vertical upward flow pattern classification of the investigated methods, it is supposed that this survey is due to the intermittent phase, which makes complex a correct prediction as reported in [16]. In all runs with every classifier, the patterns composed with a part of the intermittent regime had an expressive misrecognition, principally the Ig-Io class.

5. Conclusion

Flow pattern prediction in industrial systems that rely on complex multi-phase flows is essential for their safety, control, diagnostics, and operation. In oil extraction and oil-gas-water mixture transporting processes, the identification of flow regime is one of the essential tasks for the purpose of maintaining optimal operating and improving the performance of equipment. Usually, different flow regimes produce distinct performances of the system. However, identifying/classifying multiphase mixture continuously and precisely is still a significant and unsolved problem, due to the highly nonlinear nature of the forces which rule the flow regime transitions.

In this paper, we use the PSO/ACO2 algorithm for data mining, a hybrid method for mining classification rule. We have compared the performance of PSO/ACO2 with others data mining methods. Experimental results show that PSO/ACO2 has a predictive accuracy for untrained data (test set) greater than the others evaluated heuristic methods. Nevertheless, the *black-box* methods (neural networks and SVM) have shown a higher precision for predicting patterns than PSO/ACO2 and the others heuristic algorithms. Because the application of swarm intelligent algorithm in data mining, especially in classification rule mining, is still in infant periods, we believe that improvements in this methodology could increase the accuracy for the pattern detection, so that this kind of approach could be suitable for real world applications in Petroleum Engineering area.

Rule base methods are particularly recommended in applications where rule comprehensibility is very important, such as in engineering processes—where discovered rules should be carefully interpreted by experts before they are actually used to recognize a pattern.

Although the generation of classification rules by an expert is a feasible process, producing good results, it is an exhaustive and tedious task, mainly when a great number of parameters and samples must be analyzed. The automated classification of the data is vital when a huge volume of information is involved. The approach and methods used in this work contribute to an in-depth understanding of flow patterns dynamics, especially in the cases of multiphase flows, aiming extraction of simple rules for identifying flow pattern features. We intend to encourage researchers of the Petroleum Engineering area to use data mining to analyze real data in different domains as well.

A future research direction consists of exploring others techniques based on rule mining. Further work includes also dealing with a deeper study of the important features of three-phase flow of heavy oil in vertical pipe in order to obtain better classifier systems.

References

[1] A. Wegmann, *Multiphase flows in small scale pipes [Doctoral Dissertation]*, Federal Institute of Technology Zurich, 2005, ETH Nr. 16189.

[2] A. C. Bannwart, F. F. Vieira, C. H. M. Carvalho, and A. P. Oliveira, "Water-assisted flow of heavy oil and gas in a vertical pipe," in *Proceedings of the SPE International Thermal Operations and Heavy Oil Symposium (ITOHOS '05)*, Alberta, Canada, November 2005, Paper PS2005-SPE-97875-PP.

[3] R. S. Parpinelli, H. S. Lopes, and A. A. Freitas, "Data mining with an ant colony optimization algorithm," *IEEE Transactions on Evolutionary Computation*, vol. 6, no. 4, pp. 321–332, 2002.

[4] S. Haykin, *Neural Networks: A Comprehensive Foundation*, Prentice Hall, 2nd edition, 1999.

[5] T. Sousa, A. Silva, and A. Neves, "Particle swarm based data mining algorithms for classification tasks," *Parallel Computing*, vol. 30, no. 5-6, pp. 767–783, 2004.

[6] N. Holden and A. A. Freitas, "A hybrid particle swarm/ant colony algorithm for the classification of hierarchical biological data," in *Proceedings of the 2005 IEEE Swarm Intelligence Symposium (SIS '05)*, pp. 100–107, Pasadena, Calif, USA, June 2005.

[7] N. Holden and A. A. Freitas, "Hierarchical classification of G-protein-coupled receptors with a PSO/ACO algorithm," in *Proceedings of the 2006 IEEE Swarm Intelligence Symposium (SIS '06)*, pp. 77–84, Indianapolis, Ind, USA, 2006.

[8] J. Kennedy, R. C. Eberhart, and Y. Shi, *Swarm Intelligence*, Morgan Kaufmann, San Francisco, Calif, USA, 2001.

[9] M. Dorigo and T. Stützle, *Ant Colony Optimization*, MIT Press, Cambridge, Mass, USA, 2004.

[10] N. P. Holden and A. A. Freitas, "A hybrid PSO/ACO algorithm for classification," in *Proceedings of the 9th Annual Genetic and Evolutionary Computation Conference (GECCO '07)*, pp. 2745–2750, London, UK, July 2007.

[11] R. R. Quinlan, *C4.5: Programs for Machine Learning*, Morgan Kaufmann, San Francisco, Calif, USA, 1993.

[12] N. Landwehr, M. Hall, and E. Frank, "Logistic model trees," *Machine Learning*, vol. 59, no. 1-2, pp. 161–205, 2005.

[13] W. Cohen, "Fast effective rule induction," in *Proceedings of the 12th International Conference on Machine Learning*, pp. 115–123, Lake Tahoe, Calif, USA, 1995.

[14] I. H. Witten, M. Frank, and M. A. Hall, *Data Mining: Practical Machine Learning Tool and Technique with Java Implementation*, Morgan Kaufmann, San Francisco, Calif, USA, 3rd edition, 2011.

[15] V. Vapnik, *Statistical Learning Theory*, John Wiley & Sons, New York, NY, USA, 1998.

[16] F. Pacheco, A. C. Bannwart, J. R. P. Mendes, and A. B. S. Serapião, "Support vector ma-chines for identification of three-phase flow patterns of heavy oil in vertical pipes," *Brazilian Journal of Petroleum and Gas*, vol. 1, no. 2, pp. 95–103, 2007.

Improving the Demulsification Process of Heavy Crude Oil Emulsion through Blending with Diluent

K. K. Salam, A. O. Alade, A. O. Arinkoola, and A. Opawale

Petroleum Engineering Unit, Department of Chemical Engineering, Ladoke Akintola University of Technology (LAUTECH), PMB 4000, Ogbomoso, Nigeria

Correspondence should be addressed to K. K. Salam; kaykaysalam@yahoo.co.uk

Academic Editor: Andrea Franzetti

In crude oil production from brown fields or heavy oil, there is production of water in oil emulsions which can either be controlled or avoided. This emulsion resulted in an increase in viscosity which can seriously affect the production of oil from sand phase up to flow line. Failure to separate the oil and water mixture efficiently and effectively could result in problems such as overloading of surface separation equipments, increased cost of pumping wet crude, and corrosion problems. Light hydrocarbon diluent was added in varied proportions to three emulsion samples collected from three different oil fields in Niger delta, Nigeria, to enhance the demulsification of crude oil emulsion. The viscosity, total petroleum hydrocarbon, and quality of water were evaluated. The viscosity of the three emulsions considered reduced by 38, 31, and 18%. It is deduced that the increase in diluent blended with emulsion leads to a corresponding decrease in the value of viscosity. This in turn enhanced the rate of demulsification of the samples. The basic sediment and water (BS&W) of the top dry oil reduces the trace value the three samples evaluated, and with optimum value of diluent, TPH values show that the water droplets are safe for disposal and for other field uses.

1. Introduction

Emulsion is defined as a system in which one liquid is relatively distributed or dispersed, in the form of droplets, in another substantially immiscible liquid. Emulsions have long been of great practical interest due to their widespread occurrence in everyday life which occurs due to reliance of the behaviour of the emulsion on the magnitude and range of the surface interaction. They may be found in important areas such as food, cosmetics, pulp and paper, biological fluids, pharmaceutical, agricultural industry, and petroleum engineering. In production and flow assurance, the two commonly encountered emulsion types are water droplet dispersed in the oil phase and termed as water-in-oil emulsion (W/O) and if the oil is the dispersed phase, it is termed oil-in-water (O/W) emulsion [1].

Water-in-oil crude oil emulsions may be encountered at all stages in the petroleum production and in processing industry. With presence of water, they are typically undesirable and can result in high pumping costs and pipeline corrosions and increase the cost of transportation [2]. Reduced throughput is needed to introduce special handling equipment, contribute to plugging of gravel pack at the sand phase [3], and affect oil spill cleanup [4].

In their research work, Micheal et al. used bottle test method to simulate field condition of four emulsion samples (two Canadian and two Venezuelan emulsions) in order to determine the variables that affect emulsion stability. They were able to evaluate response to the different emulsion based on bottle test data by introducing thirty-six different demulsifiers to enable them to probe emulsion stability. Linear regression and partition tree analysis were used to analyze the effect of various variables on emulsion stability and were able to conclude that solid content significantly affects emulsion stability. Beside solid content crude oil properties, water chemistry and process condition also influence emulsion stability [5].

Christophe et al. evaluated and compared emulsion formed by different parts of the indigenous amphiphiles (the light, the intermediate, or the heavy ones) to determine their contribution to emulsion stability. The emulsions formed with the light and intermediate fractions separated immediately when the agitation stopped. The most stable emulsions were formed with the fraction of crude that distilled at temperature greater than 520°C, suggesting that the amphiphiles with the highest molecular weight, that is, resins and asphaltenes, play a major role in the protection of water droplet against coalescence [6]. This is consistent with many recent findings that the presence of these components enhanced w/o emulsion stability (Rondón et al. and Ekott and Akpabio [7, 8]). Others factors that affect emulsion stability are fine solids, temperature, size of water droplet, and brine composition [9], which is consistent with the work of previous authors [5, 6].

Despite the success of enhanced oil recovery (EOR) process, one of the problems associated with the process is emulsion problem. Efeovbokhan et al. observed that physical factors that enhance oil recovery can also greatly contribute to the formation of very stable emulsions because EOR-induced emulsions are established by surfactant/polymer (SP) and alkaline/surfactant/polymer (ASP) processes which makes breaking of emulsion different from naturally occurring emulsions which are stabilized by asphaltenes and resins [10]. Traditional demulsifiers are often not effective on emulsions created by chemical floods; therefore, the performance of demulsifier in surfactant/polymer–flooding-induced emulsion depends on the selection of the best demulsifier with respect to the system under consideration [11]. In breaking of surfactant/polymer-flooding-induced emulsion with the use of surfactant, Oseghale et al. worked on separation of oil-water emulsions expected during chemical enhanced recovery operations using crude oil from a field in Niger delta during surfactant/polymer flooding operation. Surfactant *N*-octyltrimethyammonium bromide (C_8TAB) was used as the demulsifier and a dosage between 200 and 300 ppm was the optimum dose that yielded oil and water phases with oil content reduction from 550 to 70 ppm after 4 h. Microscopy test confirmed that addition of *N*-octyltrimethyammonium bromide (C_8TAB) produced significant coalescence shortly after it was added to the emulsion, which is in agreement with an increase of the oil droplet size in the presence of the demulsifier. Their findings show that this investigation worked with the principles of using cationic surfactants as demulsifier [12].

With various problems encountered with the presence of emulsion in our system, there is need to find ways of controlling existence of emulsion or preventing it from forming in our system. One of the ways of controlling problems encountered by crude oil emulsion is the ability to predict crude oil behaviour both at the sand phase and during production by building a robust predictive model [13]. Emulsion formation or break up either for oil in water or water in oil emulsion can be characterized based on the property and type of crude oil involved in the formation or break up of emulsion which can assist in formulating method of preventing formation of such emulsion [14]. Nuraini et al. selected four groups of demulsifiers which are amine, natural, polyhydric, and alcohol demulsifier groups serving as breaking agents of stable emulsion. Their findings show that amine demulsifier group exhibited the highest efficiency to break the emulsion compared to polyhydric, alcohol, and natural groups and that demulsifier efficiency depends on two-factor solubility of demulsifier either in water or oil and molecular weight of demulsifier [15].

It has been established from the literatures that one of the ways of breaking stable emulsion is introduction of low dose of demulsifiers. For comprehensive methods of breaking emulsion, the work of Hanapi et al. treated that aspect [2]. Micheal et al. used chemical demulsifiers and statistical analysis to classify emulsion. They obtained emulsion from the field and treated the emulsion with thirty-eight chemicals that serve as demulsifiers at nine different sites. The tests were tailored towards determination of water droplet, oil dryness, and oil-water interface which were analyzed using several statistical tools. A correlation was developed for water droplet, oil dryness and oil-water interface. The results show that water droplet significantly affect oil-water interface than oil dryness [16].

Crude oil emulsions are complex and should be characterized as completely as possible. Droplet-size distribution, interfacial phenomena, and the nature of organic and inorganic components are important. The viscosity of the emulsion is affected by both the water content and droplet size distribution [17, 18]. The increase in aqueous phase of the emulsion leads to an increase in viscosity of emulsion which in turn aggravates flow of emulsion in conduct either at the sand phase or through the surface facilities [3, 19]. Stable water-in-oil emulsions have been generally found to exhibit high interfacial viscosity and/or elasticity modulus. Viscosity of crude oil emulsion was found to increase with increase in water and decreased with increase in speed of rotation of spindle when demulsifier is added [20]. The increase of the interfacial rheological parameters has been attributed to non-Newtonian nature of emulsion [20] and physical cross-links between the asphaltene particles adsorbed at the water-oil interface [21]. Demulsification of emulsion proved to be a good method of breaking emulsions but with an influence of viscosity still unaccounted for in most of the researches; this research will study the effect of adding a diluent to emulsion samples treated with diluent for three different water in oil crude emulsions collected from three different oil fields from three operators in Niger delta, Nigeria.

2. Materials and Methods

2.1. Sampling. Fresh crude oil emulsions were collected from the three oil fields flow stations operated by three different operators in the Niger delta in Nigeria, namely, Fields A, B, and C.

At the sampling points in all the three oil field mentioned above, crude oil was collected at both east and west directional sampling pipes. This to ensure that pure emulsion interface is collected and not either gas or water phase. The emulsions are collected in a tightly sealed container. The experiment was carried out after four hours from the time of sampling to avoid ageing of the crude oil. Table 1 show the initial properties of the three water-in-oil emulsion samples used for the experimental work.

Gasoline used as the diluent for this experiment was gotten from the Nigeria National Petroleum Cooperation

Table 1: Properties of crude oil emulsions used for the analysis.

	Field A	Field B	Field C
Temperature (°C)	55	60	50
Production rate (m^3/day)	11,000	32,000	41,000
Viscosity ($mPas^{-1}$)	80	100	215
Residence time (hrs)	11	7	15
API gravity (°)	23	22	21
Water cut (%)	51	8	10
Demulsifier volume used (ppm)	3	18	6

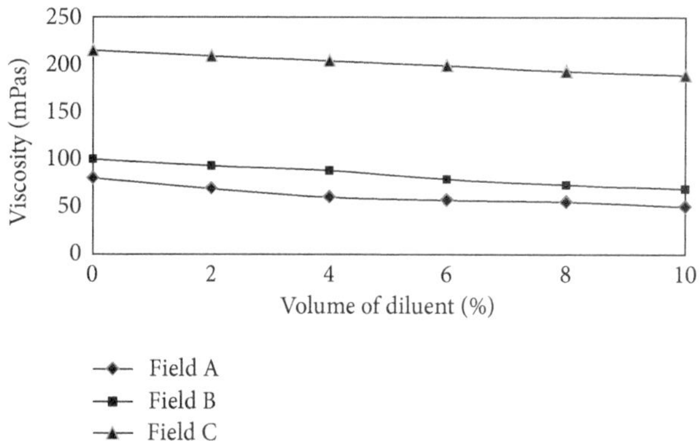

Figure 1: Volume of diluent against viscosity of emulsion.

(NNPC) Refinery, Port Harcourt, Nigeria. Two types of demulsifiers were used by the operators for breaking of the emulsions formed. Fields A and C used PhaseTreat 4633, while Field B used PhaseTreat 6074.

The equipment used for the analysis are water bath, Checktempl digital thermometer, Cannon Fenske viscometer, Model HT 5001-201, six-ounce Pyrex bottles with volume 100 mL, Beaker (100 mL), Socorex Syringe micropipette, Model: Dossy TM 174 premium, Centrifuge Machine: Robinson Centrifuge, Model T.0.2, serial no. T724, Wooden product bottle shaker, and 10 mL measuring cylinder.

2.2. Experimental Procedure. Each of the crude oil samples was analyzed in different setups; the three crude oil samples were treated according to the properties of the oilfield where they were collected. These properties vary in terms of temperature, rate of chemical injection, nature of process terminal, and time of processing, which will dictate the type of demulsifier chemical to be used. A water bath was set up and maintained at a temperature of 60°C equivalent to the average process temperature of the oil fields. This temperature was held constant to neglect the effect of temperature on the viscosity of the crude oil samples.

Six test bottles of capacity 100 mL were labeled according to their corresponding wells with A, B, and C, with suffixes 1 to 6 on each of the wells. The suffix 1 denotes 0 mL of diluent, 2 is 2 mL, 3 is 4 mL, 4 is 6 mL, 5 is 8 mL, and 6 is 10 mL of diluent. The bottles were filled with crude oil and gasoline to make up a volume of 100 mL. Prior to addition of diluent to the emulsion, demulsifier was added in a ratio of one-third of the amount used by the operators where the samples were collected. The samples (emulsion + gasoline) were placed in a bottle shaker and agitated thoroughly with 50 vertical shakes and 50 horizontal shakes to homogenize the diluent with the continuous phase of the emulsion. The bottle was returned to water bath after blending for ten minutes after which percentage-free water was recorded.

The viscosities reading of various combinations of the blend of demulsifier, emulsion, and diluent were obtained using the Cannon Fenske viscometer according to the procedure recommended by ASTM D445 (Norman) [22].

Basic sediment and water of the emulsion was determined using the method described in the published work of Sunil et al. [9]. Total petroleum hydrocarbon was measured by using TPH analyzer (Model HC-404). A sample of the effluent water was taken and fed into this analyzer and the reading recorded in parts per million, ppm.

3. Results and Discussion

3.1. Flow Assurance. Effect of diluent on viscosity is illustrated in Figure 1. Gasoline was added to the three crude oil emulsions from 2 mL to 10 mL and emulsion volume is reduced from 100 to 90 mL. There is viscosity reduction when diluent is introduced. The reduction in viscosity is proportional to the increase in the volume of gasoline. Effect of viscosity reduction is a function of initial viscosity of the emulsion because from the graph the reductions of viscosity in fields A, B, and C are 38, 31, and 17%, respectively when 10 mL of gasoline was added to them and their initial viscosity are 80, 100, and 215 mPas which means that the sample with lowest value of viscosity experienced the highest percentage reduction in viscosity value and the sample with the highest viscosity experienced the lowest percentage reduction in viscosity value.

3.2. Rate of Separation of Water. It was observed from Figures 2 and 3 that introduction of diluent affect Basic sediment and water (BS&W) of crude oil emulsion. The BS&W of Field A crude oil emulsion sample was originally 0.5% when treated with an injection rate of 1 ppm, and without blending with diluent. The value reduces as the volume of diluent increases until 8 mL when the value is zero. There is reduction in the value of BS&W of samples from fields B and C which was initially at 0.7% after the addition of 6 mL and 2 mL of demulsifiers to them to zero and 0.2% when the volume of the diluent was increased to 10 mL. Also viscosity of emulsion plays an important role in the analysis of BS&W because Field A reduces its value of BS&W to zero when the volume of diluent is 8 mL, B when the volume is 10 mL, and for C the value is at 0.2% when the volume of the diluent is at 10 mL. Field A used the lowest amount of diluent because it has the lowest viscosity while C has the the highest amount of diluent since it has the highest viscosity. Therefore, it is established that the diluent is capable of increasing reduction of BS&W in crude oil emulsions.

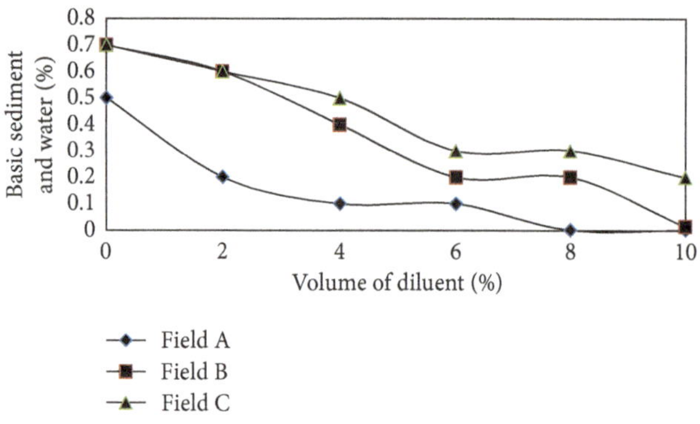

FIGURE 2: Volume of diluent against basic sediment and water.

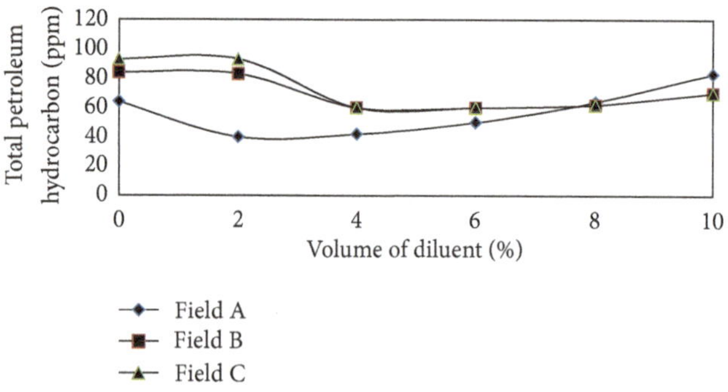

FIGURE 3: Volume of diluent against basic sediment and water.

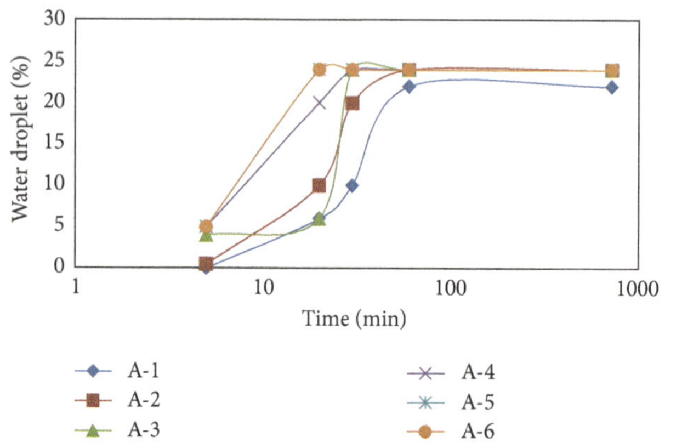

FIGURE 4: Water droplet against time at various emulsion/diluent ratio for Field A.

3.3. Total Petroleum Hydrocarbon. The TPH of each sample initially reduces with the increase in the amount of gasoline added, but later started increasing after a particular blending point. This undesirable effect is believed to be caused by excess gasoline in the mixture finding their way into the aqueous phase.

Field A crude sample maintained a good TPH of 64 ppm which reduces as the diluent value increased between 0 and 2 mL, but there is a sharp increase in the value of TPH as the volume of diluent is increased above 2.2 mL. These reduction and rising of the TPH value with the increase in the diluent volume are attributed to the relative tightness of the crude oil emulsion. Tightness is the degree at which the water droplets are held in suspension and resist separation.

Fields B and C crude emulsions demonstrated high TPH values of 84 ppm and 93 ppm. Initially, blending showed less effect of diluent on the TPH of these two crudes between 0 and 2 mL. The TPH values of B and C reduced to 60 ppm at 4 mL of diluent, which was constant till 6 mL of diluent. Above 6 mL of diluent, there is an increase in the value of TPH fields of B and C to 70 when the volume of diluent is 10 mL. Apart from tightness of the emulsion, excess diluent can penetrate aqueous phase of the emulsion which will increase the value of its TPH.

3.4. Bottle Test. The demulsification bottle test was carried out and results on water droplet are taken after 5, 20, 30, 60, and 720 minutes. Water droplet is the separation of water from the surface of emulsion formed. The effect of addition of diluent on each crude oil sample was monitored on the rate of water droplet from each of the emulsion samples. The suffixes after the fields denotations A, B, and C indicated the variation of diluent concentration added to the emulsion samples which read 1, 2, 3, 4, 5, and 6 for 0, 2, 4, 6, 8, and 10 mL of diluent concentration.

From Figure 4, depending on the amount of diluent blended with the emulsion there was a corresponding increase in the rate of water droplet with time. When 2 mL of diluent was blended with emulsion, there was no water droplet after 5 minutes; it increased to 6% at 20 minutes and 22% at the end of 60 minutes after which there was no further droplet till the end of 720 minutes which was illustrated in A-1 in Figure 4. In A-2, the trend was similar to that of A-1 but the final water droplet value is 24%. In A-3 and A-4, there was a water droplet of 4 and 5% at 5 minutes which increased to 6 and 20% after 20 minutes, 24% at the end of 30 minutes after which there was no further droplet till the end of 720 minutes.

In A-5 and A-6, water droplet was 5% at 5 minutes, 24% at the end of 20 minutes, and remained constant till the end of 720 minutes. Generally, it was observed that low amount of diluent take longer time for water to drop from the emulsion but as the volume of diluent increased, the time required for water to drop out of the emulsion decreased.

Figure 5 shows the behavior of change in diluent concentration with rate of water droplet for emulsion samples collected from Field B. When no diluent was blended with the emulsion samples obtained from Field B, there was no droplet of water until after 60 minutes with a value of 4% and progressively increased to 6% at the end of 720 minutes. In B-2 to B-4, the trend of the curve followed the trend experienced in B-1 only that the rate of water droplet was faster and higher than that of B-1 with a value of 8, 10, and 10%, respectively, for B-2, B-3, and B-4 at the end of 720 minutes. In B-5 and B-6, water droplet was experienced earlier than the previous four situations with droplet of 3 and 4% at 5 minutes; it increased to 5 and 10% at 20 minutes and was 12% from 30 to 720 minutes when the analysis was terminated.

When no diluent was blended with the emulsion obtained from Field C and when 2 mL is blended with it, the behavior of their chart was similar and illustrated in C-1 and C-2 in

TABLE 2: Effect of diluent on water quality and interface for emulsions after 720 minutes.

Volume of emulsion	Volume of diluent	Field A		Field B		Field C	
		Water quality	Interface	Water quality	Interface	Water quality	Interface
100	0	Dirty	Cloudy	Dirty	Cloudy	Dirty	Cloudy
98	2	Fair	Stained	Dirty	Stained	Dirty	Cloudy
96	4	Fair	Sharp	Fair	Stained	Dirty	Stained
94	6	Clean	Sharp	Fair	Sharp	Fair	Sharp
92	8	Clean	Sharp	Clean	Sharp	Clean	Sharp
90	10	Clean	Sharp	Clean	Sharp	Clean	Sharp

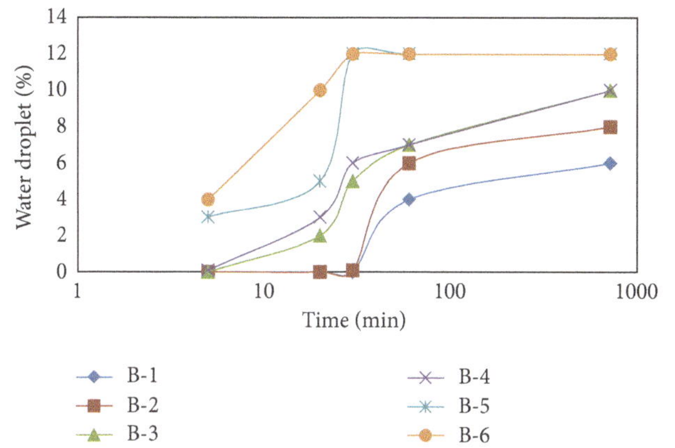

FIGURE 5: Water droplet against time at various emulsion/diluent ratio for Field B.

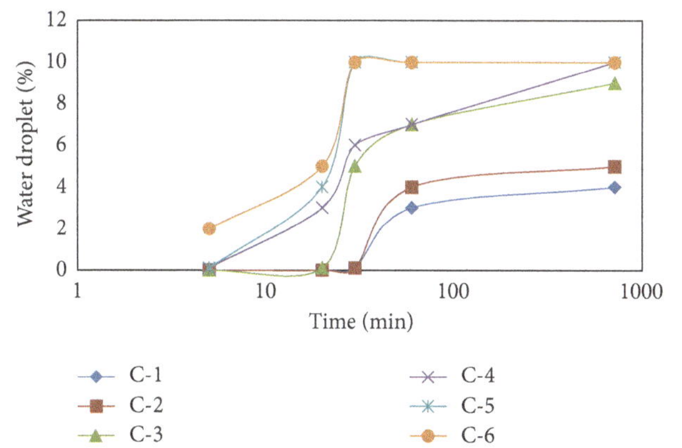

FIGURE 6: Water droplet against time at various emulsion/diluent ratio for Field C.

Figure 6. There was no water droplet in the two charts until at 30 minutes with water droplet value of 0.1% which increased to 4 and 5% at 720 minutes. C-3 and C-4 followed the trend observed in C-1 and C-2 only that water droplet rate was faster with a value higher than that of C-1 and C-2. The value of water droplet at 720 minutes in C-3 and C-4 are 9 and 10%. In C-5 and B-6, water droplet was experienced earlier than the first four situation. After 5 minutes, water droplet was 1 and 2% which increased to 4 and 5% at 20 minutes. The value of water droplet remains constant at 30 till 720 minutes at 10%.

3.5. Water Quality. The quality of water droplet and observation at the oil and water interface after separation for the three crude oil emulsions are captured in Table 2. For the three crude oil emulsion samples when diluent is not blended with the emulsion, the water quality is dirty and the interface between the water droplet and oil phase is cloudy after 720 minutes. As the diluent volume blended with emulsion increased there, is an improvement in the quality of water change from dirty to clean (i.e., there is no residual emulsion or oil in the water) and the interface between oil and dropped water changes from stained to sharp (i.e., there is a distinct different between water phase and oil phase).

4. Conclusion

Generalized conclusions are hence drawn from the observation of the three samples of crude oil used for this bottle test as follows:

(i) the viscosity of the three water-in-crude oil emulsions considered is inversely proportional to the increase in volume of diluent (gasoline) blended with emulsion. Also the effect of gasoline on the viscosity reduction was observed to be a function of the heaviness of the crude oil emulsion because the higher the viscosity of the emulsion the lower the reduction percentage in its viscosity value. Blending of emulsion reduced the viscosity of the three samples considered by 38, 31, and 17%, respectively.

(ii) Basic sediment and water (BS&W) reduces as the volume of diluent blended with the emulsion increases. This is also a function of viscosity of emulsion prior to blending because BS&W decreased with the decrease in the value of viscosity.

(iii) Total petroleum hydrocarbon (TPH) decreases with the increase in the volume of diluent until optimum concentration of diluent is reached and the TPH increases with further increase in volume of diluent. Optimum value varied for the three crude oil emulsions considered in the analysis. However, above optimum volume of diluent the TPH of the effluent increases which creates another problem when it comes to water disposal.

References

[1] D. Langevin, S. Poteau, I. Hénaut, and J. F. Argillier, "Crude oil emulsion properties and their application to heavy oil

transportation," *Oil and Gas Science and Technology*, vol. 59, no. 5, pp. 511–521, 2004.

[2] M. Hanapi, S. Ariffin, A. Aizan, and I. R. Siti, "Study on demulsifier formulation for treating malaysian crude oil emulsion," Tech. Rep., Department of Chemical Engineering, Universiti Technologi Malaysia, 2006.

[3] R. Espinoza and W. Kleinitz, "The impact of Hidden Emulsion on Oil Prooducing wells—stimulation concept and field result," in *Proceedings of the SPE European Formation Damage*, The Hague, The Netherlands, 2003, SPE paper 00082252.

[4] F. Merv and F. Ben, "Studies of the formation process of water-in-oil emulsions," *Marine Pollution Bulletin*, vol. 47, no. 9–12, pp. 369–396, 2003.

[5] K. P. Micheal, C. Shaokum, and C. M. Samuel, "The key to Predicting Emulsion stability: solid content," in *Proceedings of the SPE International Symposium on Oil Field Chemistry*, Houston, Tex, USA, 2005, SPE paper 93008.

[6] D. Christophe, A. David, S. Anne, G. Alain, and B. Patrick, "Stability of water/crude oil emulsions based on interfacial dilatational rheology," *Journal of Colloid and Interface Science*, vol. 297, no. 2, pp. 785–791, 2006.

[7] M. Rondón, J. C. Pereira, P. Bouriat, A. Graciaa, J. Lachaise, and J. L. Salager, "Breaking of water-in-crude-oil emulsions. 2. Influence of asphaltene concentration and diluent nature on demulsifier action," *Energy and Fuels*, vol. 22, no. 2, pp. 702–707, 2008.

[8] E. J. Ekott and E. J. Akpabio, "Influence of asphaltene content on demulsifiers performance in crude oil emulsions," *Journal of Engineering and Applied Sciences*, vol. 6, no. 3, pp. 200–204, 2011.

[9] K. Sunil, A. Abdullah, and N. S. Meeranpillal, "An Investigative study of potential emulsion problems before field development," in *Proceedings of the SPE Annual Technical Conference and Exhibition*, San Antonio, Tex, USA, 2007, SPE paper 102856.

[10] V. Efeovbokhan, T. Akinola, and F. Hymore, "Performance evaluation of formulated and commercially available de-emulsifiers," in *Nigerian Society of Chemical Engineers Proceedings (NSChE '10)*, vol. 40, pp. 87–99, 2010.

[11] D. T. Nguyen and N. Sadeghi, "Selection of the right demulsifier for chemical enhanced oil recovery," in *International Symposium on Oilfield Chemistry*, The Woodlands, Tex, USA, April 2011.

[12] C. I. Oseghale, E. J. Akpabio, and G. Udottong, "Breaking of oil-water emulsion for the improvement of oil recovery operations in the Niger Delta Oilfields," *International Journal of Engineering and Technology*, vol. 2, no. 11, pp. 1–7, 2012.

[13] B. Fu, "Flow assurance—a technological review of Managing fluid behaviour and solid deposition to Ensure optimum flow," in *Proceedings of the 7th Annual International Forum for deepwater Technologies (Deeptec '00)*, Aberdeen, UK, January 2000.

[14] C. Noïk, H. Malot, C. Dalmazzone, and A. Mouret, "Encapsulation of crude oil emulsions," *Oil and Gas Science and Technology*, vol. 59, no. 5, pp. 535–546, 2004.

[15] M. Nuraini, H. N. Abdurahman, and A. M. S. Kholijah, "Effect of chemical breaking agents on water-in crude oil emulsion system," *International Journal of Chemical and Environmental Engineering*, vol. 2, no. 4, pp. 1–5, 2011.

[16] K. P. Micheal, C. Shaokum, A. M. Robert, and C. M. Samuel, "Classifying crude oil emulsion using chemical demulsifiers and stastical analyses," in *Proceedings of the SPE Annual Technical Conference and Exhibition*, Denver, Colo, USA, 2003, SPE paper 84610.

[17] S. D. Taylor, "Investigations into the Electrical and rheological Behaviour of W/O—emulsions in high voltage Gradients," *Colloid & Surfaces*, vol. 29, pp. 25–51, 1988.

[18] D. G. Thompson, A. S. Taylor, and D. E. Graham, "Emulsification and demulsification related to crude oil production," *Colloid & Surfaces*, vol. 15, pp. 175–189, 1987.

[19] T. J. Jones, E. L. Neustadter, and K. P. Whittingham, "Water-in-crude oil emulsion stability and emulsion destabilization by chemical demulsifiers," *Journal of Canadian Petroleum Technology*, vol. 17, no. 2, pp. 100–108, 1978.

[20] N. H. Abdurahman and W. K. Mahmood, "Stability of water-in-crude oil emulsions: effect of cocamide diethanolamine (DEA) and Span 83," *International Journal of Physical Sciences*, vol. 7, no. 41, pp. 5585–5597, 2012.

[21] J. D. McLean and P. K. Kilpatrick, "Effects of asphaltene solvency on stability of water-in-crude-oil emulsions," *Journal of Colloid and Interface Science*, vol. 189, no. 2, pp. 242–253, 1997.

[22] J. H. Norman, *Non-Technical Guide to Petroleum, Geology, Exploration, Drilling and Production*, Penswell Corporation, Tulsa, Okla, USA, 2nd edition, 2001.

A Field Study on Simulation of CO_2 Injection and ECBM Production and Prediction of CO_2 Storage Capacity in Unmineable Coal Seam

Qin He, Shahab D. Mohaghegh, and Vida Gholami

Department of Petroleum and Natural Gas Engineering, West Virginia University, Morgantown, WV 26505, USA

Correspondence should be addressed to Shahab D. Mohaghegh; shahab.mohaghegh@mail.wvu.edu

Academic Editor: Serhat Akin

CO_2 sequestration into a coal seam project was studied and a numerical model was developed in this paper to simulate the primary and secondary coal bed methane production (CBM/ECBM) and carbon dioxide (CO_2) injection. The key geological and reservoir parameters, which are germane to driving enhanced coal bed methane (ECBM) and CO_2 sequestration processes, including cleat permeability, cleat porosity, CH_4 adsorption time, CO_2 adsorption time, CH_4 Langmuir isotherm, CO_2 Langmuir isotherm, and Palmer and Mansoori parameters, have been analyzed within a reasonable range. The model simulation results showed good matches for both CBM/ECBM production and CO_2 injection compared with the field data. The history-matched model was used to estimate the total CO_2 sequestration capacity in the field. The model forecast showed that the total CO_2 injection capacity in the coal seam could be 22,817 tons, which is in agreement with the initial estimations based on the Langmuir isotherm experiment. Total CO_2 injected in the first three years was 2,600 tons, which according to the model has increased methane recovery (due to ECBM) by 6,700 scf/d.

1. Introduction

Fossil fuels are currently playing a significant role in the whole world's energy supply. However, its damage to the environment, especially the CO_2 emission resulting in the green house effect, has gotten more and more attention. At present, several geological CO_2 sequestration technologies, such as CO_2 injection into saline aquifer, CO_2-EOR, CO_2-ECBM, and so forth, have been studied to minimize the CO_2 release into the atmosphere, and these projects have been operating all over the world [1–6]. Studies have shown that unmineable coal seams (seams too deep or too thin to be mined economically) are pretty attractive as one of the promising options for CO_2 sequestration because of their large CO_2 sequestration capacity, long time CO_2 trapping, and extra enhanced coal-bed methane (ECBM) production benefits [1, 7–10]. Field experience with CO_2 injection into coal seam is limited, although field tests are planned or are being conducted in the USA, Canada, Poland, Australia, and Japan [3].

However, unlike conventional reservoirs, gas flow in the coal seams can cause the cleat permeability and porosity variation during the injection/production process. Once gas is injected and adsorbed on the coal matrix, the matrix will swell, and correspondently decrease the cleat permeability and porosity [11, 12]. Due to its special features and the nature of gas retention in CBM reservoirs, simulating the production and injection will have more complexity compared to conventional resources.

Similar to conventional naturally fractured reservoirs, coal is characterized as a dual-porosity system consisting of matrix and cleat, in which majority of the gas is stored within the coal matrix by a process of adsorption and a small amount of free gas exists in the cleats or fractures [13]. Once CO_2 is injected into the coal seam, it will be held by coal surface because of its higher affinity to the coal matrix than methane, and then displaces the methane to boost extra natural gas production. Figure 1 shows a schematic representation of the CO_2 sequestration-ECBM process. It is estimated by

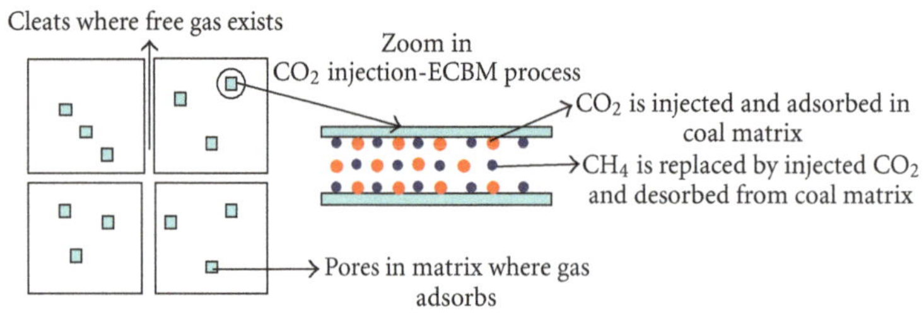

FIGURE 1: A schematic representation of CO_2 sequestration-ECBM production.

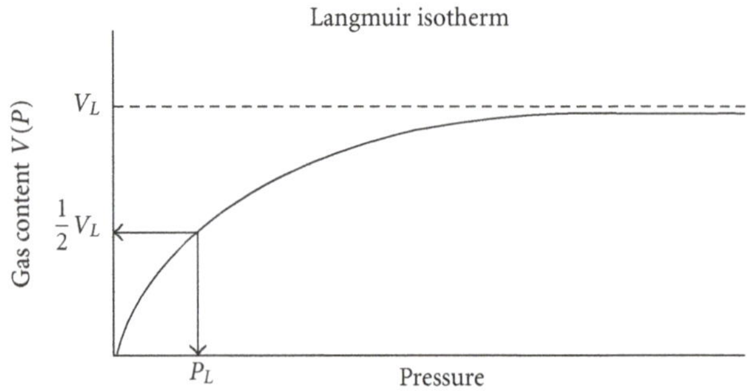

FIGURE 2: Langmuir isotherm function.

laboratory measurements that this process, known as CO_2-enhanced coal bed methane, can store twice as much CO_2 as the methane desorbed or even more [14].

The entire gas flow mechanism can be summarized in three steps: (1) desorption: once free gas or water is produced from fracture systems in coal seams, pressure starts to be released, then the adsorbed gas will be desorbed from the matrix surface, which can be described by Langmuir isotherm equation; (2) diffusion: due to the gas molecular concentration difference, gas will diffuse from matrix surface to cleats/micro-pores; (3) Darcy's flow: gas in the cleats and natural fractures will flow to the wellbore by Darcy's flow [15]. Recently, the numerical reservoir simulator have become the most popular tool to predict coal seam performance and provides a good understanding of gas flow from the reservoir to the wellbore [16].

1.1. Langmuir Isotherm. The gas adsorption/desorption process can be described by the typical formulation of Langmuir isotherm:

$$V(P) = \frac{V_L P}{P_L + P}. \quad (1)$$

As shown in Figure 2, Langmuir volume (V_L) is the maximum amount of gas that can be adsorbed on a piece of coal at infinite pressure. Langmuir pressure (P_L) is the pressure at which the Langmuir volume can be adsorbed. $V(P)$ is the amount of gas at different pressure, also known as gas content (scf/ton). Whenever the Langmuir volume and Langmuir pressure are known, the adsorbed gas amount can be calculated at any pressure.

1.2. Diffusion. Diffusion is the fact that particles move/spread from high concentration to low concentration region. Diffusion of gas out of the coal matrix can be expressed by a simple diffusion equation. The diffusion process in coal seams can be described by either diffusion coefficient or coal desorption time input in the simulator [16]:

$$\frac{\partial C}{\partial t} = \frac{1}{\tau}\left[\overline{C} - C\left(P_f\right)\right]. \quad (2)$$

1.3. Coal Shrinkage and Swelling. One of the unique characteristics of coal seam is the phenomenon of pressure dependent permeability. As the production from the reservoir takes places, two distinct phenomena occur. First, the reservoir pressure declines, which causes the pressure in the fractures to decline as well, which in turn leads to an increase in the effective stress within the cleats causing the cleats to be more compactable, so the cleat permeability will decrease. At the same time, the gas that has been desorbed is coming out of the matrix, which causes the matrix to shrink and the cleats to open-up; thereby the cleat permeability will be increased. As a function of the pressure drop, compressibility dominates in early time and shrinkage dominates in the late time [16]. Palmer and Mansoori model [17] is used to simulate the permeability change process during production/injection in this model:

$$\frac{\varnothing}{\varnothing_0} = 1 + C_f\left(\frac{P-P_0}{\varnothing_0}\right) + \frac{\varepsilon_\alpha}{\varnothing_0}\left(\frac{K}{M}-1\right)\left(\frac{P}{P+P_L} - \frac{P_0}{P_0+P_L}\right),$$
$$\frac{K}{K_0} = \left(\frac{\varnothing}{\varnothing_0}\right)^3. \quad (3)$$

2. Project Description

From 2009, the CO_2 sequestration with ECBM production project began in Marshall County, West Virginia. The objective of this project was to help mitigate climate change by providing an effective and economic way to permanently store CO_2 in un-minable coal seams. In advance of CO_2 injection, four horizontal coalbed methane wells (MH5, MH11, MH18, and MH20) were drilled into the un-minable Upper Freeport coal seam, which are 1,200 to 1,800 feet below the ground. These wells have been producing coalbed methane since 2004. The center located wells (MH18 and MH20) have been converted to CO_2 injection wells since

TABLE 1: Initial reservoir parameters used in the model.

Input parameters	Value	Unit	Input parameters	Value	Unit
Average reservoir depth	1200	ft	Poisson ratio	0.3	
Average formation thickness	4	ft	Young's Modulus	125,000	psia
Fracture spacing I/J/K	0.02	ft	CO_2 Strain	0.0065	
Perm I-Matrix	0.01	md	CH_4 Strain	0.0045	
Perm J-Matrix	0.01	md	Palmer/Mansoori exponent	3	
Perm K-Matrix	0.001	md	CO_2 Langmuir Pressure	240	psia
Perm I-Fracture	0.2	md	CO_2 Langmuir Volume	890	scf/ton
Perm J-Fracture	0.2	md	CH_4 Langmuir Pressure	402	psia
Perm K-Fracture	0.02	md	CH_4 Langmuir Volume	452	scf/ton
Porosity-Matrix	0.004		CO_2 Sorption time	100	days
Porosity-Fracture	0.001		CH_4 Sorption time	100	days
Rock compressibility-Matrix	$1.00E-06$	1/psi	Rock compressibility-Fracture	$1.00E-06$	1/psi

September 2009 [18]. 20,000 short tons are planned to be injected through well MH18 and MH20 in two years.

Several questions come with this project and need to be investigated: how much CO_2 can be stored in this coal seam? How long does the injection process take? Which parameters affect the injection and production the most? These questions could be answered by an effective coal seam model, which was represented by a dual-porosity system to show the fluid flow through both matrix and cleat under the particular conditions in this site. The following assumptions were considered for the modeling and simulation purpose.

(1) The initial seam pressure is hydrostatic pressure, which is 0.28 psi/ft after water is produced.

(2) The flow in the coal seam is single phase including only CH_4 and CO_2.

(3) The fluid flow in the cleat system is a laminar flow due to the larger pore size and it is governed by Darcy's Law, while the flow in the matrix is a diffusional flow due to smaller pore size and governed by Fick's Law.

(4) Palmer and Mansoori equation is used to allow the natural permeability and porosity to vary as a function of pressure.

In most cases, the actual in situ seam data is unavailable, which leads to the requirements of some assumptions on certain parameters, such as, in this case, matrix/cleat permeability, matrix/cleat porosity, geo-mechanical properties (Young's modulus, Poisson ratio), and so forth. Table 1 summarizes the initial physical parameters in the model.

3. History-Matching Results and Discussion

As indicated before, the CO_2 sequestration-ECBM production project went through three stages: primary methane (CBM) recovery, CO_2 injection, and secondary methane (ECBM) recovery. MH18 and MH20 were firstly performed as production wells from January 2005 to July 2007 with a following two-year shut in period; thereafter, they were transferred into CO_2 injection wells since September 2009. MH5 and MH11 keep on methane production from the all the way from beginning to present. All well productions and injection were simulated starting from the start day until the date the most updated data have been recorded and reported (August 2012 in this paper).

However, different performance of MH18 and MH20 in different time periods introduced a lot of complexity on the history matching process. A key factor should be respected in the history matching: either for initial methane production or the following CO_2 injection, well properties (MH18 or MH20) must stay the same in the model; thereby what was changed is only the operation type.

The results of sensitivity analysis were very valuable in back and forth model parameter adjustment. Sensitivity analysis is known as the study of how the variation (uncertainty) in the output of a mathematical model can be apportioned, qualitatively or quantitatively affected by the change of different variations in the input of the model [19]. Sensitivity analysis of coal modeling properties is widely studied and is addressed that it will be an important tool in future decision making [19–21]. In this case, related coal parameters, including cleat permeability, porosity, CH_4 desorption time, CO_2 desorption time, CH_4 Langmuir volume, CO_2 Langmuir volume, and Palmer and Mansoori parameters have been tested in the model. The comparison of coal physical property influences can be concluded based on the study result as: Young's modulus and Poisson ratio have little effect, while sorption time, cleat permeability, strain, and Langmuir isotherm are the key parameters that affect CH_4 production and CO_2 injection most.

The actual in-seam data for both methane production and CO_2 injection in Upper Freeport coal seam were reported daily as shown in Figure 3. The average minimum bottom hole pressure in production wells is 20 psia, and the average maximum BHP in injection wells is 900 psia. The daily injection rate is set as constraint. The trend could be observed in the production; the methane production rate has clearly increased in MH5 and MH11 after July 2009 due to the CO_2 injection. A gradual decline trend in injection rate can be noticed in the injection wells, especially in MH18, which can be a consequence of the permeability changes occurring during desorption/adsorption process on coal.

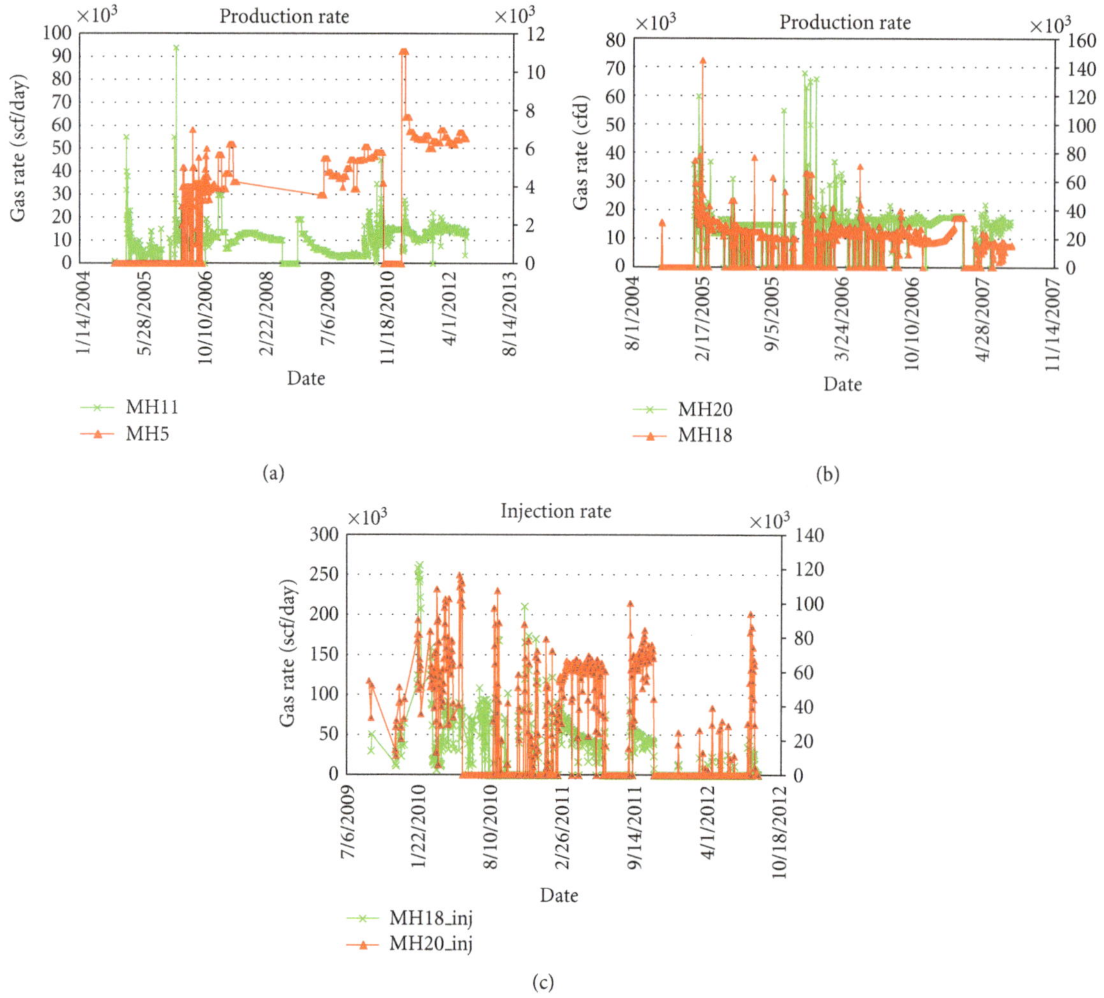

FIGURE 3: Actual CH_4 production rate/CO_2 injection rate in Upper Freeport coal seam. (a) CH_4 production rate in MH11 and MH5, (b) CH_4 production rate in MH18 and MH20, (c) CO_2 injection rate in MH18_inj and MH20_inj (MH18 and MH20 after conversion to Injection wells).

No regular tracking pattern of daily rate was observed because of frequent shut-in operations due to weather, equipment damage, or other unpredictable reasons during the injection process. Therefore, cumulative rates are considered to be the history matching target by setting bottom hole pressure as constraints in the model. History matching was performed for six wells, and final existing reservoir properties, including permeability, porosity, Langmuir isotherm parameter, sorption time, and so forth, as appropriate, were determined by history matching. The history matching results are illustrated in Figures 4 and 5 and the coal parameters are listed in Table 2. It is important to note that the degree of component isotherm and sorption time at any given in-situ condition is directly related to the rank of the coal. Values may change in a large range from different coal seams.

Figure 4 shows the fairly good history matching result of CH_4 cumulative production for all production wells. Green line and red line represents the simulated result and actual data, respectively. As shown in Figure 4(a), well5 was shut in from July 2007 to April 2009 and October 2010 to March 2011, which can be seen from two short straight lines in red cumulative curves. 7×10^6 ft^3 CH_4 could be produced from well5 by August 2012 with a stable increase. As illustrated in Figure 4(b), well11 had a short shut-in period of three months; that is why no production increase is shown in October 2005 and from July 2008 to November 2008. Totally, 2×10^7 ft^3 CH_4 were produced from well11 by August 2012, a sharp build-up could be observed after the start of large CO_2 injection on September 2009, which is because of ECBM production. Figures 4(c) and 4(d) show the cumulative CH_4 production of well18 and well20 from January 2005 to July 2007, respectively, before they were shut-in and transferred to CO_2 injection well. MH18 produced 1.6×10^7 ft^3 CH_4, while MH20 had a total of 1×10^7 ft^3 CH_4 production at the end of production period.

Figure 5 shows cumulative CO_2 injection history matching in MH18 and MH20 after they were converted into injection wells. Red dashed line represents actual CO_2 injection

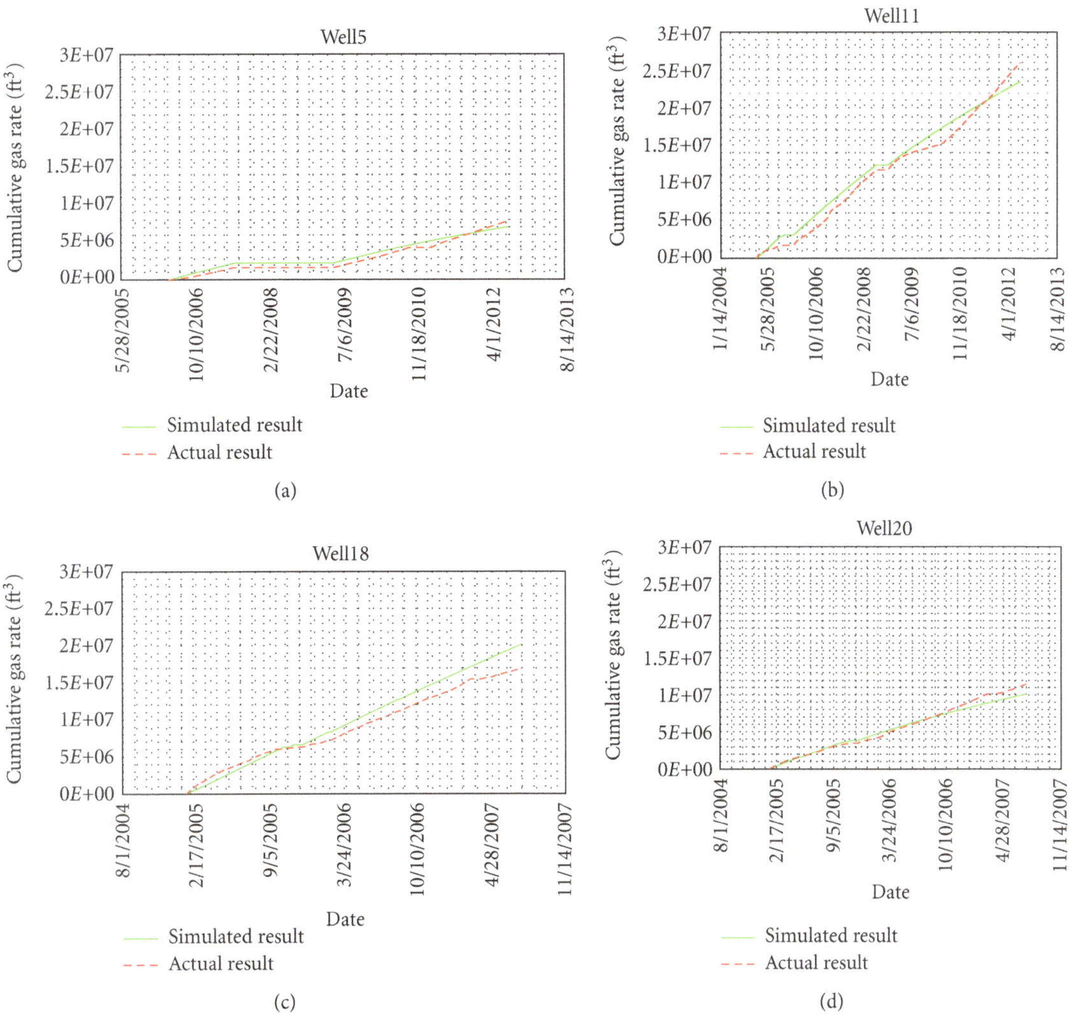

FIGURE 4: CH_4 cumulative production history matching. (a) CH_4 cumulative production in MH5, (b) CH_4 cumulative production in MH11, (c) CH_4 cumulative production in MH18, (d) CH_4 cumulative production in MH20.

data from September 2009 to August 2012, while green line shows simulation results for both wells. Certain plateaus could be seen in the curves during the whole injection periods, which is because of the shut-in times resulting from operational reasons, such as weather affects, equipment damage, and so forth. More CO_2 was injected through well18 (maximum amount of 2.5×10^7 ft^3 CO_2), compared to 2.5×10^7 ft^3 CO_2 injection in well20. The total amount of injected CO_2 through MH18 and MH20 has been almost 3,000 tons in the first three years, with an average ECBM increase of an approximation of 6,700 scf/day.

4. CO_2 Sequestration Capacity in Coal Seam

There are four main CO_2 storage mechanisms in coal seams: (a) stratigraphic and structural trapping, (b) hydrodynamic trapping, (c) mineral trapping, and (d) adsorption trapping. In un-mineable coal seams, adsorption trapping is the main sequestration method. This is the process of accumulation of injected gases which is adsorbed on the surface of micropores within the coal matrix. The adsorption capacity will mostly depend upon Langmuir isotherm factors [22]. Figure 6 illustrates the final Langmuir Isotherm in Upper Freeport coal seam in this case.

Two assumptions have been made in order to simplify the calculation here.

(1) No water production data was reported in this case; the coal reservoir was simulated with single phase production with only CH_4 and CO_2.

(2) Adsorption trapping is the main sequestration method in un-mineable coal seam, which was considered as the only storage mechanism without including free gas in the fractures in this study.

The CO_2 adsorption capacity in the coal seam can be calculated as

$$\text{OGIP} = A \times h \times \rho_b \times G_{ci} = V \times \rho_b \times G_{ci}, \quad (4)$$

where $V_L = 800$ scf/ton, $P_L = 412$ psia, $P = 0.28$ psi/ft $\times$ 1200 ft = 360 psi, and $V(P) = G_{ci} = V_L P/(P_L + P) = 800 \times$

TABLE 2: History matched reservoir parameter setting.

Input parameters	Value	Unit	Input parameters	Value	Unit
Average reservoir depth	1200	ft	Poisson ratio	0.3	
Average formation thickness	4	ft	Young's Modulus	125,000	psia
Fracture spacing I/J/K	0.015	ft	CO_2 Strain	0.0025	
Perm I-Matrix	0.01–0.02	md	CH_4 Strain	0.0045	
Perm J-Matrix	0.01–0.02	md	Palmer/Mansoori exponent	3	
Perm K-Matrix	0.001–0.002	md	CO_2 Langmuir Pressure	412	psia
Perm I-Fracture	0.2–0.4	md	CO_2 Langmuir Volume	800	scf/ton
Perm J-Fracture	0.2–0.4	md	CH_4 Langmuir Pressure	628	psia
Perm K-Fracture	0.02–0.04	md	CH_4 Langmuir Volume	652	scf/ton
Porosity-Matrix	0.002–0.004		CO_2 Sorption time	140	days
Porosity-Fracture	0.001–0.002		CH_4 Sorption time	350	days
Rock compressibility-Matrix	1.00E − 06	1/psi	Rock compressibility-Fracture	1.00E − 06	1/psi

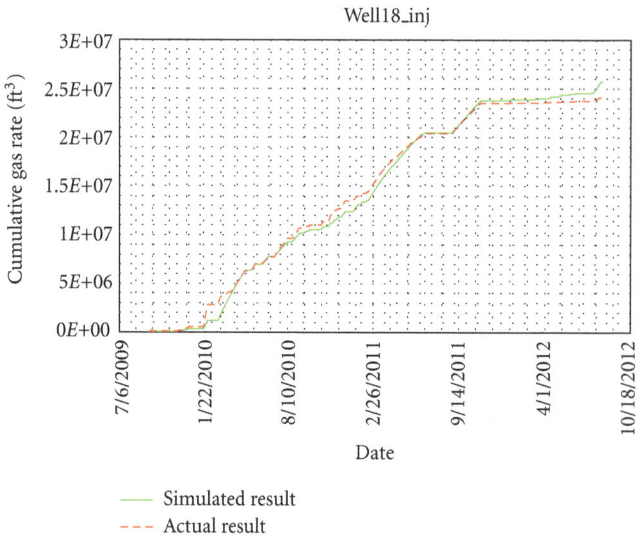

(a)

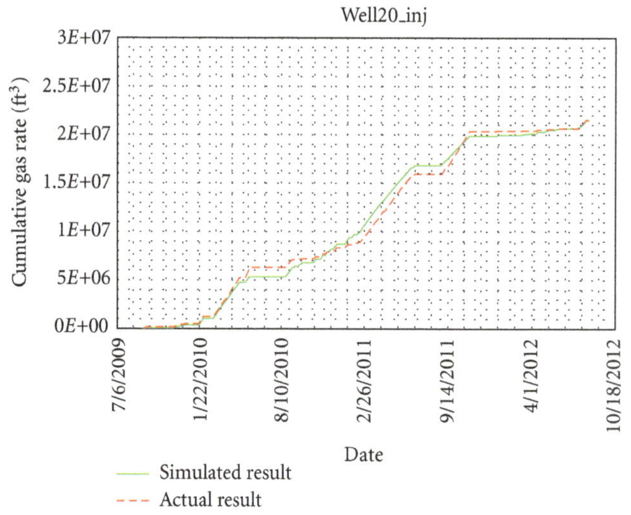

(b)

FIGURE 5: Cumulative CO_2 injection history matching. (a) Cumulative CO_2 injection in MH18_inj, (b) Cumulative CO_2 injection in MH20_inj.

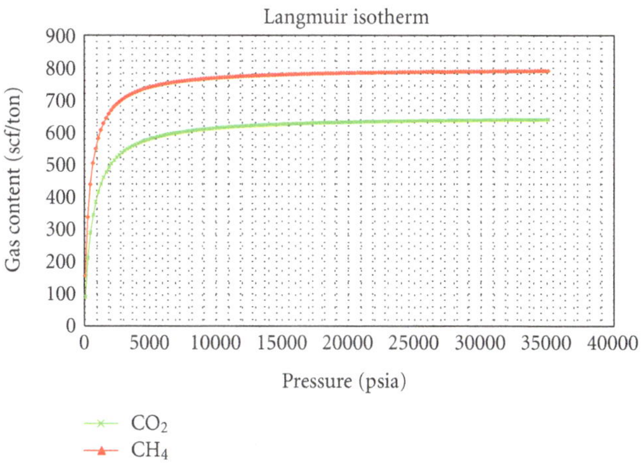

FIGURE 6: Existing Langmuir isotherm for CO_2 and CH_4 in Upper Freeport Coal seam.

360/(412 + 360) = 373 scf/ton, where, ρ_b = 85 lbs/ft^3, V = 25,193,558 ft^3, 1 ton = 2,000 lbs, Coal tonnage = 85 × 25,193,558/2,000 = 1,069,466 tons, OGIP = 1,069,466 tons × 373 scf/ton/17,483 ton/scf = 22,817 tons (coal seam volume and coal density were provided and were used directly).

5. Summary and Conclusions

The modeling and history matching process of methane production and ECBM as well as CO_2 injection in a coal bed seam was explained in this work. This process was performed using conducting actual data analysis and sensitivity analysis of related coal seam physical properties on four horizontal wells drilled in Upper Freeport coal seam. Results of history matching were compiled to show the initial and existing condition in the coal seam. CO_2 sequestration capacity prediction was completed according to the Langmuir isotherm properties obtained from the history matched reservoir model.

The simulation of CH_4 gasification and CO_2 injection process was quite complicated. The special swelling and shrinkage features and the nature of gas retention in CBM reservoirs make the modeling and history matching of production and injection data in coal bed methane more complex because of the permeability and porosity variations compared to conventional resources.

Sensitivity analysis results suggested that sorption time, cleat permeability, strain, and Langmuir isotherm are the most influential parameters during CH_4 production and CO_2 injection process. It is concluded by the Langmuir isotherm parameters from history matched model that the total CO_2 sequestration capacity is about 22,817 tons excluding the free gas part in the cleat system. The total CO_2 injection amount in the first three years was 4.5×10^7 ft^3 or 2,600 tons, which caused an increase of 6,700 scf/day in CH_4 production rate from other two wells.

Nomenclature

D: Diffusion coefficient

$\overline{C}$: Average gas concentration in the matrix

τ: Desorption time, days

$C_f((P-P_0)/\varnothing_0)$: Stress-dependent permeability term

$(K/M-1)(P/(P+P_\varepsilon)-P_0/(P_0+P_\varepsilon)(K/M-1)(P/(P+P_\varepsilon) - (P_0/(P_0 + P_\varepsilon))))$: Matrix shrinkage term

Φ_i: Initial fracture porosity, %

C_f: Pore volume compressibility, 1/psi

P: Initial pressure, psi

M: Axial modulus, psi

K: Bulk modulus, psi

ε: Langmuir strain

P_L: Langmuir pressure, psi

V_L: Langmuir volume, scf/ton

A: Drainage area, ft^2

h: Net pay, ft

ρ_b: Bulk density, lbs/ft^3

G_{ci}: Gas Content, scf/ton

$\varnothing_i$: Porosity, %

B_{gi}: Initial formation volume factor, STB/scf

OGIP: Original gas in place, tons

V: Coal volume, ft^3.

Acknowledgments

This project was funded by the Department of Energy, National Energy Technology Laboratory, Consol Energy, through a support contract with URS Energy and Construction, Inc. The authors want to acknowledge the important contributions of Consol Energy for the field data available for analysis. Acknowledgment is also extended to Tom Wilson in WVU for providing the geological maps for the studied field in this research. Thanks go to Computer Modeling Group Ltd. (CMG) for providing the software to do the PEARL research group at WVU.

References

[1] S. H. Stevens, D. Spector, and P. Riemer, "Enhanced coalbed methane recovery using CO_2 injection: worldwide resource and CO_2 sequestration potential," in *Proceedings of the 6th International Oil & Gas Conference and Exhibition in China (IOGCEC '98)*, pp. 489–501, Beijing, China, November 1998.

[2] J. Ennis-King and L. Paterson, "Engineering aspects of geological sequestration of carbon dioxide," in *Proceedings of the SPE Asia Pacific Oil and Gas Conference and Exhibition*, pp. 134–146, Melbourne, Australia, October 2002.

[3] F. M. Orr Jr., "Storage of carbon dioxide in geologic formations," *Journal of Petroleum Technology*, vol. 56, no. 9, pp. 90–97, 2004.

[4] C. Sinayuç and F. Gümrah, "Modeling of ECBM recovery from amasra coalbed in Zonguldak Basin, Turkey," in *Proceedings of the Canadian International Petroleum Conference*, Alberta, Canada, 2008.

[5] R. Petrusak, D. Riestenberg, P. Goad et al., "World class CO_2 sequestration potential in saline formations, oil and gas fields, coal, and shale: the US southeast regional carbon sequestration partnership has it all," in *Proceedings of the SPE International Conference on CO2 Capture, Storage, and Utilization*, pp. 136–153, November 2009.

[6] C. L. Liner, "Carbon capture and sequestration: overview and offshore aspects," in *Proceedings of the Offshore Technology Conference (OTC '10)*, pp. 3511–3514, May 2010.

[7] J. P. Seidle, "Reservoir engineering aspects of CO_2 sequestration in coals," in *Proceedings of the SPE/CERI Gas Technology Symposium*, Alberta, Canada, 2000.

[8] H. J. M. Pagnier, F. Van Bergen, E. Kreft, L. G. H. Van Der Meer, and H. J. Simmelink, "Field experiment of ECBM-CO_2 in the upper Silesian Basin of Poland (RECOPOL)," in *Proceedings of the 67th European Association of Geoscientists and Engineers, EAGE Conference and Exhibition, incorporating SPE (EUROPEC '05)*, pp. 3013–3015, Madrid, Spain, June 2005.

[9] G. A. Hernandez, R. O. Bello, D. A. McVay et al., "Evaluation of the technical and economic feasibility of CO_2 sequestration and enhanced coalbed-methane recovery in Texas low-rank coals," in *Proceedings of the SPE Gas Technology Symposium: Mature Fields to New Frontiers*, pp. 515–530, Alberta, Canada, May 2006.

[10] G. J. Koperna and D. Riestenberg, "Carbon dioxide enhanced coalbed methane and storage: is there promise?" in *Proceedings of the SPE International Conference on CO2 Capture, Storage, and Utilization*, pp. 183–195, November 2009.

[11] J. Q. Shi and S. Durucan, "A model for changes in coalbed permeability during primary and enhanced methane recovery," *SPE Reservoir Evaluation and Engineering*, vol. 8, no. 4, pp. 291–299, 2005.

[12] S. Mazumder and K. H. Wolf, "Differential swelling and permeability change of coal in response to CO_2 injection for ECBM," *International Journal of Coal Geology*, vol. 74, no. 2, pp. 123–138, 2008.

[13] L. Dean, "Reservoir engineering for geologists: coalbed methane fundamentals," Reservoir Issue. 2007, 11.

[14] *Storing CO_2 in Unminable Coal Seams*, IEA Greenhouse Gas R&D Programme.

[15] K. Aminian and S. Ameri, "Predicting production performance of CBM reservoirs," *Journal of Natural Gas Science and Engineering*, vol. 1, no. 1-2, pp. 25–30, 2009.

[16] I. Zulkamain, *Simulation study of the effect of well spacing, permeability, anisotropy, Palmar and Mansoori model on coalbed methane production. [M.S. thesis]*, Texas A&M University, 2005.

[17] I. Palmer and J. Mansoori, "How permeability depends on stress and pore pressure in coalbeds: a new model," *SPE Reservoir Engineering*, vol. 1, no. 6, pp. 539–543, 1998.

[18] "CO_2 storage with ECBM studybegins in West Virginia," http://www.carboncapturejournal.com/displaynews.php?NewsID=442.

[19] D. J. Remner, T. Ertekin, W. Sung, and G. R. King, "Parametric study of the effects of coal seam properties on gas drainage efficiency," *SPE Reservoir Engineering*, vol. 1, no. 6, pp. 633–646, 1986.

[20] A. N. Okeke, *Sensitivity analysis of modeling parameters that affect the dual peaking behavior in coalbed methane reservoirs [M.S. thesis]*, Texas A&M University, 2005.

[21] Q. P. Huy, K. Sasaki, Y. Sugai et al., "Numerical simulation of CO_2 enhanced coal bed methane recovery for A vietmese coal seam," *JournaL of NoveL Carbon Resource Sciences*, vol. 2, pp. 1–7, 2010.

[22] D. Jasinge and P. G. Ranjith, "Carbon dioxide sequestration in geologic formation with special reference to sequestration in deep coal seams," in *Proceedings of the 45th U.S. Rock Mechanics/Geomechanics Symposium*, 2011.

17

Well Screen and Optimal Time of Refracturing: A Barnett Shale Well

Shayan Tavassoli,[1] Wei Yu,[1] Farzam Javadpour,[2] and Kamy Sepehrnoori[1]

[1] *Department of Petroleum and Geosystems Engineering, The University of Texas at Austin, Austin, TX 78712, USA*
[2] *Bureau of Economic Geology, Jackson School of Geosciences, The University of Texas at Austin, Austin, TX 78713, USA*

Correspondence should be addressed to Farzam Javadpour; farzam.javadpour@beg.utexas.edu

Academic Editor: Jorge Ancheyta

Gas-production decline in hydraulically fractured wells in shale formations necessitates refracturing. However, the vast number of wells in a field makes selection of the right well challenging. Additionally, the success of a refracturing job depends on the time to refracture a shale-gas well during its production life. In this paper we present a numerical simulation approach to development of a methodology for screening a well and to determine the optimal time of refracturing. We implemented our methodology for a well in the Barnett Shale, where we had access to data. The success of a refracturing job depends on reservoir characteristics and the initial induced fracture network. Systematic sensitivity analyses were performed so that the characteristics of a shale-gas horizontal well could be specified as to the possibility of its candidacy for a successful refracturing job. Different refracturing scenarios must be studied in detail so that the optimal design might be determined. Given the studied trends and implications for a production indicator, the optimal time for refracturing can then be suggested for the studied well. Numerical-simulation results indicate significant improvement (on the order of 30%) in estimated ultimate recovery (EUR) after refracturing, given presented screen criteria and optimal-time selection.

1. Introduction

Shale-gas resources, predominantly lithified clays with low permeability [1], are considered unconventional gas reservoirs and important resources for the United States. However, gas production from these low-permeability resources is much greater than what is anticipated owing to non-Darcy flows and different sources of gas in their formations [2]. Gas flow is sourced from stored gas in nanopore networks and adsorbed gas on organic materials in the shale formations. However, new techniques are required for access to and economical production from these resources.

Recent advances in hydraulic-fracturing techniques have resulted in economic production from shale-gas reservoirs. Effective fracturing techniques make for successful economic production from extremely low (on the order of nanodarcies) permeability formations because they create a large, stimulated reservoir volume [3, 4]. Such a success would be attributed to the potential for developing complex fracture networks, which could significantly improve reservoir-wellbore connectivity. Refracturing is a process of improving production rates and ultimate recovery, which is an economical alternative to infill drilling. Although refracturing seems an excellent method of significantly increasing gas production, only 15% to 20% of refractured wells achieve any desired improvement in practice [5–7]. Therefore, a reliable and systematic approach is critical to an increase in the success of refracturing jobs. Several published works [8–11] have suggested a selection of methodologies for finding good shale-gas-well candidates for refracturing. However, application of these methods is limited, and results are unsatisfactory in horizontal-well or complex-fracture-network cases or when adequate completion and reservoir/geology data sets are lacking. With the aid of numerical-simulation methods, however, we are now able to study well performance properly by considering the presence of a complex fracture network, a tight matrix, and initial hydraulic fractures.

Field results demonstrate that refracturing success can be attributed to different parameters, such as the existing fracture network and reservoir properties, and a successful refracturing job should increase reservoir-wellbore connectivity, which is associated with further opening, extension, and reorientation of existing fractures. Reorientation is generally perpendicular to an initial hydraulic fracture as the result of minimum and maximum stress reversal, which occurs if induced stress changes are large enough to overcome the effect of initial horizontal stress [12]. Such a reversal could be the result of the stress created by the opening of an adjacent fracture or pore-pressure alteration due to previous production periods [13–15]. In such cases, the fracture network propagates, and, as a result, a larger portion of the reservoir is stimulated, subsequently increasing gas production and ultimate recovery.

Owing to the complexity of fracture growth in many shale-gas reservoirs, accurate prediction of fracture propagation is impossible; therefore, conductivity of the fracture network and the effectiveness of stimulation treatment are difficult to predict [16, 17]. Detailed numerical reservoir modeling is thus needed for us to better understand the mechanisms that control production in shale-gas reservoirs and to improve completion strategies and stimulation designs. First, we needed to validate our simulation model by modeling initial fractures and the refracturing process in a horizontal well in the Barnett Shale formation. Second, on the basis of our simulation model, we performed a sensitivity study on the effect of different parameters on refracturing performance. Sensitivity analyses provide insights into the dependency of the refracturing-process performance on corresponding parameters. The result would be helpful for selecting candidates for successful refracturing. We also compared results of the refracturing performance with closely spaced initial fracturing as an alternative to increasing gas recovery from the early stages of production. Note that specifying the optimal time of refracturing is crucial to maximizing performance of the process. Simulation results of refracturing at different stages of production suggest calculation of the optimal time on the basis of gas flow rate or cumulative gas-production recovery.

2. Methodology

Numerical simulation is a promising method in any design or well-performance evaluation and is commonly used to predict gas production. Specifically this approach has increased in importance for shale-gas reservoirs owing to the complexity of fracture networks and the low permeability of the formations because numerical simulations are capable of modeling fracture networks. Note that fracture-network complexity and conductivity control well productivity in shale-gas reservoirs—the more complex the fracture networks (i.e., the smaller the blocks or the denser the fracture spacing), the higher the production rates [18, 19]. Fracture networks should therefore be considered in the simulation of the refracturing process, even though their initial network fracture conductivity may be relatively low (e.g., in the Barnett Shale, it ranges from 0.5 to 5 md-ft) [18, 20].

Development of numerical simulation approaches that can properly model fluid flow in tight formations and that are capable of capturing complex fracture networks and initial hydraulic fractures is important in an evaluation of well performance and an explanation of properties that affect gas recovery. Gas flow from ultralow-permeability rock through complex fracture networks must be modeled so that stimulation designs and completion strategies can be properly evaluated. Therefore, the complex fracture network and initial hydraulic fractures must be discretely characterized in these reservoir-simulation models.

We validated our simulation methodology using field data from a refractured horizontal well in the Barnett Shale formation. We selected the Barnett Shale because of the availability of production data and because, so that economically viable production rates might be attained, hydraulic-fracture stimulation is a necessity in this formation.

The simulation model is based on a dual permeability model for two-phase (gas-water) fluid flow. Such a model considers the communication between the intergranular void spaces in contrast to the dual porosity model, which neglects this communication. The dual permeability model considers flow in the two domains of matrix and fracture, whereas the dual porosity approach assumes two domains with different hydraulic and transport properties [21]. The dual permeability model allows for simulations involving gas and water transfer between the fracture and matrix domains. Gas flow velocities in matrix and fracture are calculated on the basis of (1) and (2), respectively,

$$v_g^m = -\left(\frac{K_g^m}{\mu_g}\nabla p_g + \frac{D_g^m}{C_g^m}\nabla C_g^m\right), \tag{1}$$

$$v_g^f = -\left(\frac{K_g^f}{\mu_g}\nabla p_g + \frac{D_g^f}{C_g^f}\nabla C_g^f\right), \tag{2}$$

where v_g is gas velocity, K_g is gas permeability, D_g is gas diffusivity, P_g is gas pressure, C_g is gas concentration, and μ_g is gas viscosity. Superscripts m and f represent matrix and fracture, respectively. Water-flow velocities in matrix and fracture are calculated on the basis of (3) and (4), respectively,

$$v_w^m = -\frac{K_w^m}{\mu_w}\nabla p_w^m, \tag{3}$$

$$v_w^f = -\frac{K_w^f}{\mu_w}\nabla p_w^f, \tag{4}$$

where v_w is water velocity, K_w is water permeability, P_w is water pressure, and μ_w is water viscosity. The equations are thus simplified to represent gas flux in matrix:

$$\frac{\partial}{\partial t}\left(\frac{C_g^m P_g^m}{Z}\right) = \nabla\left(\frac{P_g^m}{Z}\frac{K_g^m}{\mu_g}\nabla P_g^m + D_g^m \nabla \frac{P_g^m}{Z} + \frac{D_g^m}{C_g^m}\frac{P_g^m}{Z}\nabla C_g^m\right) - \frac{RT}{M}\left(q_g^{mf} + q_g^m\right), \quad (5)$$

where Z is the gas compressibility factor, R is the gas constant, T is temperature, M is gas molecular weight, and q_g is gas mass-flow rate per unit matrix-block volume. Superscript mf represents the exchange between matrix and fracture. We can also write this equation for the aqueous phase in matrix as

$$\frac{\partial}{\partial t}\left(\frac{\phi_m s_w^m}{B_w}\right) = \nabla\left(\frac{K_w^m}{B_w \mu_w}\nabla P_w^m\right) - \frac{RT}{M}\left(q_w^{mf} + q_w^m\right), \quad (6)$$

where ϕ_m is matrix porosity, s_w is water saturation, and B_w is the water compressibility factor. As a result of simplification of the equations, gas flux in fracture becomes

$$\frac{\partial}{\partial t}\left(\frac{C_g^f P_g^f}{Z}\right) = \nabla\left(\frac{P_g^f}{Z}\frac{K_g^f}{\mu_g}\nabla P_g^f + D_g^f \nabla \frac{P_g^f}{Z} + \frac{D_g^f}{C_g^f}\frac{P_g^f}{Z}\nabla C_g^f\right) + \frac{RT}{M}\left(q_g^{mf} - q_g^f\right), \quad (7)$$

and also for water in fracture:

$$\frac{\partial}{\partial t}\left(\frac{\phi_f s_w^f}{B_w}\right) = \nabla\left(\frac{K_w^f}{B_w \mu_w}\nabla P_w^f\right) + \frac{RT}{M}\left(q_w^{mf} - q_w^f\right) \quad (8)$$

with auxiliary equations

$$C_g^m = \phi_m s_g^m,$$
$$C_g^f = \phi_f s_g^f, \quad (9)$$
$$s_g^m + s_w^m = 1,$$
$$s_g^f + s_w^f = 1.$$

3. History Matching

The simulation results presented in this work are obtained mainly by using the Computer Modeling Group (CMG) model [22]. Our modifications and methodology of using this model are presented in Section 2. In order to validate our simulation model, we modeled initial fractures and the refracturing process in a horizontal well in the Barnett Shale formation. The Barnett Shale is a Mississippian-age marine shelf deposit having a formation thickness that varies from 200 to 800 ft through the reservoir and ultralow permeability in the range of 70–500 nanodarcies [23]. The selected well is a cased, uncemented well in the formation.

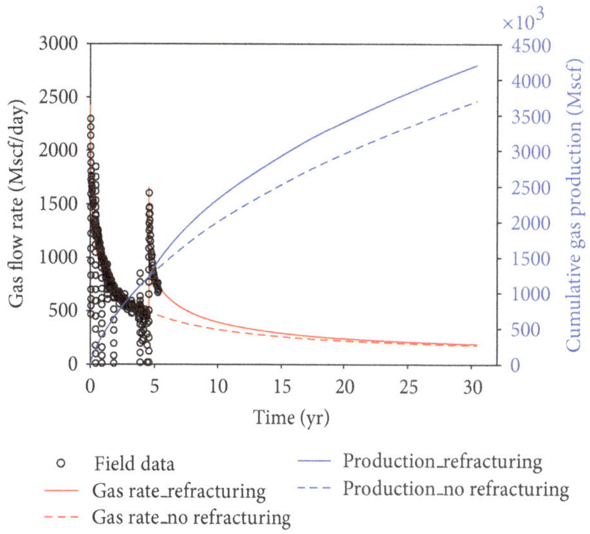

FIGURE 1: Comparison between simulation results and field data.

The well is located in the lower Barnett Shale and drilled northwest to maximize fracture-network formation. It has five equally spaced initial fractures covering a depth interval from 7,900 to 10,106 ft [23]. The well was refractured after more than 4 years in four stages in the middle of the initial fractures. Gas production from the well was monitored and plotted for its 6 years of production (4.5 years before and 1.5 years after refracturing) (Figure 1).

Input data used in the simulation model to predict gas production from the well appear in Table 1. Refracturing after 4 years of production significantly increased productivity of the horizontal well (Figure 1), and comparison of the simulation-model results and field data verifies that our model is capable of modeling initial fractures and refracturing.

Fractures postclosure occurs during production and depletion of any shale well, closure stress, and long-term degradation of proppants controlling the extent of fracture postclosure. Gas production may decrease owing to fracture-conductivity impairments, especially during later gas production. Yu and Sepehrnoori [24] recently studied the effect of reservoir depletion on fracture conductivity in the Barnett Shale. They showed that fracture conductivity reduces to about 40% of original conductivity at a limiting bottom-hole pressure of 500 psi. However, because of the great contrast between matrix permeability and fracture permeability in shale, fracture postclosure has a negligible effect (<2%) on gas production, even after 30 years [24]. This is especially true of formations with a high Young's modulus (6×10^6 psi to 10×10^6 psi), such as the Barnett Shale.

4. Results and Discussion

In this section we use our verified numerical model and the present gas production rate and cumulative gas production of the studied well. In our sensitivity analyses we present the effects of formation permeability and porosity, as well

Table 1: Input data used in simulation model.

Parameter	Value(s)	Unit: field (SI)
Model dimensions	$5000 \times 5000 \times 300$ ($1524 \times 1524 \times 91$)	ft (m)
Initial reservoir pressure	3800 (2.62×10^7)	psi (pa)
BHP	1100 (7.58×10^6)	psi (pa)
Production time	30	year
Reservoir temperature	150 (66)	°F (°C)
Gas viscosity	0.0201 (2.01×10^{-5})	cp (pa.s)
Initial gas saturation	0.70	fraction
Total compressibility	3×10^{-6} (4.35×10^{-10})	Psi^{-1} (pa^{-1})
Matrix permeability	5×10^{-4} (4.93×10^{-19})	md (m^2)
Matrix porosity	0.018	fraction
Fracture conductivity	1 (3×10^{-16})	md-ft (m^2-m)
Fracture half-length	Varied: Min = 470 (143), Max = 1870 (570)	ft (m)
Fracture spacing	550 (168)	ft (m)
Refracture half-length	970 (296)	ft (m)
Refracture spacing	275 (84)	ft (m)
Fracture height	300 (91)	ft (m)
Horizontal well length	2800 (853)	ft (m)
Number of fractures	5	number
Number of refractures	4	number

as initial induced hydraulic-fracture conductivity on well performance. These analyses are followed by a discussion on well-screening criteria and time optimization of refracturing. We used base data (Table 1) to create simulation cases by varying individual parameters for sensitivity analyses. Gas-production rate and cumulative gas production were plotted against time for all parameters to show the performance of the well. Each plot included well performance with and without refracturing to highlight refracturing effectiveness on gas-production performance.

4.1. Effects of Permeability. We created two plots to show the effects of formation permeability on gas flow rate and cumulative gas production rate (Figures 2 and 3). In each plot we considered two scenarios—performance of the well without refracturing and a case of refracturing after 4 years. Intuitively we knew that both gas flow rate and cumulative gas production would improve if permeability increased. The refractured wells showed better gas flow rates and 30% to 70% improvement in cumulative gas production for the highest (0.0005 md) and lowest (0.00005 md) permeabilities, respectively.

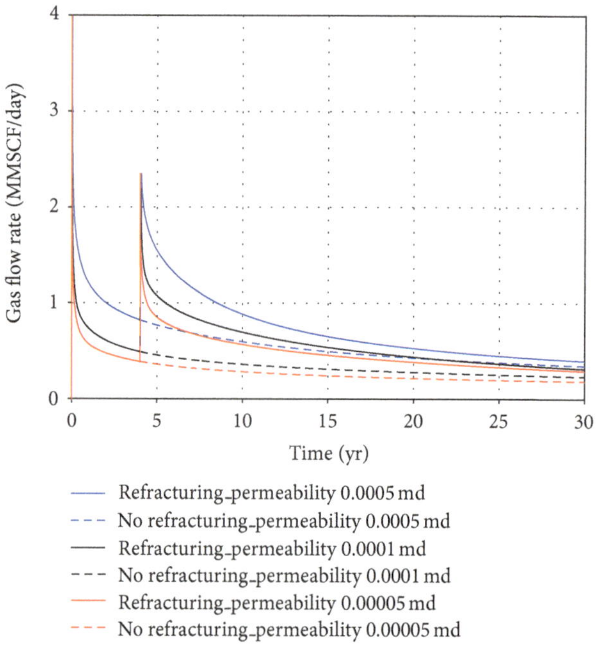

Figure 2: Effects of formation permeability on gas flow rate. Two scenarios: no refracturing and refracturing after 4 years of production.

4.2. Effects of Porosity. We created two plots to show the effects of formation porosity on gas flow rate and cumulative gas production rate (Figures 4 and 5). In each plot we considered two scenarios—performance of the well without refracturing and a case of refracturing after 4 years. The refractured wells showed better gas flow rates and 50% to 70% improvement in cumulative gas production for the lowest (0.04) and highest (0.08) porosities, respectively.

4.3. Effects of Initial Fracture Conductivity. We created two plots to show the effects of initial fracture conductivity on gas flow rate and cumulative gas production rate (Figures 6 and 7). In each plot we considered two scenarios—performance of the well without refracturing and a case of refracturing after

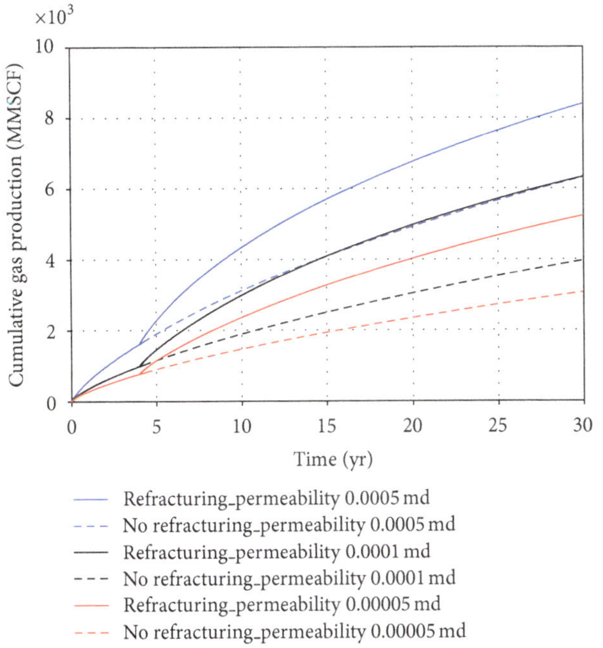

FIGURE 3: Effects of formation permeability on cumulative gas production. Two scenarios: no refracturing and refracturing after 4 years of production.

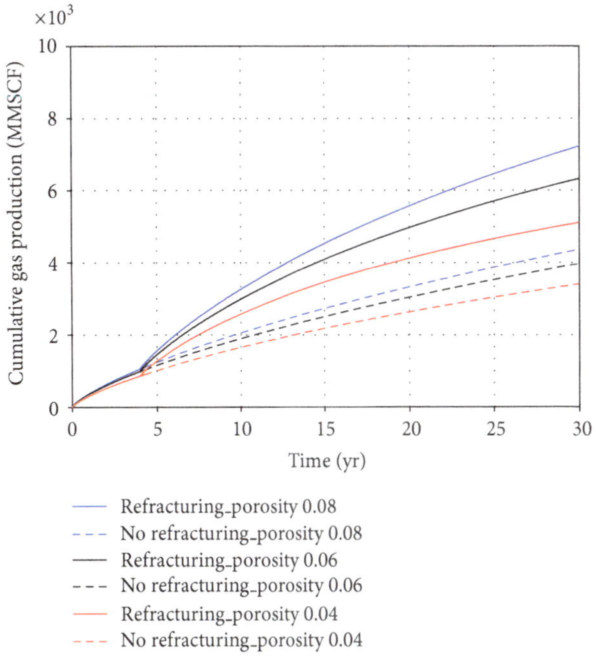

FIGURE 5: Effects of formation porosity on cumulative gas production. Two scenarios: no refracturing and refracturing after 4 years of production.

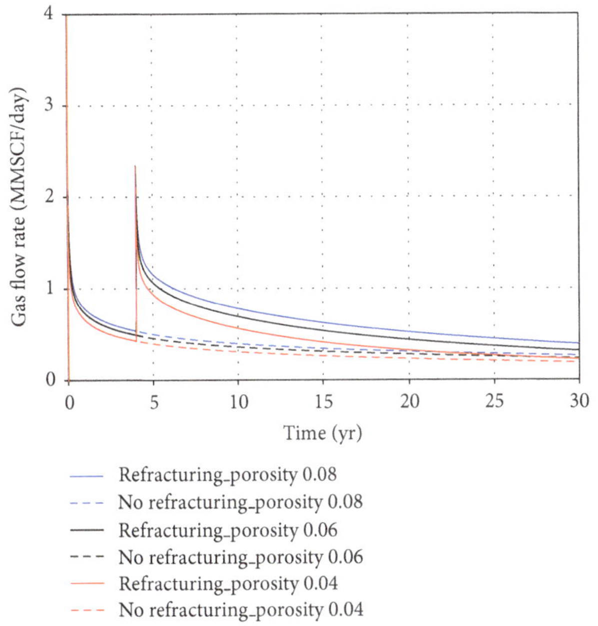

FIGURE 4: Effects of formation porosity on gas flow rate. Two scenarios: no refracturing and refracturing after 4 years of production.

4 years. Both gas-flow rate and cumulative gas production improved when initial fracture conductivity was increased. The refractured well showed better gas flow rates and 50% to 80% improvement in cumulative gas production for the highest (10 md-ft) and the lowest (0.1 md-ft) initial fracture conductivities, respectively.

4.4. Well Screen Criteria. Cumulative gas production and gas flow rate increase as reservoir permeability, porosity, and initial fracture conductivity increase (Figures 2–7). Because increase in the trends of gas production in these cases is significant in contrast to the increase in trends of gas flow rate, we are unable to refine our selection on the basis of these trends. We therefore compared our results on the basis of production indicators to select candidate shale-gas wells for refracturing. Given the correlation between monthly incremental and 5-year cumulative gas production, wells falling off the trend line indicate wells that had good initial production but for which longer-term recovery was poor [9] (Figure 8). Because all of our cases are located on the trend line, this method adds no information about range of selection.

To determine and compare production enhancement in each case study, we introduce a production indicator, known as *long-term refracturing efficiency* and defined as the ratio of cumulative gas production after refracturing to its value before refracturing in 30 years of production:

$$\text{Long-term refracturing efficiency} = \frac{\text{Total cumulative gas production after refracturing}}{\text{Total cumulative gas production with no refracturing}}. \quad (10)$$

We calculated and plotted long-term refracturing efficiency for each case and found that refracturing efficiency increases as reservoir permeability reduces to its lower limits (Figure 9). Efficiency enhancement is not that significant,

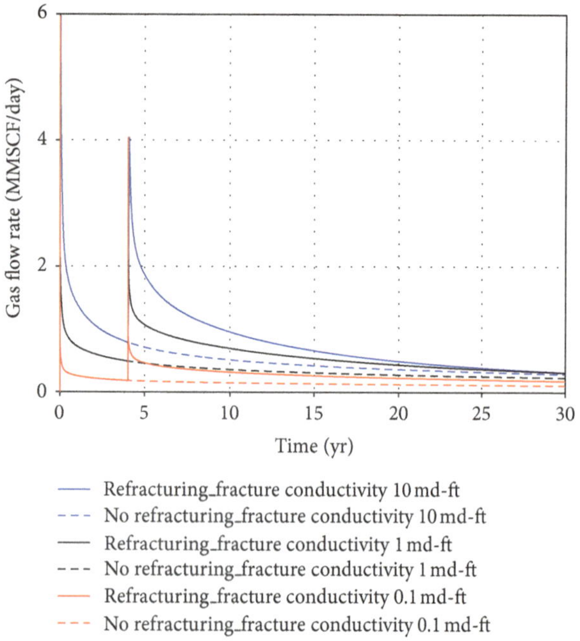

FIGURE 6: Effects of initial fracture conductivity on gas flow rate. Two scenarios: no refracturing and refracturing after 4 years of production.

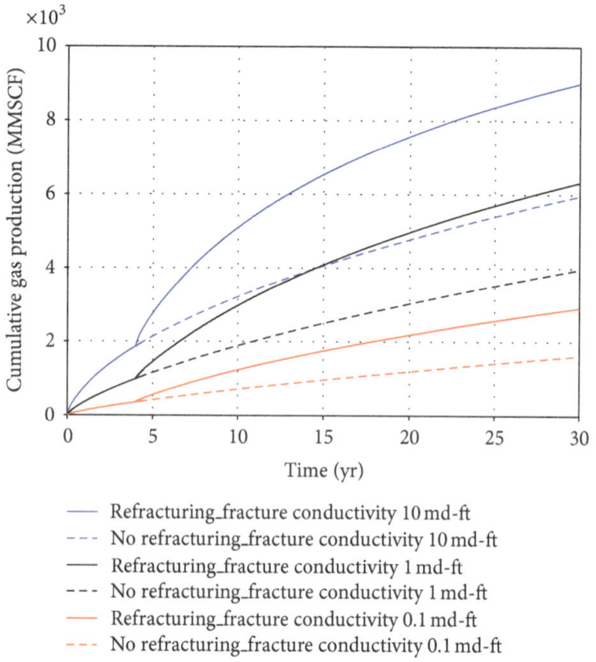

FIGURE 7: Effects of initial fracture conductivity on cumulative gas production. Two scenarios: no refracturing and refracturing after 4 years of production.

however, at permeability values greater than 0.0001 millidarcy. Refracturing efficiency increases as porosity increases (Figure 10). Efficiency enhancement due to refracturing is significant when porosity values are greater than 6%. The difference in gas-production enhancement with respect to permeability might seem to be in contradiction to its trend with respect to porosity because of the correlation between permeability and porosity. However, note that refracturing enhances gas production from low-permeability reservoirs by forming a more complex fracture network that has been depleted from a larger stimulated reservoir volume (SRV). On the other hand, larger porosity corresponds to a larger amount of initial gas in place. Hence, refracturing enhancement cannot be responsible for an increase in gas production in cases with high-porosity values. We also studied the effect of initial fracture conductivity on the performance of refracturing. Refracturing efficiency increases as fracture conductivity decreases (Figure 11), efficiency enhancement increasing in significance for values lower than 1 md-ft. Results of these simulations reveal that long-term refracturing efficiency can provide insight into the effect of each parameter and its importance. On the basis of this sensitivity study we have proposed a screening method for existing fractured horizontal wells that is based on reservoir and initial hydraulic-fracture properties (Table 2).

4.5. Initial Fracture Spacing. Here we compare results of refracturing performance with that of closely spaced initial fracturing, which can be an alternative to increasing gas recovery beginning in the early stages of production. Different values of fracture spacing (50, 100, 200, 350, and 450 ft) are considered here. Refracturing in the middle of initial

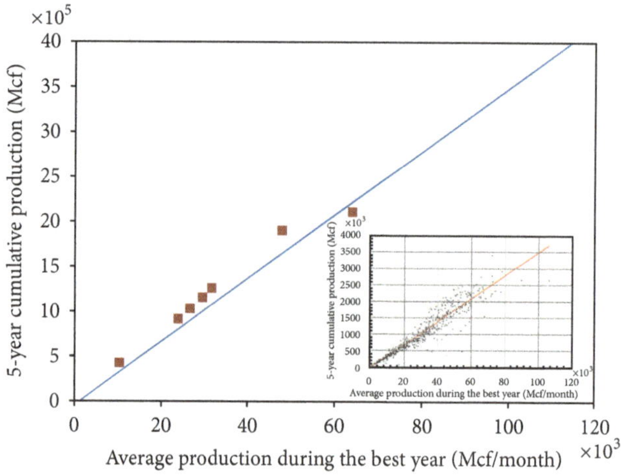

FIGURE 8: Correlation between short-term and long-term production [9], along with our simulation results.

fractures is also performed in all the cases after 4 years of production, except in cases where initial fractures are close to one another (50 ft). The total number of fractures for each case is 57 (57 initial fractures), 57 (29 initial fractures and 28 refractures), 29 (15 initial fractures and 14 refractures), 17 (9 initial fractures and 8 refractures), and 13 (7 initial fractures and 6 refractures). Cumulative gas production from refractured wells having higher initial values of fracture spacing is similar to that of closely spaced initial fractures if the ultimate fracture spacing remains the same (Figure 12). In other words, we expect the same performance in the two cases having the same value of total fracture length, including

TABLE 2: Well screening criteria.

Property	Perfect selection	Fair selection	Poor selection	Unit: field (SI)
Permeability	<0.0001 (9.8×10^{-20})	<0.0005 (4.9×10^{-19})	>0.0005 (4.9×10^{-19})	md (m^2)
Porosity	>0.06	>0.04	<0.04	fraction
Fracture conductivity	>1 (3×10^{-16})	<10 (3×10^{-15})	>10 (3×10^{-15})	md-ft (m^2-m)

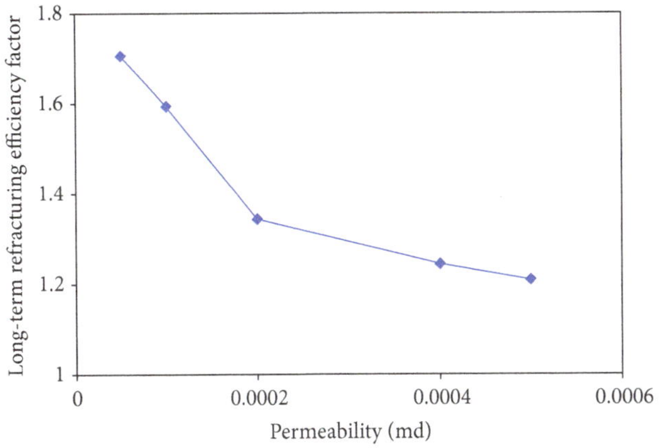

FIGURE 9: Long-term refracturing efficiency versus different permeabilities.

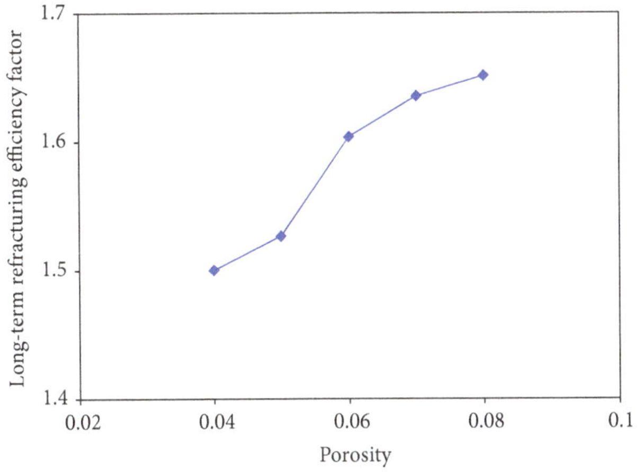

FIGURE 10: Long-term refracturing efficiency versus different porosities.

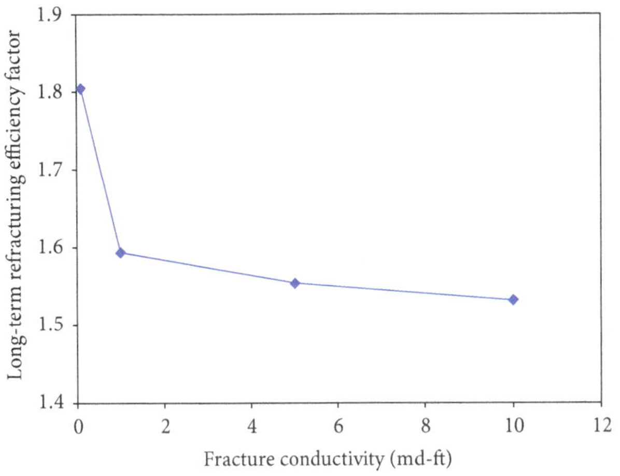

FIGURE 11: Long-term refracturing efficiency versus different fracture conductivities.

refractured lengths. Hence, refracturing at the proper time of production appears to be an attractive restimulation method for wells having large initial fracture spacing.

4.6. The Optimal Time to Refracture a Well. Finding the optimal time to refracture so as to maximize the performance of the refracturing job is crucial. On the basis of available field-production data of successful refracturing processes in Barnett Shale formations and the simulation results of different refracturing scenarios, we proposed an indicator for finding the proper time for refracturing.

To find the optimal time, we considered different times of refracturing. Cumulative gas production and gas flow rate for these different cases are obtained from simulation results (Figures 13 and 14, resp.). Because of the difficulty involved in such a task, however, we studied trends in detail and proposed an indicator for comparison between various production trends.

The *proposed indicator* suggests that if the decline rate of gas production falls below 10% to 15%, the refracturing process should be applied. The cumulative gas production result of refracturing after 2 or 3 years of production is higher than that of a well refractured after 4 years of production (Figure 13). However, the incline rate of the cumulative gas production result of refracturing after 2 or 3 years of production is still significant (more than 20%) (Figure 15). Hence, the decline rate of gas production is pronounced early in the stimulation process, but only until it reduces below 15%, at which point the initial stimulation effect has vanished and gas production levels off. Under these conditions, refracturing should be considered as a way to increase gas production from the well. Incline-decline rates can be defined as

$$\text{Decline rate of gas production} = \frac{q_r^{n-1} - q_r^n}{q_r^{n-1}},$$

$$\text{Incline rate of cumulative gas production} = \frac{Q^{n+1} - Q^n}{Q^n}, \quad (11)$$

where n is refracturing time in years, q_r is gas flow rate, and Q is cumulative gas production. Both decline rate of gas production and incline rate of cumulative gas production during the first 6 years of a project's life reduce over that time and level off (Figure 15). According to the introduced

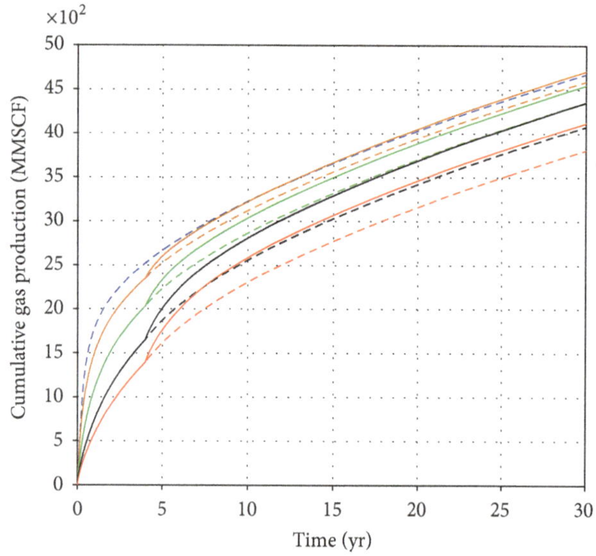

FIGURE 12: Cumulative gas production for different fracture-spacing values.

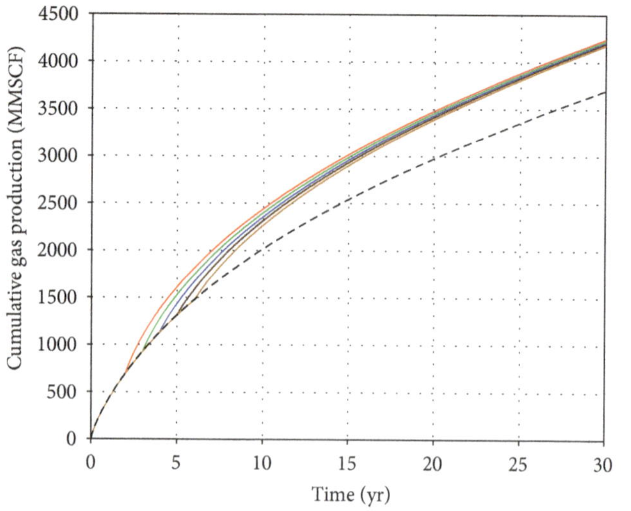

FIGURE 13: Cumulative gas production for different times of refracturing.

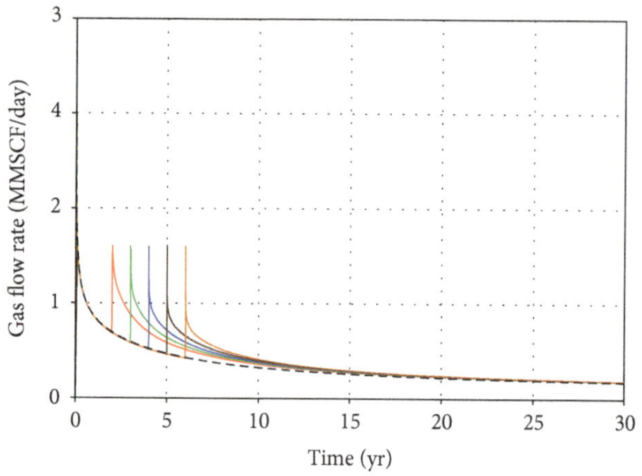

FIGURE 14: Gas flow rate for different times of refracturing.

indicator, the proper time for refracturing occurs after 4 years of production because the decline rate of gas production drops to less than 15% during that fourth year.

5. Conclusions

We used numerical simulation to study refracturing of wells in shale formations by selecting a well in the Barnett Shale formation and simulating gas production from this well. We used the verified model to perform sensitivity studies on formation properties and initial-fracture conductivity on gas-production performance. We found that refracturing is an effective restimulation technique if it is performed for a proper candidate well and at the optimal time. Numerical-simulation results indicate significant improvements on the order of 30% in estimated ultimate recovery (EUR) after refracturing. Some specific conclusions are as follows.

(1) Our simulation model results suggest that refracturing is more effective in low-permeability reservoirs. Refracturing enhances gas production by forming a more complex fracture network and producing larger stimulated reservoir volume (SRV) in these low-permeability reservoirs. Larger porosities result in higher gas production owing to an increase in the amount of gas in place.

(2) Results show that higher values of initial fracture conductivity eliminate the need for refracturing owing to the presence of conductive paths between wellbore and reservoir.

(3) Refracturing performance was compared with that of closely spaced initial fractures. Results show that cumulative gas production correlates with total fracture length, including refracture lengths.

(4) We developed a screening method for existing wells in a formation that was based on reservoir characteristics and initial hydraulic-fracture properties. The screening method provides perfect, fair, and poor selection categories that are based on ranges of the studied variables.

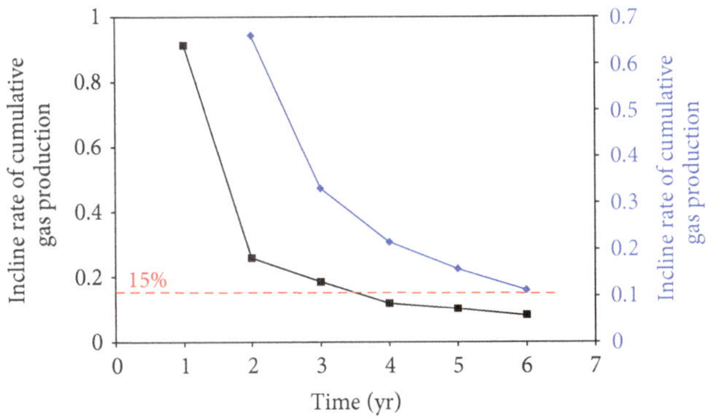

FIGURE 15: Decline rate of gas flow rate and incline rate of cumulative gas production versus time.

(5) Finding the optimal time for refracturing so as to maximize its performance is crucial. On the basis of detailed study of gas flow rate and cumulative gas production in the first 6 years of a project's life, we suggest performing refracturing at the point in the life of the project when decline of the gas flow rate falls below 10% to 15%.

Nomenclature

v: Flow velocity, m/s
μ: Viscosity, Pa·s
p: Pressure, Pa
K: Phase permeability, m^2
C: Concentration, m^3/m^3
D: Diffusivity, m^2/s
Z: Gas-compressibility factor
ϕ: Porosity
B: Compressibility factor, Pa^{-1}
R: Gas constant, $Pa{\cdot}m^3/(mol{\cdot}K)$
M: Molar mass, kg/mol
T: Temperature, K
q: Mass flow rate per unit matrix-block volume, $kg/(m^3{\cdot}s)$
q_r: Gas flow rate, MMSCF/D
Q: Cumulative gas production, MMSCF
s: Phase saturation.

Superscripts

m: Matrix
f: Fracture
mf: Exchange between matrix and fracture.

Subscripts

g: Gas
w: Water.

Acknowledgments

The authors would like to thank Computer Modeling Group Ltd. for use of the CMG-IMEX software. They would also like to express their gratitude for financial support from the Hilcorp Energy Company and NanoGeosciences lab at the Bureau of Economic Geology (UT-Austin) and to thank Ms. L. Dieterich for editing the paper.

References

[1] F. Javadpour, D. Fisher, and M. Unsworth, "Nanoscale gas flow in shale sediments," *Journal of Canadian Petroleum Technology*, vol. 46, no. 10, pp. 55–561, 2007.

[2] F. Javadpour, "Nanopores and apparent permeability of gas flow in mudrocks (shales and siltstone)," *Journal of Canadian Petroleum Technology*, vol. 48, no. 8, pp. 16–21, 2009.

[3] N. R. Warpinski, M. J. Mayerhofer, M. C. Vincent, C. L. Cipolla, and E. R. Lolon, "Stimulating unconventional reservoirs: maximizing network growth while optimizing fracture conductivity," *Journal of Canadian Petroleum Technology*, vol. 48, no. 10, pp. 39–51, 2009.

[4] M. Y. Soliman and C. S. Kabir, "Testing unconventional formations," *Journal of Petroleum Science and Engineering*, vol. 92-93, pp. 102–1109, 2012.

[5] S. R. Reeves, "Restimulation technology for tight gas sand wells," in *Proceedings of the SPE Annual Technical Conference*, SPE-56482, Houston, Tex, USA, 1999.

[6] D. G. Hill and S. R. Reeves, "Restimulation research to target low cost, incremental gas reserves," *Gas Tips*, vol. 4, no. 3, 1998.

[7] T. Lantz, "Refracture treatments proving successful in horizontal Bakken wells," *SPE Production & Operations*, vol. 23, no. 3, pp. 373–378, 2008.

[8] L. P. Moore and H. Ramakrishnan, "Restimulation: candidate selection methodologies and treatment optimization," in *Proceedings of the SPE Annual Technical Conference*, SPE-102681, San Antonio, Tex, USA, 2006.

[9] S. Sinha and H. Ramakrishnan, "A novel screening method for selection of horizontal refracturing candidates in shale gas reservoirs," in *Proceedings of the North American Unconventional Gas Conference*, SPE-144032, The Woodlands, Tex, USA, 2011.

[10] N. P. Roussel and M. M. Sharma, "Selecting candidate wells for refracturing using production data," in *Proocedings of the SPE Annual Technical Conference*, SPE-146103, Denver, Colo, USA, 2011.

[11] A. A. Ketter, J. L. Daniels, J. R. Heinze, and G. Waters, "A field study optimizing completion strategies for fracture initiation in barnett shale horizontal wells," in *Proocedings of the SPE Annual Technical Conference and Exhibition (ATCE '06)*, Focus on the Future, SPE-103232, pp. 4531–4536, San Antonio, Tex, USA, September 2006.

[12] E. Siebrits, J. L. Elbel, R. S. Hoover et al., "Refracture reorientation enhances gas production in Barnett Shale tight gas wells," in *Proceedings of the SPE Annual Technical Conference*, SPE-63030, Dallas, Tex, USA, October 2000.

[13] M. C. Vincent, "Refracs—why do they work, and why do they fail in 1000 published field studies?" in *Proceedings of the SPE Annual Technical Conference*, SPE-134330, Florence, Italy, 2010.

[14] M. C. Vincent, "Restimulation of unconventional reservoirs: when are refracs beneficial?" in *Proceedings of the Canadian*

Unconventional Resources and International Petroleum Conference, SPE-136757, pp. 407–443, Alberta, Canada, October 2010.

[15] M. K. Fisher, J. R. Heinze, C. D. Harris, B. M. Davidson, C. A. Wright, and K. P. Dunn, "Optimizing horizontal completion techniques in the barnett shale using microseismic fracture mapping," in *Proceedings of the SPE Annual Technical Conference*, SPE-90051, Houston, Tex, USA, September 2004.

[16] C. L. Cipolla, E. P. Lolon, M. J. Mayerhofer, and N. R. Warpinski, "Fracture design considerations in horizontal wells drilled in unconventional gas reservoirs," in *Proceedings of the SPE Hydraulic Fracturing Technology Conference*, SPE-119366, pp. 366–375, The Woodlands, Tex, USA, January 2009.

[17] C. L. Cipolla, E. P. Lolon, and B. Dzubin, "Evaluating stimulation effectiveness in unconventional gas reservoirs," in *Proceedings of the SPE Annual Technical Conference and Exhibition 2009, ATCE 2009*, SPE-124843, pp. 3397–3417, New Orleans, La, USA, October 2009.

[18] M. J. Mayerhofer, E. P. Lolon, N. R. Warpinski et al., "What is stimulated reservoir volume?" *SPE Production & Operations*, vol. 25, no. 1, pp. 89–898, 2006.

[19] C. L. Cipolla, E. P. Lolon, J. C. Erdle, and B. Rubin, "Reservoir modeling in shale-gas reservoirs," *SPE Reservoir Evaluation & Engineering*, vol. 13, no. 4, pp. 638–6653, 2010.

[20] C. L. Cipolla, N. R. Warpinski, M. J. Mayerhofer, E. P. Lolon, and M. C. Vincent, "The relationship between fracture complexity, reservoir properties, and fracture treatment design," in *Proceedings of the SPE Annual Technical Conference and Exhibition*, SPE-115769, Denver, Colo, USA, 2008.

[21] T. Vogel, H. H. Gerke, R. Zhang, and M. T. Van Genuchten, "Modeling flow and transport in a two-dimensional dual-permeability system with spatially variable hydraulic properties," *Journal of Hydrology*, vol. 238, no. 1-2, pp. 78–89, 2000.

[22] CMG, *IMEX User's Guide*, Computer Modeling Group, 2011.

[23] D. I. Potapenko, S. K. Tinkham, B. Lecerf et al., "Barnett Shale refracture stimulations using a novel diversion technique," in *Proceedings of the SPE Hydraulic Fracture Technology Conference*, SPE-119636, The Woodlands, Tex, USA, 2009.

[24] W. Yu and K. Sepehrnoori, "Simulation of gas desorption and geomechanics effects for unconventional gas reservoirs," in *Proceedings of the SPE Western Regional Meeting*, SPE-165377, Monterey, Calif, USA, April 2013.

18

Review of Sand Production Prediction Models

Hossein Rahmati,[1] Mahshid Jafarpour,[1] Saman Azadbakht,[1] Alireza Nouri,[1] Hans Vaziri,[2] Dave Chan,[1] and Yuxing Xiao[2]

[1] *Department of Civil and Environmental Engineering, University of Alberta, Edmonton, AB, Canada T6G 2W2*
[2] *BP America Inc., Houston, TX 77079, USA*

Correspondence should be addressed to Hossein Rahmati; hrahmati@gmail.com

Academic Editor: Jorge Ancheyta

Sand production in oil and gas wells can occur if fluid flow exceeds a certain threshold governed by factors such as consistency of the reservoir rock, stress state and the type of completion used around the well. The amount of solids can be less than a few grams per cubic meter of reservoir fluid, posing only minor problems, or a substantial amount over a short period of time, resulting in erosion and in some cases filling and blocking of the wellbore. This paper provides a review of selected approaches and models that have been developed for sanding prediction. Most of these models are based on the continuum assumption, while a few have recently been developed based on discrete element model. Some models are only capable of assessing the conditions that lead to the onset of sanding, while others are capable of making volumetric predictions. Some models use analytical formulae, particularly those for estimating the onset of sanding while others use numerical models, particularly in calculating sanding rate. Although major improvements have been achieved in the past decade, sanding tools are still unable to predict the sand mass and the rate of sanding for all field problems in a reliable form.

1. Introduction

A significant proportion of the world oil and gas reserves is contained in weakly consolidated sandstone reservoirs and hence is prone to sand production. Material degradation is a key process leading to sanding. Drilling operations, cyclic effects of shut-in and start-up, operational conditions, reservoir pressure depletion, and strength-weakening effect of water may gradually lead to sandstone degradation around the perforations and boreholes. High pressure gradient due to fluid flow also facilitates the detachment of sand particles. In addition, fluid flow is responsible for the transport and production of cohesionless sand particles or detached sand clumps to the wellbore.

Sand production is the cause of many problems in the oil industry and it affects the completion adversely. These problems include, but are not limited to, plugging the perforations or production liner, wellbore instability, failure of sand control completions [1], collapse of some sections of a horizontal well in unconsolidated formations, environmental effects, additional cost of remedial and clean-up operations, and pipelines and surface facilities erosion, in case the sand gets out of the well. The mechanical prevention of sanding is costly and leads to low productivity/injectivity. Therefore, there is always a cost benefit if sand management and modeling is implemented early before well completions.

Sand production takes place if the material around the cavity is disaggregated and additionally, the operation of the well generates sufficient seepage force to remove the sand grains. It is a complex phenomenon which depends on various parameters such as the stress distribution around the wellbore, the properties of the rock and fluids in the reservoir, and the completion type. Therefore, capturing all the factors and mechanisms in the numerical models is difficult and the models have many limitations.

Due to the importance of the sand production prediction in the oil industry, considerable efforts have been made in developing robust numerical methods for sand production prediction. In this paper, the techniques, advances, problems, and the likely future development in numerical models for the prediction of sand production are presented.

2. Common Techniques Used in Sand Management Decisions

A number of approaches have been developed to predict or help to understand the sand production problem using physical model testing, analytical and empirical relationships, and numerical models. Routine laboratory tests can only predict the onset of sand production [2]. More sophisticated physical model could predict volumetric sand production [3]. They are also time-consuming and expensive. In addition, because of the small sizes of the laboratory setup, the results are usually influenced by boundary effects. Analytical models are fast and easy to use but they are only suitable to predict the onset of sand production and they have limitations. Most of them are only valid for capturing a single mechanism of sanding and under simplified geometrical and boundary conditions which are not usually the case in complicated field-scale problems. Numerical models are by far the most powerful tools for predicting sand production. They can be combined with analytical correlations to obtain the results more efficiently. Experimental results are also utilized to calibrate or validate a numerical model. Yet, numerical models have their own limitations and extensive efforts have been made to improve them.

Modeling of sand production requires coupling of two mechanisms. The first mechanism is mechanical instability and degradation around the wellbore and the second one is hydromechanical instability due to flow-induced pressure gradient on degraded material surrounding the cavity (e.g., perforation and openhole). In general, numerical methods in the mechanical modeling are categorized under continuum and discontinuum approaches.

In the continuum approach, matters are treated as continuous in deriving the governing differential equations. The assumption of continuity implies that the material cannot be separated or broken into smaller pieces. In the case when there is a discontinuity, the magnitudes of deformation along or across the discontinuity are about the same as the rest of the continuum [4].

Discrete element method (DEM) is a useful tool to simulate sand production especially to understand the mechanism of sanding. However, it cannot be used for large-scale problems because of large computational time required. The calibration of the model is also difficult and involves several uncertainties as it is not possible to create a model with the exact particle arrangement as the real material. Further, methodologies to directly measure sandstone micro properties have not been developed yet. Presently, the micro-properties are obtained in calibrating against the actual sand behavior [5, 6]. Therefore, continuum-based models are more popular especially for field-scale problems. However, there can be advanced models which couple continuum and discontinuum models together to take advantage of both methods. These are known as hybrid models and are discussed later.

A comprehensive sand management may require the use of some or all of the above mentioned techniques.

2.1. Numerical Models Based on Continuum Approach. Developments of continuum models are based on various assumptions, constitutive laws, sanding criteria, and numerical procedures with different levels of complexity to capture the physical behavior of the material.

Table 1 summarizes the majority of continuum-based sanding models. Initially many researchers first calculated the onset of sand production or the initiation of mechanical failure until Vardoulakis et al. [7] proposed the basic theory for hydrodynamic erosion of sandstone which is based on filtration theory without solving the equilibrium equation. Later Papamichos and Stavropoulou [8] combined the evolution of localized deformation with hydrodynamic erosion. This was the beginning for many researches that adopt full strength hardening/softening behavior of sandstone in their models [9–17]. The results are highly mesh-dependent for strain softening material and hence a regularization method is necessary. Regularization methods include an internal length, which has been related to the grain size, in the formulation. Papanastasiou and Vardoulakis [18] applied Cosserat microstructure method [18, 19] for cavity failure around boreholes. Zervos et al. [20] considered Gradient elastoplasticity [21] for thick-walled cylinders. Fracture energy regularization technique [22] was also applied by Nouri et al. [16], Wang et al. [23], and Rahmati et al. [24] in sand production modeling.

2.1.1. Constitutive Models Used in Continuum Approaches. Rock failure and/or degradation in commonly accepted as prerequisite for sanding. Failure of geomaterials is usually associated with formation of shear bands which are narrow zones of concentrated plastic deformation. This phenomenon, which is known as "deformation localization" or simply "localization," is one of the key parameters in sanding prediction models. Further details about this concept can be found in Sulem et al. [40], Nouri et al. [16], and Jafarpour et al. [41].

In the simplest form, an elastic brittle failure model has been implemented in sand production models by Nordgren [42], Coates and Denoo [43], Risnes et al. [44], and Edwards et al. [45]. Most of these models predict the onset of sand production by considering failure of the sand matrix. Elastic brittle failure rock behavior leads to excessive stress concentrations at the borehole wall and therefore results in overestimation of initial sand production conditions. This model may be used as a quick estimate of sanding onset in relation to production parameters and in situ stresses. The predictions of the elastic models are cumbersome unless they are combined with apparent strength models.

An elastoplastic material model can simulate the material behavior more realistically. However, it requires more computational effort and more input data. Many researchers have implemented elastoplastic models in sand production analysis (e.g., Morita et al. [25]; Antheunis et al. [46];

Table 1: Summary of the numerical works on sand production in the literature (continuum approach).

Model	Geometry and solution method	Yield	Hardening/ softening	Coupling	Phases	Sanding criteria	Permeability alteration in the sanded zone	Other features
Morita et al. [25], Morita et al. [26], Burton et al. [27]	3D; finite element (FE) (SAND3D software)	Kinematic model with a cap	yes (flow friction)	Iteratively coupled	(1) Fluid (2) Solid	Maximum plastic strain limit	No change	(1) Only the onset of sanding (2) Burton applied it for gas reservoirs
(Vardoulakis et al. [7])	1D; finite difference (FD)	N/A	N/A	Fluid flow and erosion are coupled	(1) Fluid (2) Fluidized solid	Erosion	$k/k_0 = \varphi^3/(1-\varphi^2)$ (Carman-Kozeny)	(1) Only hydromechanical effects; equilibrium eqn. is not solved (2) Sand deposition is neglected in modeling
Skjaerstein et al. [28]	1D; FD	N/A	N/A	Fluid flow and erosion are coupled.	(1) Fluid (2) Fluidized solid	Erosion	$1/k = B\varphi + A$ (found experimentally)	Forcheimer's law was used instead of Darcy's law to account for turbulence
Papamichos and Malmanger [29]	2D Axial symmetry and 3D; FE; Newoton-Paphson (NR) iterations	MC	Yes	Fully coupled	(1) Fluid (2) Solid (3) Fluidized solid	Erosion	$k/k_0 = \varphi^3/\overline{\rho}(1-\varphi^2)$ $\overline{\rho} = \rho_f + c(\rho_s - \rho_f)$	Tension cut-off: function of both plastic strain and porosity; by the factor $(1-\varphi)/(1-\varphi_i)$
Papamichos et al. [9]	2D Axial symmetry; FEM; NR iterations	MC	Yes	Fully coupled	(1) fluid (2) solid	Erosion	Carman-Kozeny	Tension cut-off and Young mod changed by the factor $(1-\varphi)/(1-\varphi_i)$
Yi [30]	2D axial symmetry; FD	MC	No		(1) Fluid (2) Solid (3) Fluidized sand	Erosion	$k/k_0 = \gamma(\varphi^3(1-\varphi_0^2)/\varphi_0^3(1-\varphi^2)) \times (1/(1+\beta m_d/\rho_s))$	Sand deposition in porous media is considered
Vaziri et al. [31]	2D axial symmetry; FE; fully-implicit	Modified MC with tensile failure	Yes	Fully coupled	(1) Fluid (2) Solid	Tensile failure	$k/k_0 = \exp[-2.88 \times 10^{-3}(\sigma_i' - \sigma_0')]$	Zero stiffness, compressibility and high k for the liquefied tensile zone

TABLE 1: Continued.

Model	Geometry and solution method	Yield	Hardening/ softening	Coupling	Phases	Sanding criteria	Permeability alteration in the sanded zone	Other features
Wang and Xue [32]	FE; Crank-Nicholson for time integration	MC	No	Fully coupled	(1) Oil (2) Water (3) Solid (4) Fluidized sand	Erosion	Kozeny-Poiseuille law and Carman-Kozeny	
Chin and Ramos [33]	2D and 3D; FE; explicit; NR iteration	Drucker-Prager	No	Coupled	(1) Fluid (2) Solid	Shear dilation	Power law with porosity (exponent = 5.6)	Porosity is changed as a function of plastic volumetric strain
Nouri et al. [34]	2D plane strain; FD	Bilinear MC	Yes	Fully coupled	(1) Fluid (2) Solid	Tensile failure or shear-failed element falls in tension	0	Capillary is considered as a residual cohesion
Wang et al. [35]	2D; FE	Drucker-Prager	No	Fully coupled	(1) Solid (2) Fluidized solid (3) Oil (4) Water (5) Gas		$k = k_0 \exp\left[A\left((\varphi - \varphi_0)/(\varphi_{\max} - \varphi_0)\right)\right]$	Cohesion and friction drop linearly with porosity
Servant et al. [36]	2D; FE	MC	No	Iteratively coupled	(1) Fluid (2) Solid (3) Fluidized solid	Yielding	Not mentioned	Failed material is treated as a Poiseuille fluid. Constant viscosity for the slurry
Detournay et al. [11]	2D; FD	MC	Yes	Iteratively coupled	(1) Fluid (2) Solid	Erosion	$k/k_0 = \varphi^3(1-\varphi_0^2)/\varphi_0^3\left(1-\varphi^2\right)$	Bulk mod. Change by φ_0/φ
Nouri et al. [12]	2D axial symmetry; FD	Bilinear MC	Yes	Iteratively coupled	(1) Fluid (2) Solid	Tensile failure or a shear-failed element falls in tension	$k = C\varphi^3/\gamma_f(1-\varphi^2)$	
Nouri et al. [37]	FE; NR iterations	MC	Yes	Fully coupled	(1) Fluid (2) Solid	Tension	0	Adaptive mesh is used
Nouri et al. [38], Vaziri et al. [14]	2D plane strain and axial symmetry; FD	Bilinear MC	Yes	Iteratively coupled	(1) Fluid (2) Solid	Complete degradation and tensile mean effective stress	High permeability is assigned to infill materials (elements that satisfy sanding criteria)	WH pulsing is included in the model, and stiffness changes with sanding (Vaziri et al. [14])

Table 1: Continued.

Model	Geometry and solution method	Yield	Hardening/ softening	Coupling	Phases	Sanding criteria	Permeability alteration in the sanded zone	Other features
Detournay [15]	2D plane strain; FD	Double-yield	Yes	Iteratively coupled	(1) Fluid (2) Solid	Same as (Detournay et al. [11])	Not mentioned	Only cap yielding is considered
Nouri et al. [16]	All the features are the same as (Nouri et al. [38]) but fracture energy regularization is applied to remove mesh dependency and the calibration of hardening/softening behavior is discussed thoroughly							
Kim et al. [17]	3D; FD	MC	Yes	Not mentioned	(1) Fluid (2) Solid	Calculated from force balance on the element	N/A (element removal)	No calibration parameter to match the experiment is used and yet a good match is observed
Wang et al. [23]	2D; FE (ELFEN software)	Soft rock model	Yes	Fully coupled	(1) Fluid (2) Solid	Displacement (the materials passing the wellface are removed)	K changes as a function of dilation or volumetric strain: $k = \exp(A\varepsilon_v)$	(1) Adaptive mesh (2) Arbitrary Largrangian Eulerian formulation (3) Fracture Energy Regularization
Azadbakht et al. [39]	2D, FE for fluid and FD for solid	Bilinear MC	Yes	Sequentially coupled	(1) Fluid (2) Solid	Erosion	K changes linearly with porosity	Water effect is considered by reducing the cohesion Bulk moduli change linearly with porosity After critical porosity, residual properties are assigned to the element as infill material

Peden et al. [47]; Papamichos and Vardoulakis [9]; Detournay et al. [11]; Wan and Wang [48]; Servant et al. [36]; Wang et al. [35]; Wan et al. [49]; Detournay [15]; Wan and Wang [50]; Vaziri et al. [14]; Nouri et al. [13]; Nouri et al. [51]; Nouri et al. [34]; Rahmati et al. [24]; Azadbakht et al. [39]). These works will be discussed in more detail in the next section.

Implementation of classical elastoplastic continuum models and the abovementioned failure criteria are inefficient in modeling the localization phenomena, which are of discontinuum nature. Therefore, employment of such models in simulation of material degradation leads to inability in recovering size and dependency of the results on numerical mesh design. This problem would be addressed by pursuing two distinct solutions: using discontinuum models that we discuss in the next sections and enriching the material model by an appropriate regularization strategy. Papanastasiou and Zervos [52] used Cosserat continuum and gradient elastoplasticity theories to predict the localization phenomenon. These models also proved efficient in capturing the size effect often observed in laboratory tests performed on thick-walled cylinder samples [53].

Vardoulakis et al. [54] used bifurcation theory to predict the failure mode and the critical value of the stress at infinity. They showed that failure depends on stress path and boundary conditions. Tronvoll et al. [55] performed finite element modeling using bifurcation theory to solve the axisymmetric problem for the trivial solution and checked the condition for non-axisymmetric bifurcation modes. Van Den Hoek et al. [56] and Papanastasiou and Vardoulakis [57] used bifurcation theory in a Cosserat continuum which takes into account the material microstructure. In a Cosserat continuum individual points possess, in addition to their translational degrees of freedom, independent rotational degrees of freedom, which result in an internal characteristic length in the classical elastoplastic constitutive laws.

The models based on bifurcation theory require special numerical techniques to capture size effects, localization mechanisms, and so forth, which make it computationally demanding and hard to apply in solving field problems.

As shown in Table 1, the most basic improvement in sand models is the yield function. Mohr-Coulomb (MC) model is the most common model being used. Vaziri et al. [10] modified the MC model using a bilinear yield function to differentiate sand behavior under low and high confining stresses. The model was later used by Nouri et al. [12], Nouri et al. [38], and Vaziri et al. [14]. The theory is based on Sulem et al. [40] and is described more thoroughly by Nouri et al. [16] and Jafarpour et al. [41]. Haimson and Lee [58] showed that the slit mode of cavity development is related to the formation of compaction bands, which are thin bands of localized compressive deformation in high porosity rocks. Therefore, Detournay [15] extended Detournay et al. [11] work by accounting for the compaction mode of failure. The model used Double Yield cap constitutive law to capture compaction bands. The simulation results show that the slit mechanism develops as a combination of volumetric collapse and transport of failed material by seepage forces. It was found out that pore collapse is the responsible mechanism for slit mode of cavity failure associated with sand production. Although it was considered as one of the main sanding mechanisms, many researchers have avoided the incorporation of compression yielding to simplify their models.

It appears that the optimum constitutive models are those that are based on the critical state theory and use a combined isotropic and kinematic hardening model which allows capturing all kinds of failure (shear, tensile, and compressional). In addition, it would be ideal to capture the effect of hysteresis to simulate fatigue in cyclic start-up and shut-in conditions in the wellbore.

2.1.2. Sanding Criteria Used in Continuum Approaches. Several mechanisms are recognized as responsible for sand production. They are mainly based on shear and tensile failure, critical pressure gradient, critical drawdown pressure, critical plastic strain, and erosion criteria. Table 2 summarizes the main sanding criteria used in sand models.

When the effective minimum principal stress is equal to the tensile strength of the formation rock, tensile failure may occur. This mode of failure is responsible for rock degradation. It can occur as a standalone degradation mechanism or in combination with shear failure [22]. Tensile mode is also believed to be the responsible mechanism for particle removal after the degradation during production. In this case, tensile failure relates to seepage forces on the degraded sand particles.

Shear failure may occur when some planes in the vicinity of the wellbore are subjected to higher stress than they can sustain. This is the dominant mechanism in cemented sands and when it is combined with tensile cracks and high compressive stress, it can lead to buckling at the wellbore wall [18].

When effective hydrostatic stresses are increased due to the reservoir pressure depletion, pore collapse may occur which can lead to sand production. Plastic volumetric compression can be captured using a compression yield cap in the failure envelope. This mainly occurs in high porosity sandstones.

Risnes and Bratli [59] proposed a tensile failure criterion for perforation tunnel inner shell collapse. Bratli and Risnes [60] and Weingarten and Perkins [61] proposed sanding criteria in terms of pressure gradient. Morita et al. [25] proposed a sand production model that can be triggered by either shear failure or tensile failure.

Dynamic seepage drag forces lead to internal and surface erosion that result in releasing and transporting sand particles. Internal erosion may be related to micromechanical impacts imposed on solid skeleton by gas bubbles, water droplets, and so forth. Surface erosion may be related to parallel flow scouring the surface and normal flow over the surface [28]. Numerous authors studied surface erosion criterion in sand production modeling. Vardoulakis et al. [7] proposed a model that was based on mass balance of the produced solids and radial flowing flow conditions, the constitutive law for particle erosion, and Darcy's law but neglecting skeleton deformation. Based on sand production experiments, Tronvoll et al. [62] showed that in addition to the radial flow, axial flow parallel to the perforation

Table 2: Sand criteria used in continuum-based sanding models.

Sanding criterion		Formulation	Further developments	Assumptions
	Vardoulakis et al. [7]	$d\varphi/dt = \dot{m}/\rho_s = \lambda(1-\varphi)c\|q_i\|$ And with sand deposition: $= \lambda(1-\varphi)(c - c^2/c_{\mathrm{cr}})\|q_i/q_0\|^n$	Sand deposition is also considered in some models [28]	It is assumed that the sand in place is fully degraded from the beginning and the production is only due to the hydrodynamic forces. Equilibrium eqn. for solid phase is often ignored To initiate the process a very small solid concentration is given as a boundary condition. The results are insensitive to this value as long as it is small
	Skjaerstein et al. [28]	$d\varphi/dt = \dot{m}/\rho_s = \lambda c(\varphi - \varphi^2/\varphi_{\mathrm{cr}})q_x$		
	Papamichos and Malmanger [29]	$\dot{m}/\rho_s = \lambda(1-\varphi)c\sqrt{q_iq_i}$ $\lambda = \lambda(\varepsilon_p)$ if $\varepsilon_p \geq \varepsilon_p^{\mathrm{peak}}$		The critical flow rate is ignored Softening is not simulated
	Yi [30]	$\partial\varphi/\partial t =$ $\eta\lambda(1-\varphi-(m_d/\rho_s))v^2 - \lambda_d v c_s$ $\dot{m}_d/\rho_s = \lambda_d v c_s + (1-\eta)(\dot{m}/\rho_s)$		
	Papamichos et al. [9]	$\varepsilon_p > \varepsilon_p^{\mathrm{cr}}$ $\dot{m}/\rho_s = \lambda(1-\varphi)\|q_i\|$ if $\varphi < \varphi_{cr}$, else pore collapse or sudden element removal. $\lambda = \lambda(\varepsilon_p)$	Later models defined $\varepsilon_p^{\mathrm{cr}}$ as a function of stresses (Wang and Xue [32]). Critical plastic strain $\varepsilon_p^{\mathrm{cr}}$ was later linked to the state of zero cohesion (Azadbakht et al. [39])	Porosity is assumed to increase until it reaches unity. Usually, it is assumed that eroded elements collapse once their porosities reach a critical level less than unity (say 0.5)
(1) Erosion criterion	Chin and Ramos [33]	$d\varphi/dt = (1-\varphi)\nabla\cdot v_s$		
	Wang [32, 35]	$d\varphi/dt = \dot{m}/\rho_s = \lambda(1-\varphi)c\sqrt{q_iq_i}$	Wang et al. [35] assumed $\lambda = \lambda(\varepsilon_p)$	The resisting effect of cohesion was ignored resulting in conservative sanding assessment Softening is not considered
	Servant et al. [36]	After failure, new material properties are assigned to the failed material (such as zero cohesion). The infill is assumed to behave as a Poiseuille fluid with a viscosity depending on the sand concentration		
	Detournay et al. [11], Detournay [15]	$q > q_{\mathrm{cr}}$ $q_{\mathrm{cr}} = q_{\mathrm{cr}}(k, \varphi, R_p, \varnothing, C)$ if $\varphi < \varphi_{\mathrm{cr}}$, $\partial\varphi/\partial t = \lambda(1-\varphi)(q - q_{\mathrm{cr}})(S_L/V_L)$ else element removal		
	Azadbakht et al. [39]	$\varepsilon_p \geq \varepsilon_p^{\mathrm{cr}}$ and $q > q_{\mathrm{cr}}$ if $\varphi < \varphi_{\mathrm{cr}}$ $\dot{m}/\rho_s = \lambda(1-\varphi)(q_i - q_{\mathrm{cr}})$ $\lambda = \lambda(\varepsilon_p)$ $q_{\mathrm{cr}} = q_{\mathrm{cr}}(k, \varphi, R_p, \varnothing)$ else pore collapse or element removal	Some models consider sand production from elements whose cohesion is already reduced but not equal to zero yet (Detournay et al. [11]): $q_{\mathrm{cr}} = q_{\mathrm{cr}}(k, \varphi, R_p, C, \varnothing)$	

TABLE 2: Continued.

Sanding criterion		Formulation	Further developments	Assumptions
(2) Tensile criterion	Vaziri et al. [10], Nouri et al. [12], Nouri et al. [13], Vaziri et al. [14]	Complete degradation ($c = 0$) and tensile failure	Tensile failure has been replaced by either tensile effective stress or tensile mean effective stress, assuming cohesionless sand has no tensile strength A more comprehensive one is: shear-failed element with zero cohesion falls in tension or sand fails in tension (Nouri et al. [12])	The element that satisfies sanding criteria is removed from the mesh (assuming cavity can grow adjacent to the wellbore)
(3) Force balance on elements	Kim et al. [17]	$A_r p \|_{r=L} - A_r \|_{r=0} p_w > A_r\sigma_r\|_{r=L}^{r=0} - A_\theta\sigma_{\theta r}\|_{\theta=0}^{\theta=L} - A_z\sigma_{zr}\|_{z=0}^{z=L} + \mu_f(A_\theta\sigma_\theta\|_{\theta=0} + A_\theta\sigma_\theta\|_{\theta=L} + A_z\sigma_z\|_{z=0} + A_z\sigma_z\|_{z=L})$ For illustration purposes the boundary is shown from 0 to L in all directions. This can be considered as an elaboration of tensile criterion considering the effect of friction coefficient		The element is removed suddenly without allowing the porosity to grow gradually

channels is important in sand production and it may result in surface erosion of the perforation channels. Consequently, Vardoulakis et al. [63] extended Vardoulakis et al. [7] work to account for axial flow conditions. In the governing equations, they included Brinkman's extension of Darcy's law, which accounts for a smooth transition between channel flow and Darcian flow. The results show that erosion progresses in time at the front of high transport concentration.

First Stavropoulou et al. [64] and later Papamichos et al. [9] advanced a purely hydromechanical model proposed by Vardoulakis et al. [7] by coupling the poromechanical behavior of the solid fluid system with the erosion behavior of the solids due to fluid flow. Papamichos et al. [65] extended their own work by using a porosity diffusion type law that results in a sand rate that decreases over time when the process of erosion zone enlargement takes place. The model is based on nonlinear elastoplasticity, nonlinear stress dependent elasticity, friction hardening and cohesion softening, and single-phase flow fully coupled with geomechanics.

Wang et al. [35] also performed a fully coupled single-phase flow analysis based on erosion mechanics using the FEM. They applied the model in 2D plain strain geometry for both open hole and perforated casing completion.

Based on laboratory experiments, Haimson [66] and Papamichos [67] observed the slit cavity evolution pattern during sand production. Subsequently Detournay et al. [11] proposed a model to predict the formation of slit channels based on a critical flow rate. They modified the erosion law used by Vardoulakis et al. [7] with the addition of a critical flow rate which is a function of the grain size. They also assumed that sand continuously is produced until a critical porosity is reached beyond which the material suddenly collapses. This process can be responsible for the periodic sand bursts observed in experiments. The model was applied to 2D plain strain geometry for a long wellbore or perforation using a finite difference software. The model can predict different erosion features such as surface spalling and small burst events.

Kim et al. [17] calculated sanding conditions and utilized a sanding criterion based on the force balance on each element and achieved a good match with experimental data reported by Nouri et al. [51]. In their criterion, the forces leading to sand production are hydrodynamic forces generated by the pore pressure gradient between element faces in the flow direction. The resistance forces are the forces that retain the elements in place and are generated by vertical and tangential stresses and the friction coefficient. The advantage of using this method is that no calibration parameter for sanding (such as sand production coefficient) is necessary. The friction coefficient is an empirical parameter which depends on the grain size and mineralogy.

However, the abovementioned coupled hydromechanical erosion models have been criticized due to the following conflicting assumptions. Material mass balance equations are based on rigid porous media while equilibrium equations are based on deformable porous media. Therefore in order to establish a proper coupled mechanical erosion model in a consistent manner, the porosity changes can be split into two parts: one related to volume changes as a result of erosion, and the other one due to deformation in the matrix subjected to stress changes [48].

To sum up, a realistic sanding model should ideally account for all failure mechanisms (shear, tensile, and compressional) and must also consider the effect of fluid flow. Therefore, a suitable sand erosion model consists of a combination of erosion criterion, tensile criterion, and compressional criterion. Considering the physics of sand production, erosion seems more suitable for weak rock where full decementation and degradation to small particles is more likely [68]. On the other hand strong rocks are more prone to localized failure that result in larger chunks of sandstone which are not easily eroded away. Lastly, compressional failure is more dominant in highly porous weak materials where void spaces collapse easily under high loading.

2.1.3. Phases Involved in Continuum Approaches. The models can be categorized into two groups based on the phases involved. In the first group, mass balance equation is solved for only fluid and solid phases while the second group recognizes fluidized solid as a phase and solves for solid concentration in the fluid. Fluidized solids are the particles in suspension that move with the fluid. Any other loose particle which is trapped inside the void space is seen as part of the solid phase. However, these models use a constant viscosity for the slurry. It is notable over time that researches tend to use the simpler approach (solid and fluid phases) and couple the equations with an erosion model. This is mainly because good agreement between the model and field observations were obtained when combined with a suitable sanding criterion.

Multiphase fluid flow may also affect both the onset of sand production and sand rate. Tronvoll et al. [69] and Vaziri et al. [10] observed water cut effects on the onset of sand production. Water inflow changes the relative permeability and capillary pressure. It also can dissolve cement bonds and weaken the strength of the porous media.

Wang et al. [48] presented an integrated modular approach to predict volumetric sand production and cavity growth under two-phase flow (oil and water) and 3D geometry. In this work, the effect of water on rock strength reduction is reflected in material properties such as cohesion. The results show that water contact has a significant impact on the sand rate.

Gas flow also may hasten the instability process in sand production. When gas comes out the oil phase due to pressure drop and flows towards the wellbore at high velocities, it applies additional drag forces on sand particles and increases sand production. Wan and Wang [49] studied the gas effects in 1D model using the finite element method. They assumed that eroded mass in erosion law is proportional to the total fluid flux, which is referred to the oil and gas flux. The effect of multiphase fluid flow was also considered by Wang and Xue [32] for oil-water phases and later by Wang et al. [35] for oil-water-gas phases.

The only available numerical works in the literature that incorporate the water contact effect on sanding were performed by adjusting the cohesion or lowering the rock strength [17, 39].

2.1.4. Treating the Sanded Elements. Different strategies have been used for dealing with those elements that satisfy sanding criteria. The first one is to remove such elements from the model assuming cavities and wormholes grow as a result of sanding. This seems to be a suitable approach in stronger rocks in which stable cavities can form. The other approach is to keep the element in place but alter the material properties to residual values. We believe that stable cavities cannot develop in weak sandstone. Rather, the space is occupied by infill material, or cohesionless sand particles. In this approach, sandstone properties are altered to those of degraded cohesionless sand. The property change should also be applied in erosion models as a function of increasing porosity in the eroding elements.

It is evident that the moduli, tension cut off and permeability vary with the production of sand and increase of the porosity. However, the correct method to apply these changes requires experimental data. Most researchers use arbitrary choice of permeability change with porosity or with volumetric strain or even with mean stress. Wang and Xue [32] used two permeability correlations and found that the permeability relationship plays an important role. There is also disagreement on whether permeability should increase or decrease. It is reported that for high permeable sand formations, the permeability of the disaggregated sand is much less than that of the intact sand. This is mainly attributed to the sand deposition and plugging of the pore space, which is not the case for less permeable sand [26].

The moduli of the elements will also vary with porosity to the values of loose cohesionless sand (about 6.9 MPa for bulk modulus and 4.14 MPa for shear modulus). These numbers are the lowest values reported for loose sand [70].

2.1.5. Model Design. Openhole completion is often treated with axisymmetrical models. Strictly speaking, this is correct only when horizontal stresses are equal. However, most of the times, if not always, principal horizontal stresses, σ_h and σ_H, are not equal. In such cases a plane strain model can be a suitable choice for 2D analysis. Plane strain may not always be an appropriate assumption as vertical deformation may not necessarily be negligible. Papanastasiou and Zervos [52] suggested that generalized plane strain could be an appropriate assumption in the modeling of vertical and inclined wellbores.

As sand production model is commonly used in cased and perforated (C&P) completions, it is important to consider the sand behavior under such geometrical and boundary conditions. For instance the frequency of shots, the length of perforations, and their orientations may lead to a more intense commingling of the failed zones and finally higher sanding rate. One solution is using 3D simulation but it is computationally demanding because very fine mesh is required around perforations. In addition, creating the hollow geometry in weak rocks may not be reasonable as in reality the perforation can collapse upon creation and be filled with infill degraded sand.

Some perforation simulations have used axisymmetric models in which the perforation is assumed to be ring-shaped rather than conical or cylindrical [39]. Such an assumption influences the pressure gradient around the perforations and also impacts the mechanical response. These models are also unable to capture the perforation direction effect, which can play a significant role in perforation stability and failure as demonstrated by Papanastasiou and Zervos [71]. They performed a 3D numerical simulation to study the effect of orientation on stability and failure of perforation. Based on results of this work, it is recommended to avoid perforating the wellbore parallel to the minimum horizontal stress direction as perforations in this direction suffer more compressive stress and hence more chance of failure and sand production. These models require calibration against field and/or physical model testing before application to real-world sanding problems.

2.1.6. Other Factors for Continuum-Based Numerical Models. Sand production is a moving-boundary problem. As sand is produced, a sanded zone is formed around the perforations. Adaptive meshing can be very useful in processes where the geometry or the boundary is changing. Yet, there are only two applications of adaptive meshing in the literature [23, 37].

The current models are unable of predicting the sand bridging and fines retention in the rock. Sand particles can aggregate at the perforation cavity and form a stable entity called *sand bridge* and act as a filter which may reduce further sand production as long as the flow rate remains constant. The theory for modeling sand deposition was proposed by Vardoulakis et al. [7] but it has been applied only in one model [30] using a similar but not exactly the same approach. Yi [30] considered a part of the degraded sand as to be deposited in the porous media. The difficult part about modeling sand deposition is the calibration of the critical porosity or the critical sand concentration after which sand deposition initiates.

An important issue about the current sand models is that almost all of them are applied in modeling production wells. A few researchers [14, 30] performed numerical studies to predict sanding in injector wells. Sanding mechanisms in injectors have not been investigated thoroughly and may be quite different such as the effects of water hammer (WH) waves. The observation in injection wells is often described as the cases with high sand production within a short duration. Few papers [72–75] tried to explain the sanding problems in injectors. It is stated that sand liquefaction due to WH pressure pulses is the most likely mechanism for massive sand production. Liquefaction is defined as the process by which saturated sand loses shear strength and stiffness in response to dynamic loading [76]. WH is a general term describing generation, propagation, and damping of pressure waves in pipes. It occurs due to sudden velocity changes such as quick shutting of the well [72]. Nevertheless, no work has been published which investigates liquefaction around the wellbore and the conditions leading to liquefaction. As a result, sand particles can flow easily like a liquid. As no investigation has been performed on liquefaction in sand production, it is difficult to confirm its role in massive sand production.

2.2. Numerical Models Based on Discontinuum Approach. Sand production is a continuous and dynamic process that occurs at the microscopic scale and the rock becomes a discontinuum in nature. As mentioned before, conventional continuum approaches cannot capture local discontinuous phenomena. Therefore, discontinuum approach is promising to simulate phenomena such as detachment of individual particles from the rock matrix.

Cundall [83] first introduced the Discrete Element Method (DEM). The method can be used to simulate the disintegration of granular media subjected to loading. Each particle of the granular media is considered as an individual entity with a geometric representation of its surface topology and a description of its physical state. Particle bonds are modeled with a spring-dashpot in the normal direction and a spring-dashpot-frictional slider in the tangential direction. In the DEM, the interaction of the particles is treated as a dynamic process and a state of equilibrium is reached whenever the internal forces are equal to the external forces. The contact forces and displacements of a stressed assembly of particles are found by tracing the movements of the individual particles [84].

Table 3 summarizes some of the discontinuum-based sanding models. At first, O'Connor et al. [77] introduced the application of DEM to model the mechanics of sand production during oil recovery. Using laser scanning and sieve testing they developed techniques to consistently represent particles with irregular geometries. They used particle bonding scheme to mimic the cement and cohesion due to capillary forces. The bond also incorporated spring stiffness and a nominal tensile breaking strength assuming a bond dimension proportional to the size of the particles it connects. They incorporated the fluid flow calculations by combining the continuity equation and Darcy's law using the finite element method. Darcy flow is formulated with a measure for effective permeability in the solid medium based on the porosity and average diameter of the solid particles. Their 2D model provides an understanding of the fundamental physics involved in sand production and the relative importance of various rock and fluid properties.

Jensen and Preece [78] explored the coupling of 2D DEM and finite element implementation of the 2D continuity equation for Darcy flow to assess the sanding potential. The basic particle shape used by the model was an n-sided polygon and only the tensile mode for bond failure was considered. They concluded that lower strength of the cohesive bonds increased the number of particles breaking free from the solid matrix.

Li et al. [79] used commercial DEM code PFC2D to simulate hollow cylinder tests with fluid flow to study sanding. PFC2D simulates an assembly of circular disks with the bonds inserted between them. The disks are assumed to be rigid but they can overlap. The bonds have normal and shear stiffness and strength. In the standard PFC2D code, a bond fails when the tensile or shear stress in the bond exceeds its strength. Bond breakage may be interpreted as microfailure in the real rock. The growth of such microfailures eventually leads to macroscopic failure of the rock. Li and Holt [80] showed that DEM model may not result in realistic macroscopic friction coefficients if only circular or spherical grain shapes are used.

Li et al. [79] improved the prediction of macroscopic friction coefficient by setting the bond strength so high that no bonds in the model would fail due to the stress in the bond. Instead, all bonds associated with a given disk break when the stresses inside the disk satisfy a failure criterion. They used a failure criterion composed of tensile failure, shear failure, and compressive failure. A simple approach was

Table 3: Summary of the numerical works on sand production in the literature (Discontinuum approach).

Model	Geometry and software	Particle shape	Failure criterion of the particle	Fluid flow analysis
O'Connor et al. [77]	2D	Irregular particles	Tensile failure	2D, FE, Darcy's law
Jensen and Preece [78]	2D, MIMES software	n-sided polygon	Tensile failure	2D, FE, Darcy's law
Li et al. [79]	2D, PFC2D software	Circular	(i) Tensile failure (ii) Shear failure (iii) Compressive failure	2D, Explicit FD, Darcy's law
Li and Holt [80]	2D, PFC2D software	Circular	(i) Tensile failure (ii) Shear failure	2D, fluid flow networks, Darcy's law
Cheung [81]	3D, PFC3D software	Spherical	(i) Tensile failure (ii) Shear failure	1D, Navier-Stokes equations
Zhou et al. [82]	3D, PFC3D software	Spherical	(i) Tensile failure (ii) Shear failure	3D, Navier-Stokes equations (CFD software)

used to calculate and couple the fluid flow. They found three typical failure patterns in the simulations similar to those observed in laboratory experiments. The slit-like breakout failure pattern was observed when the material is prone to localized compressive failure due to grain crushing. For those cases where the material was weak and tensile strength was low, uniform failure around the borehole was observed along with a rather uniform hole enlargement. In those cases with relatively competent rock properties, which were unlikely to fail in localized compaction, the failure pattern was observed to be in the form of dog-eared breakouts.

Several researchers (e.g., [80, 85]) have modeled fluid flow in 2D DEM codes by introducing fluid flow networks and simulating flow along the flow paths connecting the voids which are referred to as pipes. The fluid velocities that flow through the pipes and the pore pressures were computed based on Darcy's theory. The forces arising from the pore fluid were then calculated and applied to the particles in the DEM model. Li and Holt [80] implemented this type of fluid-solid coupling system in the PFC2D codes. While the geometrical limitations of using a 2D model to investigate breakout geometry are obvious, the fluid flow simulation through flow networks is computationally expensive.

The entire coupling techniques described above used Darcy's law to calculate fluid flux or pressure. Darcy's law has been derived from the Navier-Stokes equations via homogenization that is only valid for slow and viscous flow. The conditions may not be met around the wellbore where breakout forms [81]. Chan and Tipthavonnukul [86] proposed a method to couple continuum and discontinuum flow to simulate granular particles movement in a flowing fluid. The fluid flow is modeled using the 2D Navier-Stokes equations solved by a finite volume method. The coupling is achieved by detecting the presence of the solid in the flow domain and altering the flow resistance accordingly. The method is computationally expensive and therefore it is not applicable in large-scale problems.

In 2D DEM models, only two force components and one moment component are considered. However three force components and three moment components exist in a 3D DEM model. Since sand production is a three-dimensional problem with complex geometry, where cement bonds and arching of particles are important, there is a need to model this problem using 3D DEM. The 2D DEM models overestimate the effect of fluid flow on the integrity of the assembly of particles as the resistance to particle dislodgment due to contact forces and bonds normal to the fluid flow is neglected in 2D models. In addition, 2D models cannot represent three-dimensional pore flow networks.

Cheung [81] used a coupled fluid-solid 3D DEM model for a perforation test simulation to study the sand production problem. They used the Navier-stokes equations assuming radial flow. Although this scheme is computationally inexpensive, it has two major problems. First, by assuming radial flow, it is not suitable for investigating the impact of the flow at the tip of the perforation where the fluid flow is in all directions. Second, the fluid flow scheme does not consider the presence of the cement between particles. The magnitude of the radial pore fluid velocity in each fluid cell is calculated, considering only the presence of the particles. The presence of cement can highly affect the rock conductivity and therefore the fluid velocity.

Later, Zhou et al. [82] employed DEM with computational fluid dynamics (CFD) and showed that the main features of sand erosion can be captured by the CFD-DEM approach.

The main advantage of the DEM models is that it captures the motion and interaction of individual sand grains and its failure micro mechanism in a dynamic process. It enables the model to predict many real behaviors such as continuously nonlinear stress strain response, behavior that changes in character according to stress state, memory of previous stress or strain excursion in both magnitude and direction, dilatancy that depends on history, mean stress, and initial states, hysteresis at loading/unloading among others.

To the best of our knowledge, there is no existing continuum constitutive model that reproduces all of these behaviors. However, since the DEM involves many individual particles and interactions between them, it is computationally expensive and therefore it is not applicable to large-scale problems.

Another disadvantage of the DEM model is the lack of a systematic method for an objective determination of micro material parameters. As opposed to the continuum-based models for which the strength and elastic properties can be determined directly from laboratory testing, the micro properties cannot be determined by direct measurements of the macro responses on the laboratory specimens. It could be found by means of a calibration process in which a particular assembly of particles with a set of micro parameters is used to simulate a set of material tests and the micro parameters then are evaluated to reproduce the macro responses measured in such tests [84]. Several researchers (e.g., [5, 6, 84, 87]) have proposed calibration procedures relating the micro parameters to macro properties of the material. However, the calibration procedure is challenging for the 3D DEM models. There may be several variations in the parameters and it is difficult to conclude which set of parameters is most appropriate for the material.

2.3. Hybrid Approaches. Considering the advantages and disadvantages of the continuum- and discontinuum-based approaches, a hybrid model that combines them can be practical and efficient in sand production modeling. Continuum-based approaches can be used in far field where the deformation is small hence continuum assumption is still valid and computationally efficient. On the other hand, a discontinuum-based approach can be used to describe large deformation or discontinuity near the wellbore or the perforations. In this manner, accurate and descriptive simulation of field-scale problems becomes possible with nowadays computational power available.

Some researchers have used this approach to analyze geomechanical problems. For example, El Shamy and Elmekati [88] and Elmekati and Shamy [89] combined a FEM code with a DEM code to analyze soil-structure interaction problems. Also, Azevedo and Lemos [90] used the same approach to study fracture growth in tensioned columns. In a similar work, Zeghal and El Shamy [91] coupled a continuum fluid model with discontinuum particle model to analyze the dynamic liquefaction of granular soils.

To the best of our knowledge, the hybrid scheme has not been used in sand production modeling.

3. Conclusions

Despite the numerous efforts in sand production and modeling, there are still some fundamental deficiencies which require to be addressed. Considering the works reported in the literature, there is still significant room for improvement in sanding models. Some are listed below.

(i) Critical-state-based constitutive models are expected to provide a more realistic representation of sand production. Such models can also predict compressive yielding which is the main mechanism for the creation of slit mode of degradation that may precede sanding. Yet, researchers have tended to avoid them due, perhaps, to several calibration parameters.

(ii) Sanding criteria which are based on erosion mechanics are most popular among researchers for describing and simulating the sanding phenomenon. These criteria usually involve a constitutive law for mass generation that requires calibration against laboratory tests. The erosion criteria are then coupled with hydromechanical behavior of the fluid-rock system for the purpose of numerical simulation. Erosion sanding criteria are more applicable for weak rocks where degradation of the rock into small particles is more probable. For strong rocks, other sanding criteria such as those that are based on tensile or shear failure or their combination with the erosion criteria could be more appropriate.

(iii) The most appropriate constitutive law may be a combined isotropic and kinematic hardening model, which allows capturing of the principal mechanisms of failure (shear, tensile, and compressional) as well as capturing the hysteresis effect in multiple shut-in and start-up scenarios.

(iv) Simulation of perforated wellbores requires a very fine mesh around perforations which makes the analysis computationally demanding especially is 3D modeling. 2D models of perforation using plane strain and axisymmetric assumptions do not fully capture the accurate mechanical and fluid flow responses. Yet, the majority of present models are simulated in 2D. A possible solution may be developing special numerical elements that best represent the perforation geometry and response.

(v) In capturing sanding in weak rocks, it is more realistic to keep the elements that satisfy sanding criteria in place and assign residual properties than to remove them. Element removal makes more sense when treating stronger rocks.

(vi) The assessment of the properties of infill materials requires more investigations and experimental data. The choice of these properties plays a significant role in sanding prediction. More accurate correlations for the changes of rock and flow properties with sand production and increase of porosity are essential.

(vii) Methods to capture sand arching have not been explicitly developed as such require complex interaction between the geometry of the opening in the completion and the disaggregated rock mass characteristics under prevailing stress state. Sand particles aggregate at the perforation cavity and form a stable entity called sand bridge and acts as a filter that prevents further sand particles to produce as long as the flow rate remains constant.

(viii) Sand liquefaction around injection wells has not been investigated fully yet. Since there is no recorded measurement of how sanding occurs in an injector, it is not yet clear whether it is occurs suddenly or gradually over many cycles, and if it is due to waterhammer or cross and back flow, or whether the produced material is sand or shale/clay.

(ix) The calibration procedure in the DEM model for determining micro material parameters needs further research.

(x) In order to have more accurate analysis in the DEM model, the fluid flow scheme needs to be modified. For instance, in current models, the permeability is usually related to porosity changes due to particles removal. However, cement debonding and wash out could also affect the pore-network and therefore the permeability.

(xi) Hybrid model combining DEM for the rocks around the wellbore and continuum approach for far-field rocks is expected to provide a more realistic and yet practical representation of a number of critical factors governing sanding.

Nomenclature

a: Dimensionless erosion onset coefficient
A: Constant for permeability equation
B: Constant for permeability equation
c: Transport concentration
C: Constant for permeability equation
c_r: Critical value of c for which the two competing phenomena, erosion and deposition, balance each other
k: Permeability
k_0: Initial permeability
$\dot{m}$: Mass rate of sand production
$\dot{m}_d$: Mass rate of sand deposition
p: Pore pressure
p_w: Pore pressure at wellbore
q: Fluid flow rate (L/T)
q_{cr}: Critical specific discharge
q_i: Specific discharge in the ith direction
r_w: Wellbore radius
R_p: Average grain size
S_L: The boundary surface, $2\pi r_w$, (unit length along wellbore axis)
t: Time
v: Flow velocity
v_s: Solid velocity
V_L: Volume of boundary layer, $\pi L(2r_w + L)$
β: Plugging permeability reduction coefficient
ε_v: Volumetric strain
$\dot{\varepsilon}_v$: Volumetric strain rate
σ_h: Minimum horizontal stress
σ_H: Maximum horizontal stress
σ_r: Radial stress
σ_θ: Tangential stress
σ_z: Vertical stress
σ_t': Effective stress acting parallel to the boundary
φ: Porosity
φ_i: Initial porosity
ϕ: Friction angle
λ: Sand production coefficient (must be calibrated experimentally)
λ_d: Sand deposition coefficient
γ: Permeability reduction coefficient
γ_f: Specific (unit) weight of fluid
μ: Fluid viscosity
μ_f: Friction coefficient
ρ_f: Fluid density
ρ_s: Solid density
ρ: Density
$\overline{\rho}$: Equivalent density, $\rho_f + c(\rho_s - \rho_f)$
η: Proportion of particles with size less than average pore size
σ_0': Effective initial stress
$\|\|$: Notation representing the norm of a vector.

Conflict of Interests

The authors of this paper are not involved with those software companies whose products have been mentioned in this paper that may give rise to conflict of interests.

Acknowledgments

The authors would like to acknowledge the research funding for this study provided by NSERC through a Collaborative Research Development program supported by BP.

References

[1] S. M. Willson, Z. A. Moschovidis, J. R. Cameron, and I. D. Palmer, "New model for predicting the rate of sand production," in *SPE/ISRM Rock Mechanics Conference*, pp. 152–160, Irving, Tex, USA, October 2002.

[2] Y. Xiao and H. H. Vaziri, "Import of strength degradation process in sand production prediction and management," in *Proceedings of the 45th U.S. Rock Mechanics/Geomechanics Symposium*, San Francisco, Calif, USA, June 2011.

[3] E. Papamichos, P. Cerasi, J. F. Stenebråten et al., "Sand production rate under multiphase flow and water breakthrough," in *Proceedings of the 44th U.S. Rock Mechanics Symposium and the 5th US/Canada Rock Mechanics Symposium*, Salt Lake City, Utah, USA, June 2010.

[4] L. Jing and O. Stephansson, *Fundamentals of Discrete Element Methods For Rock Engineering, Theory and Applications*, Elsevier, 2007.

[5] N. Belheine, J. P. Plassiard, F. V. Donzé, F. Darve, and A. Seridi, "Numerical simulation of drained triaxial test using 3D discrete element modeling," *Computers and Geotechnics*, vol. 36, no. 1-2, pp. 320–331, 2009.

[6] P. H. S. W. Kulatilake, B. Malama, and J. Wang, "Physical and particle flow modeling of jointed rock block behavior under uniaxial loading," *International Journal of Rock Mechanics and Mining Sciences*, vol. 38, no. 5, pp. 641–657, 2001.

[7] I. Vardoulakis, M. Stavropoulou, and P. Papanastasiou, "Hydromechanical aspects of the sand production problem," *Transport in Porous Media*, vol. 22, no. 2, pp. 225–244, 1996.

[8] E. Papamichos and M. Stavropoulou, "An erosion-mechanical model for sand production rate prediction," *International Journal of Rock Mechanics and Mining Sciences and Geomechanics Abstracts*, vol. 35, no. 4, pp. 531–532, 1998.

[9] E. Papamichos, I. Vardoulakis, J. Tronvoll, and A. Skjrstein, "Volumetric sand production model and experiment," *International Journal for Numerical and Analytical Methods in Geomechanics*, vol. 25, no. 8, pp. 789–808, 2001.

[10] H. Vaziri, B. Barree, Y. Xiao, I. Palmer, and M. Kutas, "What is the magic of water in producing sand?" in *SPE Annual Technical Conference and Exhibition*, pp. 2973–2985, San Antonio, Tex, USA, October 2002.

[11] C. Detournay, C. Tan, and B. Wu, "Modeling the mechanism and rate of sand production using FLAC," in *Proceedings of the of 4th International FLAC Symposium on Numerical Modeling in Geomechanics*, pp. 8–10.

[12] A. Nouri, H. Vaziri, H. Belhaj, and R. Islam, "Sand-production prediction: a new set of criteria for modeling based on large-scale transient experiments and numerical investigation," *SPE Journal*, vol. 11, no. 2, pp. 227–237, 2006.

[13] A. Nouri, E. Kuru, and H. Vaziri, "Enhanced modelling of sand production through improved deformation and stress analysis," in *Proceedings of the Canadian International Petroleum Conference*, pp. 12–14, Calgary, Canada, June 2007.

[14] H. Vaziri, A. Nouri, K. Hovem, and X. Wang, "Computation of sand production in water injectors," *SPE Production and Operations*, vol. 23, no. 4, pp. 518–524, 2008.

[15] C. Detournay, "Numerical modeling of the slit mode of cavity evolution associated with sand production," in *Proceedings of the SPE Annual Technical Conference and Exhibition (ATCE '08)*, pp. 3422–3431, Denver, Colo, USA, September 2008.

[16] A. Nouri, E. Kuru, and H. Vaziri, "Elastoplastic modelling of sand production using fracture energy regularization method," *Journal of Canadian Petroleum Technology*, vol. 48, no. 4, pp. 64–71, 2009.

[17] A. S. Kim, M. M. Sharma, and H. Fitzpatrick, "A Predictive model for sand production in poorly consolidated sands," in *Proceedings of the International Petroleum Technology Conference*, pp. 15–17, Bangkok, Thailand, November 2011.

[18] P. C. Papanastasiou and I. G. Vardoulakis, "Numerical treatment of progressive localization in relation to borehole stability," *International Journal for Numerical & Analytical Methods in Geomechanics*, vol. 16, no. 6, pp. 389–424, 1992.

[19] H. B. Muehlhaus and I. Vardoulakis, "The thickness of shear bands in granular materials," *Geotechnique*, vol. 37, no. 3, pp. 271–283, 1987.

[20] A. Zervos, P. Papanastasiou, and I. Vardoulakis, "Modelling of localisation and scale effect in thick-walled cylinders with gradient elastoplasticity," *International Journal of Solids and Structures*, vol. 38, no. 30-31, pp. 5081–5095, 2001.

[21] A. Zervos, P. Papanastasiou, and I. Vardoulakis, "A finite element displacement formulation for gradient elastoplasticity," *International Journal for Numerical Methods in Engineering*, vol. 50, no. 6, pp. 1369–1388, 2001.

[22] T. Crook, S. Willson, J. G. Yu, and R. Owen, "Computational modelling of the localized deformation associated with borehole breakout in quasi-brittle materials," *Journal of Petroleum Science and Engineering*, vol. 38, no. 3-4, pp. 177–186, 2003.

[23] J. Wang, D. P. Yale, and G. R. Dasari, "Numerical modeling of massive sand production," in *Proceedings of the SPE Annual Technical Conference and Exhibition*, Denver, Colo, USA, October 2011.

[24] H. Rahmati, A. Nouri, D. Chan, and H. Vaziri, "Validation of predicted cumulative sand and sand rate against physical-model test," *Journal of Canadian Petroleum Technology*, vol. 51, no. 5, pp. 403–410, 2011.

[25] N. Morita, D. L. Whitfill, I. Massie, and T. W. Knudsen, "Realistic sand-production prediction: numerical approach," *SPE Production Engineering*, vol. 4, no. 1, pp. 15–24, 1989.

[26] N. Morita, E. Davis, and L. Whitebay, "Guidelines for solving sand problems in water injection wells," in *Proceedings of the SPE International Symposium on Formation Damage Control*, pp. 189–200, Lafayette, La, USA, February 1998.

[27] R. C. Burton, E. R. Davis, N. Morita, and H. O. McLeod, "Application of reservoir strength characterization and formation failure modeling to analyze sand production potential and formulate sand control strategies for a series of North Sea gas reservoirs," in *Proceedings of the SPE Technical Conference and Exhibition*, pp. 159–168, New Orleans, La, USA, September 1998.

[28] A. Skjaerstein, M. Stavropoulou, I. Vardoulakis, and J. Tronvoll, "Hydrodynamic erosion: a potential mechanism of sand production in weak sandstones," *International Journal of Rock Mechanics and Mining Sciences*, vol. 34, no. 3-4, pp. 292.e1–292.e18, 1997.

[29] E. Papamichos and E. M. Malmanger, "A sand erosion model for volumetric sand predictions in a north sea reservoir," in *Proceedings of the SPE Latin American and Caribbean Petroleum Engineering Conference*, Caracas, Venzuela, 1999.

[30] X. Yi, "Water injectivity decline caused by sand mobilization: simulation and prediction," in *Proceedings of the SPE Permian Basin Oil and Gas Recovery Conference*, pp. 202–209, Midland, Tex, USA, May 2001.

[31] H. H. Vaziri, Y. Xiao, R. Islam, and A. Nouri, "Numerical modeling of seepage-induced sand production in oil and gas reservoirs," *Journal of Petroleum Science and Engineering*, vol. 36, no. 1-2, pp. 71–86, 2002.

[32] Y. Wang and S. Xue, "Coupled reservoir-geomechanics model with sand erosion for sand rate and enhanced production prediction," in *Proceedings of the SPE International ymposium on Formation Damage Control*, pp. 373–383, Lafayette, La, USA, February 2002.

[33] L. Y. Chin and G. G. Ramos, "Predicting volumetric sand production in weak reservoirs," in *Proceedings of the SPE/ISRM Rock Mechanics Conference*, pp. 161–170, Irving, Tex, USA, October 2002.

[34] A. Nouri, M. M. Al-Darbi, H. Vaziri, and M. R. Islam, "Deflection criteria for numerical assessment of the sand production potential in an openhole completion," *Energy Sources*, vol. 24, no. 7, pp. 685–702, 2002.

[35] J. Wang, R. G. Wan, A. Settari, and D. Walters, "Prediction of volumetric sand production and wellbore stability analysis of a well at different completion schemes," in *Proceedings of the 40th U.S. Symposium on Rock Mechanics (USRMS '05)*, Anchorage, Alaska, USA, June 2005.

[36] G. Servant, P. Marchina, Y. Peysson, E. Berner, and J. F. Nauroy, "Sand erosion in weakly consolidated reservoirs: experiments and numerical modeling," in *Proceedings of the 15th SPE-DOE Improved Oil Recovery Symposium: Old Reservoirs New Tricks a Global Perspective*, pp. 949–956, Tulsa, Okla, USA, April 2006.

[37] A. Nouri, H. Vaziri, and E. Kuru, "Numerical investigation of sand production under realistic reservoir/well flow conditions," in *Proceedings of the Canadian International Petroleum Conference*, pp. 13–15, Calgary, Canada, June 2006.

[38] A. Nouri, H. Vaziri, H. Belhaj, and M. R. Islam, "Comprehensive transient modeling of sand production in horizontal wellbores," *SPE Journal*, vol. 12, no. 4, pp. 468–474, 2007.

[39] S. Azadbakht, M. Jafarpour, H. Rahmati, A. Nouri, H. Vaziri, and D. Chan, "A numerical model for predicting the rate of sand production in injector wells," in *Proceedings of the SPE Deep Water and Completions Conference and Exhibition*, pp. 20–21, Galveston, Tex, USA, June 2012.

[40] J. Sulem, I. Vardoulakis, E. Papamichos, A. Oulahna, and J. Tronvoll, "Elasto-plastic modelling of Red Wildmoor sandstone," *Journal of Mechanics of Cohesive-Frictional MaterialsNo*, vol. 4, no. 3, pp. 215–245, 1999.

[41] M. Jafarpour, H. Rahmati, S. Azadbakht, A. Nouri, D. Chan, and H. Vaziri, "Determination of mobilized strength properties of degrading sandstone," *Journal of Soils and Foundations*, vol. 52, no. 4, pp. 658–667, 2012.

[42] R. P. Nordgren, "Strength of well completions," in *Proceedings of the 18th US Symposium on Rock Mechanics (USRMS '77)*, Golden, Colo, USA, June 1977.

[43] G. R. Coates and S. A. Denoo, "Mechanical properties program using borehole analysis and mohr's circle," in *Proceedings of the 22nd Annual Logging Symposium (SPWLA '81)*, pp. 23–26, 1981.

[44] R. Risnes, R. K. Bratili, and P. Horsrud, "Sand stresses around a wellbore," *SPE Journal*, vol. 22, no. 6, pp. 883–898, 1982.

[45] D. P. Edwards, Y. Sharma, and A. Charron, "Zones of sand production identified by log-derived mechanical properties: a case study," in *Proceedings of the 8th European Formation Evaluation Symposium (SPWLA '83)*, London, UK, 1983.

[46] D. Antheunis, P. B. Vriezen, B. A. Schipper, and A. C. van der Vlis, "Perforation collapse: failure of perforated friable sandstones," in *Proceedings of the SPE European Spring Meeting*, pp. 8–9, Amsterdam, The Netherlands, April 1976.

[47] J. M. Peden, U. Heriot Watt, and A. A. M. Yassin, "The determination of optimum completion and production conditions for sand-free oil production," in *Proceedings of the SPE Annual Technical Conference and Exhibition*, New Orleans, La, USA, October 1986.

[48] R. G. Wan and J. Wang, "Analysis of sand production in unconsolidated oil sand using a coupled erosional-stress-deformation model," *Journal of Canadian Petroleum Technology*, vol. 43, no. 2, pp. 47–53, 2004.

[49] J. Wang, D. Walters, A. Settari, and R. G. Wan, "An integrated modular approach to modeling sand production and cavity growth with emphasis on the multiphase flow and 3D effects," in *Proceedings of the 41st U.S. Symposium on Rock Mechanics Symposium*, Golden, Colo, USA, June 2006.

[50] R. G. Wan and J. Wang, "Modeling of sand production and wormhole propagation in oil saturated sand pack using stabilized finite element methods," *Journal of Canadian Petroleum Technology*, vol. 42, no. 12, pp. 1-8, 2003.

[51] A. Nouri, H. Vaziri, H. Belhaj, and R. Islam, "Effect of volumetric failure on sand production in Oil-wellbores," in *Proceedings of the SPE Asia Pacific Oil and Gas Conference and Exhibition*, pp. 86–93, Jakarta, Indonesia, April 2003.

[52] P. Papanastasiou and A. Zervos, "Wellbore stability analysis: from linear elasticity to postbifurcation modeling," *International Journal of Geomechanics*, vol. 4, no. 1, pp. 2–12, 2004.

[53] E. Papamichos, J. Tronvoll, A. Skjærstein, and T. E. Unander, "Hole stability of Red Wildmoor sandstone under anisotropic stresses and sand production criterion," *Journal of Petroleum Science and Engineering*, vol. 72, no. 1-2, pp. 78–92, 2010.

[54] I. Vardoulakis, J. Sulem, and A. Guenot, "Borehole instabilities as bifurcation phenomena," *International Journal of Rock Mechanics and Mining Sciences and Geomechanics Abstracts*, vol. 25, no. 3, pp. 159–170, 1988.

[55] J. Tronvoll, E. Papamichos, and N. Kessler, "Perforation cavity stability: investigation of failure mechanisms," in *Proceedings of the International Symposium on Geotechnical Engineering of Hard Soils—Soft Rocks*, pp. 1687-1693, Balkema, Rotterdam, The Netherlands, 1993.

[56] P. J. Van Den Hoek, G. M. M. Hertogh, A. P. Kooijman, P. De Bree, C. J. Kenter, and E. Papamichos, "New concept of sand production prediction: theory and laboratory experiments," *SPE Drilling and Completion*, vol. 15, no. 4, pp. 261–273, 2000.

[57] P. C. Papanastasiou and I. G. Vardoulakis, "Bifurcation analysis of deep boreholes: II. Scale effect," *International Journal for Numerical & Analytical Methods in Geomechanics*, vol. 13, no. 2, pp. 183–198, 1989.

[58] B. Haimson, "Micromechanisms of borehole instability leading to breakouts in rocks," *International Journal of Rock Mechanics and Mining Sciences*, vol. 44, no. 2, pp. 157–173, 2007.

[59] R. Risnes and R. K. Bratli, "Stability and failure of sand arches," in *Proceedings of the 54th Annual Technical Conference and Exhibition of the Society of Petroleum Engineers of AIME*, Las Vegas, Nev, USA, September 1979.

[60] R. K. Bratli and R. Risnes, "Stability and failure of sand arches," *SPE Journal*, vol. 21, no. 2, pp. 236–248, 1981.

[61] J. S. Weingarten and T. K. Perkins, "Prediction of sand production in gas wells: methods and gulf of Mexico case studies," *Journal of Petroleum Technology*, vol. 47, no. 7, pp. 596–600, 1995.

[62] J. Tronvoll, A. Skjaerstein, and E. Papamichos, "Sand production: mechanical failure or hydrodynamic erosion," *International Journal of Rock Mechanics and Mining Sciences and Geomechanics Abstracts*, vol. 34, no. 3-4, pp. 291.e1–291.e17, 1997.

[63] I. Vardoulakis, P. Papanastasiou, and M. Stavropoulou, "Sand erosion in axial flow conditions," *Transport in Porous Media*, vol. 45, no. 2, pp. 267–281, 2001.

[64] M. Stavropoulou, P. Papanastasiou, and I. Vardoulakiz, "Coupled wellbore erosion and stability analysis," *International Jounral for Numerical and Analytical Methods in Geomechanics*, vol. 22, no. 9, pp. 749–769, 1998.

[65] E. Papamichos and I. Vardoulakis, "Sand erosion with a porosity diffusion law," *Computers and Geotechnics*, vol. 32, no. 1, pp. 47–58, 2005.

[66] B. Haimson and H. Lee, "Borehole breakouts and compaction bands in two high-porosity sandstones," *International Journal of Rock Mechanics and Mining Sciences*, vol. 41, no. 2, pp. 287–301, 2004.

[67] E. Papamichos, "Sand production physical and experimental evidence," *Journal of Geomechanics in Energy Production*, vol. 10, no. 6-7, pp. 803–816, 2006.

[68] J. Tronvoll and E. Fjær, "Experimental study of sand production from perforation cavities," *International Journal of Rock Mechanics and Mining Sciences and Geomechanics*, vol. 31, no. 5, pp. 393–410, 1994.

[69] J. Tronvoll, M. B. Dusseault, F. Sanfilippo, and F. J. Santarelli, "The tools of sand management," in *Proceedings of the SPE Annual Technical Conference and Exhibition*, pp. 3101–3115, New Orleans, La, USA, October 2001.

[70] M. F. Trentacoste, Evaluation of LS-DYNA soil material model 147, no. FHWA-HRT-04-094, 2004.

[71] P. Papanastasiou and A. Zervos, "Three-dimensional stress analysis of a wellbore with perforations and a fracture," in *Proceedings of the SPR/ISRM Rock Mechanics In Petroleum Engineering (Eurock '98)*, pp. 347–355, Trondheim, Norway, July 1998.

[72] F. J. Santarelli, E. Skomedal, P. Markestad, H. I. Berge, and H. Nasvig, "Sand production on water injectors: just how bad can it get?" in *Proceedings of the SPR/ISRM Rock Mechanics In Petroleum Engineering (Eurock '98)*, pp. 107–115, Trondheim, Norway, July 1998.

[73] F. J. Santarelli, E. Skomedal, P. Markestad, H. I. Berge, and H. Nasvig, "Sand production on water injectors: how bad can it get?" *SPE Drilling and Completion*, vol. 15, no. 2, pp. 132–139, 2000.

[74] F. J. Santarelli, F. Sanfilippo, J. Embry, M. White, and J. B. Turnbull, "The Sanding mechanisms of water injectors and their quantification in terms of sand production: example of the Buzzard Field (UKCS)," in *Proceedings of the SPE Annual Technical Conference and Exhibition*, Denver, Colo, USA, October 2011.

[75] A. Hayatdavoudi, "Formation sand liquefaction: a mechanism for explaining fines migration and well sanding," in *Proceedings of the SPE Mid-Continent Operations Symposium*, Oklahoma City, Okla, USA, March 1999.

[76] P. K. Robertson and C. E. Fear, "Liquefaction of sands and its evaluation," in *Proceedings of the 1st International Conference on Earthquake Geotechnical Engineering*, 1997.

[77] R. M. O'Connor, J. R. Torczynski, D. S. Preece, J. T. Klosek, and J. R. Williams, "Discrete element modeling of sand production," *International Journal of Rock Mechanics and Mining Sciences and Geomechanics Abstracts*, vol. 34, no. 3-4, pp. 231.e1–231.e15, 1997.

[78] R. P. Jensen and D. S. Preece, *Modeling of Sand Production with Darcy's Flow Coupled with Discrete Elements*, OSTI, 2000.

[79] L. Li, E. Papamichos, and P. Cerasi, "Investigation of sand production mechanisms using DEM with fluid flow," in *Proceedings of the International Symposium of the International Society for Rock Mechanics (Eurock '06)*, pp. 241–247, Liège, Belgium, May 2006.

[80] L. Li and R. M. Holt, "Particle scale reservoir mechanics," *Oil and Gas Science and Technology*, vol. 57, no. 5, pp. 525–538, 2002.

[81] L. Y. G. Cheung, *Micromechanics of Sand Production in Oil Wells [Ph.D. thesis]*, Imperial College of London, 2010.

[82] Z. Y. Zhou, A. B. Yu, and S. K. Choi, "Numerical simulation of the liquid-induced erosion in a weakly bonded sand assembly," *Powder Technology*, vol. 211, no. 2-3, pp. 237–249, 2011.

[83] P. A. Cundall, "A computer model for simulating progressive large scale movement in blocky rock systems," in *Proceedings of the Symposium International Society of Rock Mechanics*, 1971.

[84] D. O. Potyondy and P. A. Cundall, "A bonded-particle model for rock," *International Journal of Rock Mechanics and Mining Sciences*, vol. 41, no. 8, pp. 1329–1364, 2004.

[85] S. Thallak, L. Rothenburg, and M. Dusseault, "Hydraulic fracture (parting) simulation in granular assemblies using the discrete element method," *AOSTRA Journal of Research*, vol. 6, pp. 141–153, 1990.

[86] D. Chan and S. Tipthavonnukul, "Numerical simulation of granular particles movement in fluid flow," *International Journal of Nonlinear Sciences and Numerical Simulation*, vol. 9, no. 3, pp. 229–248, 2008.

[87] J. Yoon, "Application of experimental design and optimization to PFC model calibration in uniaxial compression simulation," *International Journal of Rock Mechanics and Mining Sciences*, vol. 44, no. 6, pp. 871–889, 2007.

[88] U. El Shamy and A. Elmekati, "An efficient combined DEM/FEM technique for soil-structure interaction problems," in *International Foundation Congress and Equipment Expo*, pp. 238–245, Orlando, Fla, USA, March 2009.

[89] A. Elmekati and U. E. Shamy, "A practical co-simulation approach for multiscale analysis of geotechnical systems," *Computers and Geotechnics*, vol. 37, no. 4, pp. 494–503, 2010.

[90] N. M. Azevedo and J. V. Lemos, "Hybrid discrete element/finite element method for fracture analysis," *Computer Methods in Applied Mechanics and Engineering*, vol. 195, no. 33–36, pp. 4579–4593, 2006.

[91] M. Zeghal and U. El Shamy, "A continuum-discrete hydromechanical analysis of granular deposit liquefaction," *International Journal for Numerical and Analytical Methods in Geomechanics*, vol. 28, no. 14, pp. 1361–1383, 2004.

Effect of Bed Deformation on Natural Gas Production from Hydrates

Mohamed Iqbal Pallipurath

Mechanical Engineering Department, TKM College of Engineering, Kollam, Kerala 691005, India

Correspondence should be addressed to Mohamed Iqbal Pallipurath; mohamediqbalp@gmail.com

Academic Editor: Jorge Ancheyta

This work is based on modelling studies in an axisymmetric framework. The thermal stimulation of hydrated sediment is taken to occur by a centrally placed heat source. The model includes the hydrate dissociation and its effect on sediment bed deformation and resulting effect on gas production. A finite element package was customized to simulate the gas production from natural gas hydrate by considering the deformation of submarine bed. Three sediment models have been used to simulate gas production. The effect of sediment deformation on gas production by thermal stimulation is studied. Gas production rate is found to increase with an increase in the source temperature. Porosity of the sediment and saturation of the hydrate both have been found to significantly influence the rate of gas production.

1. Introduction

Energy demand is on the rise globally but the production rates of major fossil fuels are going down. Several analysts predict a drastic reduction in energy production due to diminishing reserve of fossil fuels. The major result from the global analysis is that world oil production peaked in 2006. Production has started to decline at a rate of several percentages per year. This necessitates a search for commercially viable and clean source of energy capable of meeting future energy demands. Natural gas hydrate (NGH) is one of the possible energy sources to meet these requirements. It is a highly condensed form of natural gas formed by capture of natural gas molecules in a cage of water molecules: each cubic meter of natural gas hydrate yields about 160 cubic meter of gas at STP.

A large amount of natural gas hydrate exists on our planet. Such deposits are found both on land (in the permafrost region), and offshore (in the submarine sediment). Over 230 gas hydrate deposits have been found globally. Gas hydrates have also been located in the coastal regions of India [1]. Needless to say, the vastness of gas hydrates has attracted global attention for its exploration and exploitation for future energy supply. It is predicted that utilization of even 17% to 20% of this resource could meet the energy demands for next 200 years [2].

Methods suggested for the production of natural gas from gas hydrate include depressurization, thermal stimulation, chemical inhibitor injection, and CO_2 sequestration. Among these, depressurization and thermal stimulation have been considered to be the most economical, though other methods are under investigation. The type of method depends on the reservoir characteristics. Due to less energy input for depressurization, this method has been studied more than the thermal stimulation. However, the efficacy of latter method needs to be studied in more detail. In recent times, during December 2001 to March 2002, field tests were conducted at a permafrost region located at Mallik gas hydrate site (Canada) in which both depressurization and thermal stimulation were conducted. Each of these methods has its relative merits and demerits, and for the development of commercial technology it is essential to have a good understanding of each of the proposed methods.

2. Current Scenario

The study of any of the methods for gas production from natural gas hydrates needs not only a proper knowledge of the thermodynamics, kinetics, and heat and mass transfer effects, but also that of sediment response to any change in its morphology during hydrate dissociation. The presence of hydrate has a cementing effect on the sediment structure and hence any loss of hydrate tends to weaken the sediment. Sediment deformation has profound influence on the gas productivity. Moreover, a reduction in the sediment strength may destabilize sediment matrix to an extent that could cause serious damage to the production operation and uncontrolled release of methane to the environment.

The earlier modeling on gas production from gas hydrates considered an undeformed bed as the main objective was to study the effectiveness of a given method in the gas production. Some of these works considered the kinetics of hydrate dissociation (e.g., [3]), and others assumed instantaneous dissociation and hence equilibrium condition during hydrate dissociation (e.g., [4]). Also, the heat transfer effect was not considered in all the studies (e.g., [3] assumed isothermal operation). Analytical solutions were obtained by Selim and Sloan [4], Yousif et al. [3], Tsypkin [5], Ji et al. [6], and many others. However, such solutions were based on assumptions which were found not to be valid in reality. Numerical approach to the problem has been attempted to address the real conditions existing in the reservoir and other operational issues. Gas production from hydrate reservoir by the combination of warm water flooding and depressurization was proposed by Bai and Li [7] which can overcome the deficiency of single production method. Sun et al. [8] developed a one-dimensional model of hydrate depressurization in porous media.

Recently, a research group led by Professor Kimoto [9, 10] from Kyoto University, Japan, reported studies on gas production from gas hydrates considering bed deformation. They proposed a model based on chemothermomechanically coupled analysis which could predict the deformation of sediment; they did not report the gas production under this condition. Gas production from gas hydrate was also studied by some researchers by using some of the reservoir simulators such as CMG-STARS, Hydrate Res Sim, MH-21 HYDRES, STOMP-HYD, and TOUGH-HYDRATE.

Kinetics and thermodynamics of hydrate formation and dissociation dictate the choice of operating conditions and hence the gas production. Pioneering work on methane hydrate kinetics was done by Professor Bishnoi and his research team from University of Calgary (e.g., [11–13]). Other studies on methane hydrate kinetics were performed with different additives such as polymeric inhibitor [14], electrolyte solution [15], and promoter [16] A kinetic rate model was proposed by Kim et al. [13] for methane hydrate decomposition. Kinetics of hydrate formation of gases other than CH_4, such as C_2H_6, and natural Gas, was studied by Kaschiev and Firoozabadi [17].

The gas hydrate thermodynamics dictates the pressure-temperature relationship to predict the zone of hydrate stability. Also, the heat of dissociation of hydrate is obtained from such studies. Selim and Sloan [4] reported a thermodynamic relationship based on Antoine equation and also an equation to determine the heat of hydrate dissociation. Lu and Sultan (2008) summarized the studies on hydrate stability and proposed a correlation to relate pressure and temperature at hydrate equilibrium which gave a good match with the previous experimental data.

The effects of some polymers and surfactants on methane hydrate formation were investigated by Karaaslan and Parlaktuna (2004) in a high-pressure system.

Results of field test on gas production by thermal stimulation of hydrated sediments performed at Mallik 5L-38 gas hydrate production research well were reported by Hancock et al. [18]. No other field data on thermal stimulation is available.

A review of the reported studies in the literature shows that the following issues need to be investigated in more detail.

(i) The type of soil mechanical model that can best predict the production behaviour during thermal stimulation has not previously been investigated. The previous work available in the literature on the effect of soil deformation indicates that soil model may have a significant bearing on the gas production.

(ii) Till now, experimental data are available only from one field test on thermal stimulation in permafrost region. It is necessary to conduct experiments to study this method under submarine conditions.

2.1. Description of the Model Used. Numerical modelling can be an effective tool that enables understanding mechanisms leading to wellbore instability in oceanic hydrate bearing sediment. To assess deformations caused by hydrate dissociation and the effect of these deformations on the gas generation, numerical techniques are essential. Thermal dissociation of hydrated sediment by a pumped hot fluid is modeled. A radial heat flow from the hot pipe is assumed. The coordinate system is cylindrical. Four components (soil, hydrate, gas (methane), and water) and three phases (hydrate, gas, and aqueous-phase) are considered in the simulator. The intrinsic kinetics of hydrate formation or dissociation is considered using the Kim-Bishnoi model. Mass transport and heat transfer involved in formation or dissociation of hydrates are included in the governing equations. The arrangement of heat source is shown in Figures 1 and 2.

Factors to be evaluated are as follows.

(i) Changing the stresses and pore pressures.

(ii) Impact of the selected constitutive model for pore pressures.

(iii) Effect of heating the formation on thermodynamic stability of the hydrates.

(iv) Deformation of the sediment as a result of dissociation of hydrate.

(v) Effect of this deformation on the gas production rates.

(vi) Effect of different saturations of hydrate on gas production.

(vii) Effect of source temperature on gas production.

(viii) Effect of porosity on gas production.

Mechanisms to be considered are as follows.

(i) Kinetic rate, heat and mass transfer equilibrium, and fluid flow relations for gas hydrate dissociation/reformation with the change in pressure and temperature.

(ii) Resultant changes in the mechanical and petrophysical properties of the sediments.

(iii) Representative constitutive equation and yield criterion for the mechanical behavior of HBS of various hydrate concentrations.

(iv) Different soil models and their responses to deformation.

The generation rates of gas and water are dictated by stoichiometry of the hydrate. The transition between gas, water, and hydrates can be represented as a chemical reaction

$$g\left(\frac{G}{A}\right) + n \cdot w\left(A\right) \Longleftrightarrow h\left(H\right), \quad (1)$$

where g is the gas component, existing as free gas (G) or dissolved in water (A), h is the hydrate component present only in the hydrate phase (H), and n is the hydration number.

2.2. Assumptions

(i) Gas hydrate bearing zone at a total depth of 3000 to 3300 meters (~700 m below sea floor and 2400 m below sea level) is considered the default range of study (Yun et al., 2010).

(ii) We neglect the adsorption of any component by the rock phase; that is, the rock phase is inactive in mass transfer.

(iii) Momentum (fluid flow) and heat transfer are axisymmetric.

(iv) The water phase is incompressible.

(v) The gas follows the Peng Robinson equation of state.

(vi) The porous medium (rock) is nondeformed.

(vii) Gas can occur only in gaseous and hydrate states since CH_4 solubility under conditions of model is negligible.

(viii) Water can occur only in liquid and hydrate states; that is, ice and water vapour formation are neglected since the sediment conditions preclude its formation.

Equivalent thermal conductivity of hydrated sediment is given by the equation

$$\lambda = \left(1 - \varphi\right)\lambda_s + \sum_{j=h,g,w} \phi S_j \lambda_j. \quad (2)$$

For a porosity of 0.47, let us consider two extreme cases of hydrate saturation, namely, 0.8 and 0.1. For hydrate saturation of 0.8 and water and gas saturations of 0.1 each, the equivalent thermal conductivity is 3.968 W m^{-1} K^{-1} whereas for a hydrate saturation of 0.1 and water and gas saturations of 0.45 each, the equivalent thermal conductivity is 3.868 W m^{-1} K^{-1} which is a difference of just 2.53%. This shows that an assumption of invariant equivalent thermal conductivity is valid.

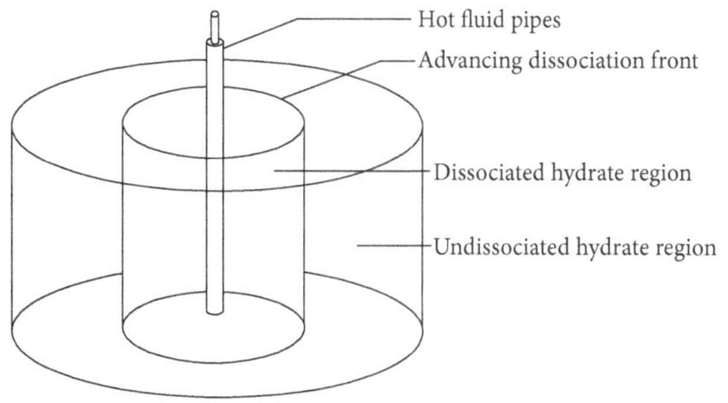

FIGURE 1: Schematic of a hydrate reservoir heated with pipes carrying hot fluid.

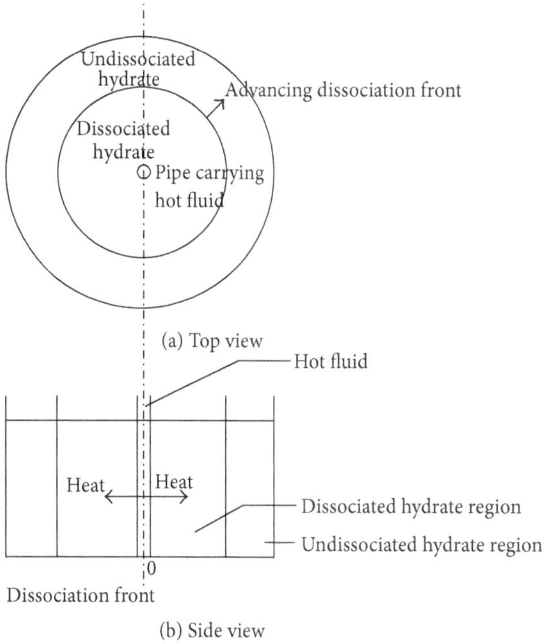

FIGURE 2: Schematic representation of hydrated sediment to study gas hydrate dissociation by thermal stimulation.

2.3. Flow through Porous Media Applied to Hydrate Bearing Sediment. A porous medium is modelled in Abaqus/Standard by a conventional approach that considers the medium as a multiphase material and adopts an effective stress principle to describe its behaviour. The porous medium modelling provided considers the presence of two fluids in the medium. One is the "wetting liquid," which is assumed to be relatively (but not entirely) incompressible. Often the other is a gas, which is relatively compressible. An example of such a system is marine hydrated sediment containing sea water and gas. When the medium is partially saturated, both fluids exist at a point; when it is fully saturated, the voids are completely filled with the wetting liquid.

The porous medium is modelled by attaching the finite element mesh to the solid phase; fluid can flow through this mesh.

2.4. Coupled Flow and Heat Transfer through Porous Media. Optionally, heat transfer due to conduction in the soil skeleton and pore fluid, as well as convection in the pore fluid, can also be modeled. This capability represents an enhancement to the basic pore fluid flow capabilities discussed in the earlier paragraphs and requires the use of coupled temperature-pore pressure elements that have temperature as an additional degree of freedom in addition to the pore pressure and the displacement components. When you use the coupled temperature-pore pressure elements, Abaqus solves the heat transfer equation in addition to and in a fully coupled manner with the continuity equation and the mechanical equilibrium equations. Only linear brick, first-order axisymmetric, and second-order modified tetrahedrons are available for modeling coupled heat transfer with pore fluid flow and mechanical deformation. Coupled temperature-pore pressure elements are not supported in Abaqus/CAE.

2.5. Total and Excess Pore Fluid Pressure. The coupled pore fluid diffusion/stress analysis capability can provide solutions either in terms of total or "excess" pore fluid pressure. The excess pore fluid pressure at a point is the pore fluid pressure in excess of the hydrostatic pressure required to support the weight of pore fluid above the elevation of the material point. The difference between total and excess pore pressure is relevant only for cases in which gravitational loading is important. Total pore pressure solutions are provided when the gravity distributed load is used to define the gravity load on the model. Excess pore pressure solutions are provided in all other cases, for example, when gravity loading is defined with body force distributed loads.

2.6. Transient Analysis. In this transient coupled pore pressure/effective stress analysis, the backward difference operator is used to integrate the continuity equation and the heat transfer equation: this operator provides unconditional stability so that the only concern with respect to time integration is accuracy.

For fully saturated flow analyses in which heat transfer is also modelled, the contributions to the model's stiffness matrix arising from convective heat transfer due to pore fluid flow are unsymmetric.

2.7. Partially Saturated Flow. In gas hydrate sediment analysis, we shall be dealing with partially saturated flow. In partially saturated flow cases, the corresponding guideline for the minimum time increment is

$$\Delta t > \frac{\gamma_w n^0 \left(1 + \beta v_w\right)}{6 k_s k} \frac{ds}{du_w} (\Delta \ell)^2, \tag{3}$$

where s is the saturation, k_s is the permeability-saturation relationship, ds/du_w is the rate of change of saturation with respect to pore pressure, n^0 is the initial porosity of the material, Δt is the time increment, γ_w is the specific weight of the wetting liquid, k is the permeability of the soil, v_w is the magnitude of the velocity of the pore fluid, β is the velocity coefficient in Forchheimer's flow law ($\beta = 0$) in the case of Darcy flow, and $\Delta\ell$ is a typical element dimension.

2.8. Automatic Incrementation. Since automatic time incrementation is left to Abaqus, three tolerance parameters are chosen. The accuracy of the time integration of the flow continuity equations is governed by the maximum wetting liquid pore pressure change, $\Delta u_w^{\max}$, allowed in an increment. Abaqus/Standard restricts the time increments to ensure that this value is not exceeded at any node (except nodes with boundary conditions) during any increment in the analysis.

Since heat transfer is modelled, the accuracy of time integration is also governed by the maximum temperature change, $\Delta\theta_{\max}$, allowed in an increment. Abaqus/Standard restricts the time increments to ensure that this value is not exceeded at any node (except nodes with boundary conditions) during any increment of the analysis.

2.9. Mechanical Constitutive Models. The constitutive library provided in Abaqus contains a range of linear and nonlinear material models for all of these categories of materials. In general, the library has been developed to provide those models that are most usually required for practical applications. There are several distinct models in the library, and for the more commonly encountered materials, several ways of modeling the material are provided, each suitable to a particular type of analysis application. But the library is far from comprehensive: the range of physical material behavior is far too broad for this ever to be possible. If there is no model in the library that is useful for a particular case, Abaqus/Standard contains a user subroutine UMAT. In these routines the user can code a material model (or call other routines that perform that task). This "user subroutine" capability proved to be a powerful resource for the modelling of gas hydrate bearing sediment.

From a numerical viewpoint, the implementation of a constitutive model involves the integration of the state of the material at an integration point over a time increment during a nonlinear analysis (the implementation of constitutive models in Abaqus assumes that the material behavior is entirely defined by local effects, so each spatial integration point can be treated independently). Since Abaqus/Standard is most commonly used with implicit time integration, the implementation must also provide an accurate "material stiffness matrix" for use in forming the Jacobian of the nonlinear equilibrium equations.

2.10. FEM Formulation of Gas Production from Sediment Bed. The Abaqus package was customized with user subroutines and modified input files to solve the model equations. Some typical results obtained for submarine sediment are given. Figure 3 is a comparison of the three developed soil models for the base case using simulation of cumulative gas production for comparison. The three models of soil dynamical system give comparable results as shown in Figure 3.

In Figures 4 and 5, the temperature profiles for the viscoplastic and viscoelastic soil models are compared, and the figures indicate that, by around 60 hours, the temperature profile steadies and is unvarying with time thereafter. It may be inferred that, for a given radius of reservoir, there exists

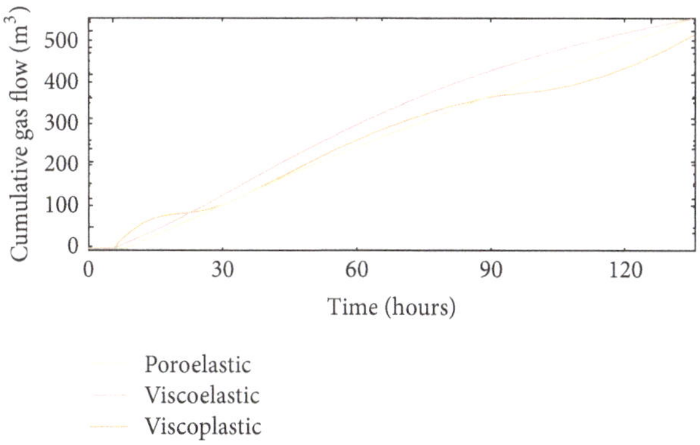

FIGURE 3: Three soil models compared in cumulative gas production rate.

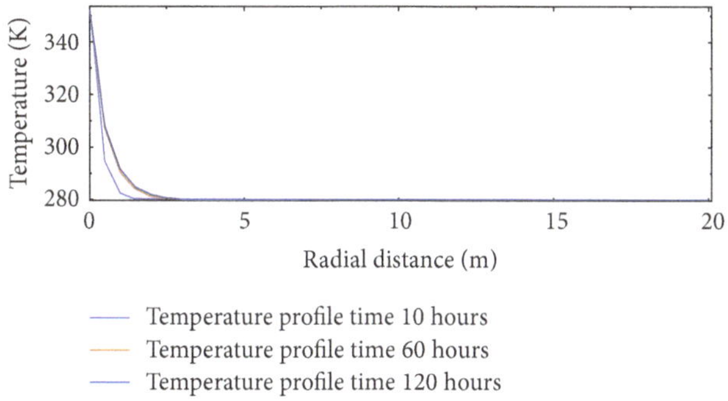

FIGURE 4: Temperature profile in sediment at various times using viscoplastic model.

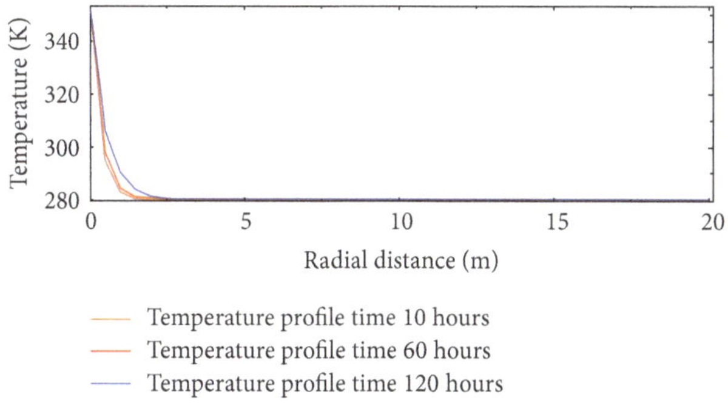

FIGURE 5: Temperature profile in sediment at various times using viscoelastic model.

a constant period of time after which the temperature profile can be said to have achieved steady state.

It appears from the curves in Figures 4 and 5 that the steady state temperature distribution is not dependant on the soil model, provided the thermal properties of the sediment are more or less constant, as they are in these two soil models. But, as can be seen, the rates at which the steady states are achieved can vary marginally with the soil models used.

The effects of deformation on gas production are shown in Figures 6 and 7. Figure 6 shows cumulative gas production without deformation. When the plots are compared side by

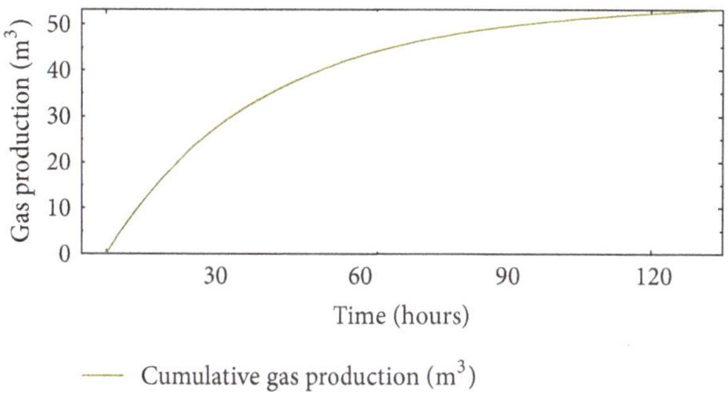

FIGURE 6: Cumulative gas production without deformation.

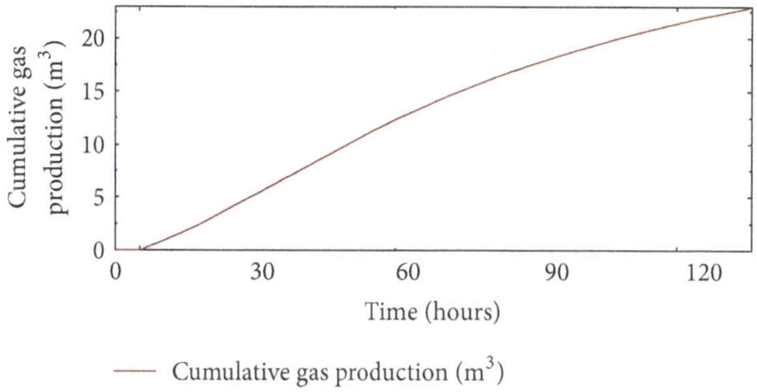

FIGURE 7: Cumulative gas production with deformation (viscoelastic soil model).

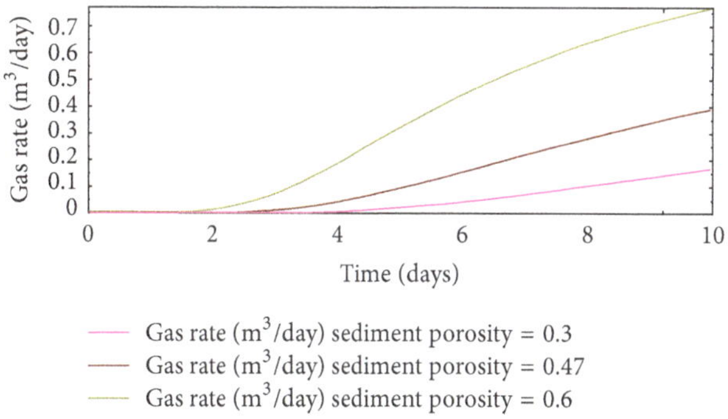

FIGURE 8: Effect of initial bed porosity on gas production rate.

side, it appears that the amount of gas produced is reduced by more than 50% when we allow for deformation of the sediment. This could be explained by the reduced permeability of the sediment after dissociation and subsequent subsidence.

Figure 7 shows cumulative gas production with deformation of sediment bed using the viscoelastic soil model. The gaps between the sediment grains are reduced after subsidence, which could account for the drop in gas/water permeability, which in turn reduces the gas production by more than half.

2.11. Parametric Study. A few representative results are shown of parametric studies conducted to gauge the impact of porosity and hydrate saturation upon gas production rate. The effect of sediment porosity upon gas production is graphically presented in Figure 8. Higher porosity seems to aid

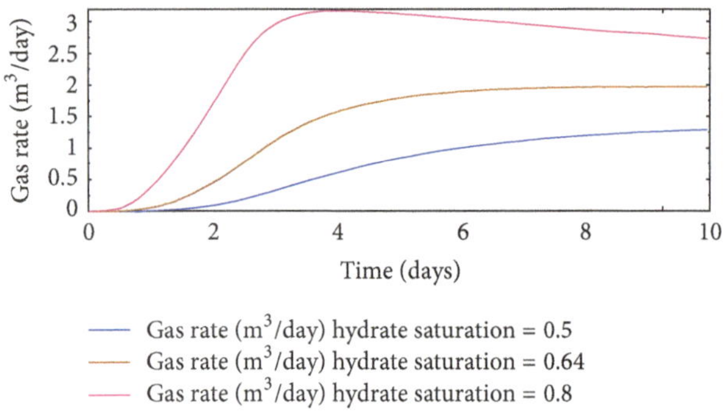

FIGURE 9: Effects of initial hydrate saturation on gas production rate.

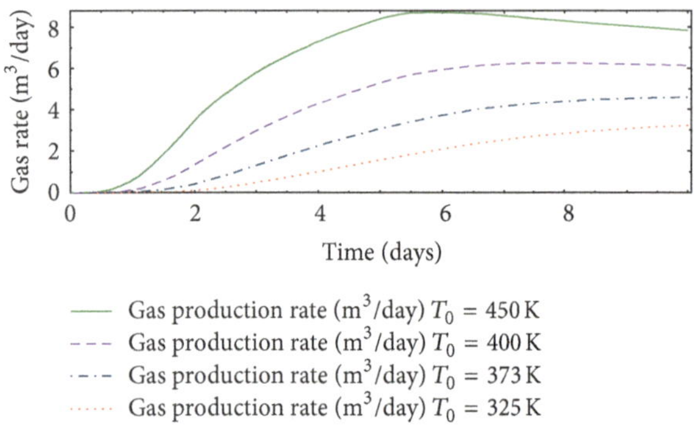

FIGURE 10: Effect of heat source temperature upon the gas production rate.

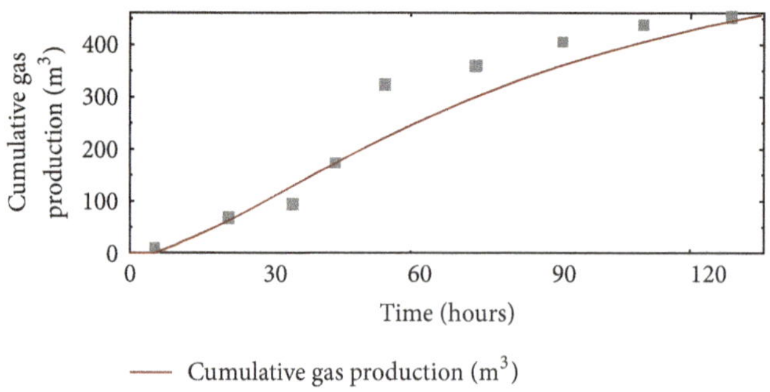

FIGURE 11: Poroelastic model of sediment validated with Mallik 5L-38 well data.

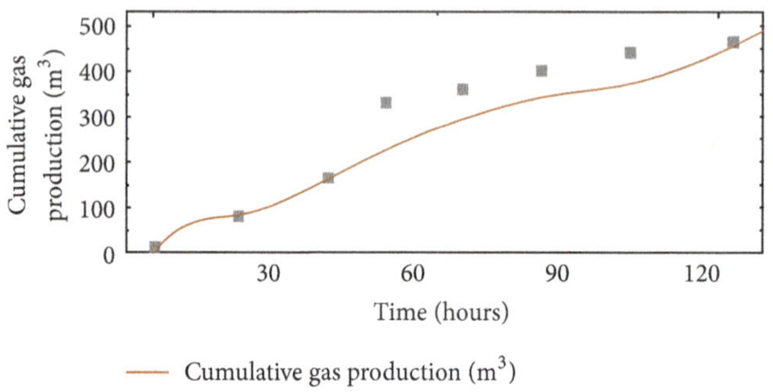

FIGURE 12: Viscoplastic model of sediment validated with Mallik 5L-38 well data.

the production rate due to greater mobility for gas in higher porosity sediment (with higher permeability).

Gas production rates at different initial hydrate saturations are shown in Figure 9. The production rates are dependent on initial hydrate saturation in a way that is expected, with higher saturations yielding greater amount of gas produced per day.

For a constant temperature of the heat source, a higher saturation of hydrate is expected to correspond to larger volumes of produced CH_4 because of larger hydrate abundance. The substantial increase in the volume of the released gas when hydrate saturation increases from 0.5 to 0.8 as shown in Figure 9 confirms this expectation. Higher saturations of hydrate mean richer sediment and higher production rates. The effect of gas production rate on the heat source temperature is shown below in Figure 10.

The heat source temperature has a marked influence upon the rate of production of gas, with a factor of ~50% increase in production with 12.5% increase in temperature. But maintaining the heat source temperature at say 450 K is a costly proposition. A more economically sustainable temperature would be around 350–360 K.

2.12. Validation of Model. The three soil models were validated with the well data from JAPEX/JNOC/GSC and others, Mallik 5L-38, gas hydrate production research well [18] as shown in Figures 11 and 12.

2.13. FEM Profiles. Figures 13, 14, 15, 16, and 17 show the raw Abaqus screen output for parameters such as nodal temperature, temperature, and porosity across the width and depth of sediment.

3. Conclusions

A commercial package Abaqus was customized to simulate the gas production from natural gas hydrate by considering the deformation of submarine bed. The effect of sediment deformation on gas production by thermal stimulation is studied. The effect of three soil dynamic models, which are generally used for submarine sediments, on the prediction of the gas production has been studied. Gas production rate is found to increase with an increase in the source temperature. Porosity of the sediment and saturation of the hydrate both have been found to significantly influence the rate of gas production.

The economics of the method of gas production from gas hydrates using thermal stimulation has been evaluated in the literature prior to this work. But such evaluations (Table 1) have always assumed that the gas production rate cannot be increased significantly by raising the temperature of the heating fluid. This study has found evidence to the contrary (Figure 10) which, though not conclusive, opens up new possibilities in the economic extraction of natural gas from ocean gas hydrate deposits worldwide in general and the KG and KK basins in India in particular.

Table 1: Economics of gas production from hydrates (modified from [19]).

	Gas extraction methods		
	Thermal stimulation	Depressurization	Conventional gas production
Investment in millions of ₹*	254200	166000	157500
Annual cost in millions of ₹	160000	125500	100000
Total production (million m^3/year)	1274.26	1557.43	1557.43
Production cost (₹/million m^3)	6356.6	4025.9	3213.6
Break-even wellhead price (₹/million m^3)	7945.8	5032.3	3972.9

*Calculated for currency rates at 2004 levels.

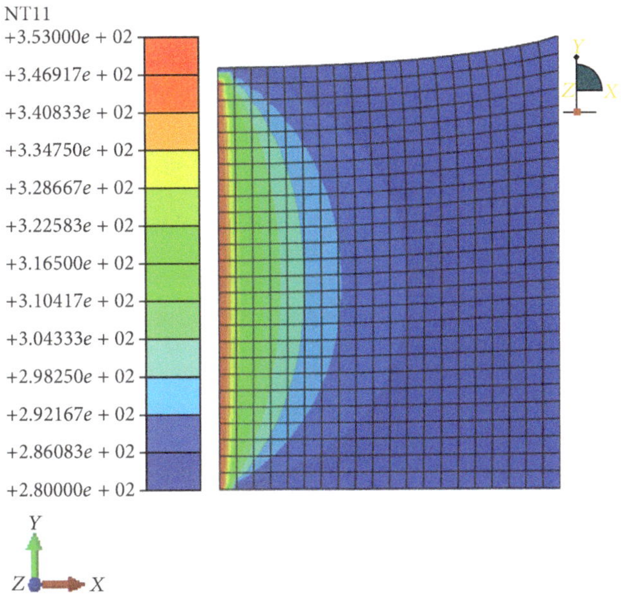

Figure 13: Nodal temperature profile of the sediment bed (with deformation). Radius of bed is 10 m, and depth is 13 m. Initial temperature was 280 K.

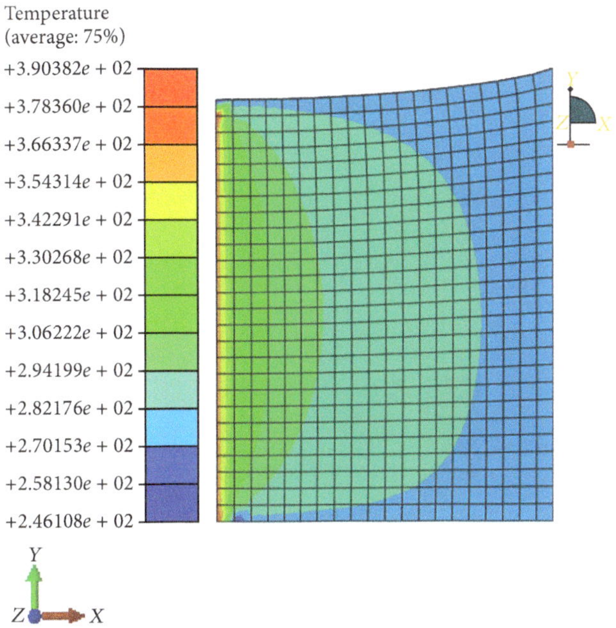

Figure 14: Temperature distribution after 50 days (with deformation).

The higher overall annual average temperature of the regions where these basins are located do mitigate to some extent the energy overhead required to maintain a high temperature flow to the sediment bed.

The earlier techniques of hot fluid injection have many limitations as compared to the method studied in this work. In the earlier method, the heat losses to adjacent sediments are too big to permit economic extraction of gas from hydrate reservoirs by steam injection, even if the steam can be introduced at high rates (15 MW) into impenetrable hydrate reservoirs.

So also, low injection temperatures involve very big volumetric flow rates to carry useful amounts of heat into the reservoir. Injection of approximately 3600 m^3 per day of 66°C water is required if a heat flux of 15 MW is to be maintained. The limitations of disproportionate heat losses on the one hand and unrealistically high injection flow rates on the other hand will probably limit injection temperatures to between 66 and 120°C.

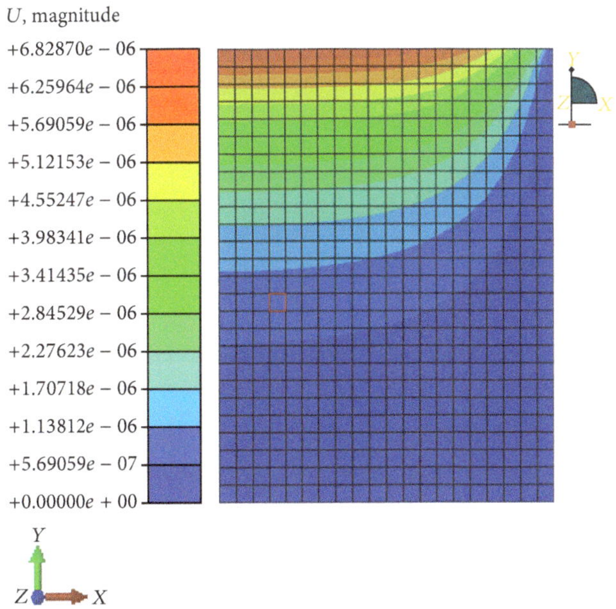

Figure 15: Magnitude of deformation (in m) after 6 hours.

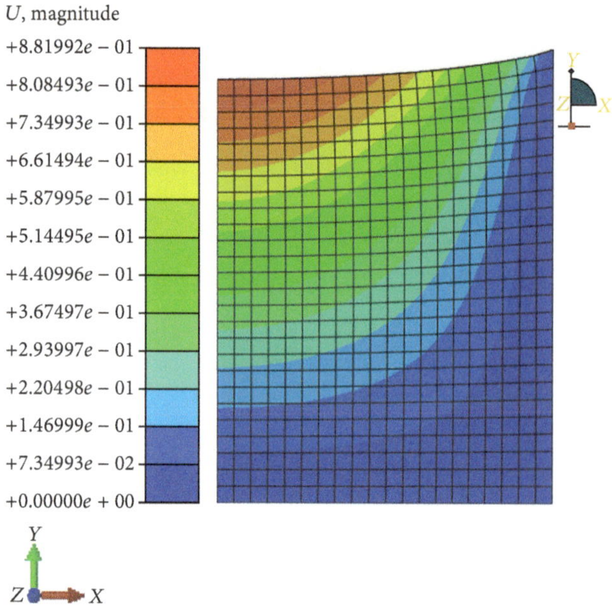

FIGURE 16: Magnitude of deformation after 50 days (m).

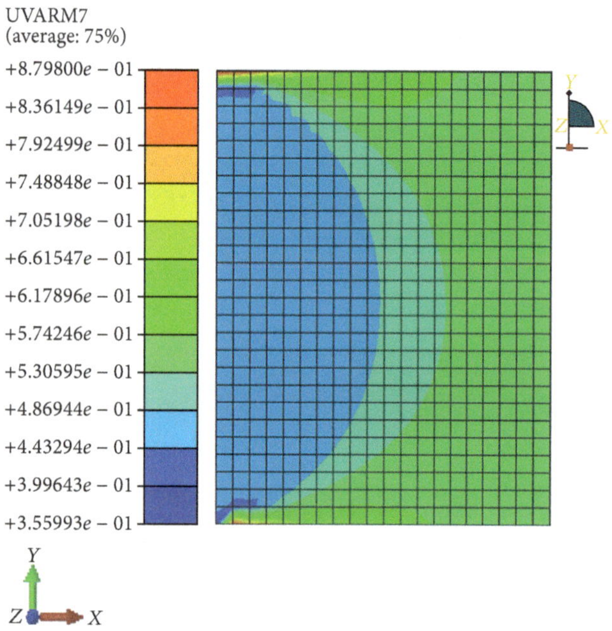

FIGURE 17: Porosity of sediment after 50 days.

Similarly, unless the porosity is at least 0.15, the heat lost in raising the soil matrix temperature will render thermal stimulation (by injection) ineffective in producing useful quantities of gas. The relative importance of porosity in determining gas production is illustrated in Figure 8.

Scope for future work could be as follows.

(i) Consideration of inhomogeneity and anisotropy of the hydrate bed.

(ii) Study of the interaction between the multiphase flow and the porous bed.

(iii) Study of hybrid and alternate techniques for gas production.

(iv) Further experimental validation of the above simulated results.

References

[1] "DGH Annual Activity Report 2010-2011," Hydrocarbon exploration and production activities.

[2] Y. F. Makogon, "Natural gas hydrates: a promising source of energy," *Journal of Natural Gas Science and Engineering*, vol. 2, no. 1, pp. 49–59, 2010.

[3] M. H. Yousif, H. H. Abass, M. S. Selim, and E. D. Sloan, "Experimental and theoretical investigation of methane gas hydrate dissociation in porous media," in *Proceedings of the SPE Annual Technical Conference & Exhibition*, pp. 571–18320, October 1988.

[4] M. S. Selim and E. D. Sloan, "Heat and mass transfer during the dissociation of hydrate in porous media," *AIChE Journal*, vol. 35, pp. 1049–1052, 1989.

[5] G. G. Tsypkin, "Mathematical model for dissociation of gas hydrates coexisting with gas in strata," *Doklady Physics*, vol. 46, no. 11, pp. 806–809, 2001.

[6] C. Ji, G. Ahmadi, and D. H. Smith, "Natural gas production from hydrate decomposition by depressurization," *Chemical Engineering Science*, vol. 56, no. 20, pp. 5801–5814, 2001.

[7] Y. Bai and Q. Li, "Simulation of gas production from hydrate reservoir by the combination of warm water flooding and depressurization," *Science China Technological Sciences*, vol. 53, no. 9, pp. 2469–2476, 2010.

[8] X. Sun, N. Nanchary, and K. K. Mohanty, "1-D modeling of hydrate depressurization in porous media," *Transport in Porous Media*, vol. 58, no. 3, pp. 315–338, 2005.

[9] S. Kimoto, F. Oka, T. Fushita, and M. Fujiwaki, "A chemo-thermo-mechanically coupled numerical simulation of the subsurface ground deformations due to methane hydrate dissociation," *Computers and Geotechnics*, vol. 34, no. 4, pp. 216–228, 2007.

[10] S. Kimoto, F. Oka, and T. Fushita, "A chemo-thermo-mechanically coupled analysis of ground deformation induced by gas hydrate dissociation," *International Journal of Mechanical Sciences*, vol. 52, no. 2, pp. 365–376, 2010.

[11] A. Vysniauskas and P. R. Bishnoi, "A kinetic study of methane hydrate formation," *Chemical Engineering Science*, vol. 38, no. 7, pp. 1061–1072, 1983.

[12] P. Englezos, N. Kalogerakis, P. D. Dholabhai, and P. R. Bishnoi, "Kinetics of formation of methane and ethane gas hydrates," *Chemical Engineering Science*, vol. 42, no. 11, pp. 2647–2658, 1987.

[13] H. C. Kim, P. R. Bishnoi, R. A. Heidemann, and S. S. H. Rizvi, "Kinetics of methane hydrate decomposition," *Chemical Engineering Science*, vol. 42, no. 7, pp. 1645–1653, 1987.

[14] J. S. Pic, J. M. Herri, and M. Cournil, "Experimental influence of kinetic inhibitors on methane hydrate particle size distribution during batch crystallization in water," *Canadian Journal of Chemical Engineering*, vol. 79, no. 3, pp. 374–383, 2001.

[15] P. D. Dholabhai, N. Kalogerakis, and P. R. Bishnoi, "Kinetics of methane hydrate formation in aqueous electrolyte solutions," *Canadian Journal of Chemical Engineering*, vol. 71, no. 1, pp. 68–74, 1993.

[16] N. Gnanendran and R. Amin, "Modelling hydrate formation kinetics of a hydrate promoter-water-natural gas system in a semi-batch spray reactor," *Chemical Engineering Science*, vol. 59, no. 18, pp. 3849–3863, 2004.

[17] D. Kashchiev and A. Firoozabadi, "Driving force for crystallization of gas hydrates," *Journal of Crystal Growth*, vol. 241, no. 1-2, pp. 220–230, 2002.

[18] S. H. Hancock, T. S. Collett, and S. R. Dallimore, "Overview of thermal-stimulation production test results for the JAPEX/JNOC/GSC et al. Mallik 5L-38 gas hydrate production research well," 2005.

[19] B. S. Pierce and T. S. Collett, "Energy resource potential of natural gas hydrates," in *Proceedings of the 5th Conference Exposition on Petroleum Geophysics*, pp. 899–903, Hyderabad, India, 2004.

Permissions

The contributors of this book come from diverse backgrounds, making this book a truly international effort. This book will bring forth new frontiers with its revolutionizing research information and detailed analysis of the nascent developments around the world.

We would like to thank all the contributing authors for lending their expertise to make the book truly unique. They have played a crucial role in the development of this book. Without their invaluable contributions this book wouldn't have been possible. They have made vital efforts to compile up to date information on the varied aspects of this subject to make this book a valuable addition to the collection of many professionals and students.

This book was conceptualized with the vision of imparting up-to-date information and advanced data in this field. To ensure the same, a matchless editorial board was set up. Every individual on the board went through rigorous rounds of assessment to prove their worth. After which they invested a large part of their time researching and compiling the most relevant data for our readers. Conferences and sessions were held from time to time between the editorial board and the contributing authors to present the data in the most comprehensible form. The editorial team has worked tirelessly to provide valuable and valid information to help people across the globe.

Every chapter published in this book has been scrutinized by our experts. Their significance has been extensively debated. The topics covered herein carry significant findings which will fuel the growth of the discipline. They may even be implemented as practical applications or may be referred to as a beginning point for another development. Chapters in this book were first published by Hindawi Publishing Corporation; hereby published with permission under the Creative Commons Attribution License or equivalent.

The editorial board has been involved in producing this book since its inception. They have spent rigorous hours researching and exploring the diverse topics which have resulted in the successful publishing of this book. They have passed on their knowledge of decades through this book. To expedite this challenging task, the publisher supported the team at every step. A small team of assistant editors was also appointed to further simplify the editing procedure and attain best results for the readers.

Our editorial team has been hand-picked from every corner of the world. Their multi-ethnicity adds dynamic inputs to the discussions which result in innovative outcomes. These outcomes are then further discussed with the researchers and contributors who give their valuable feedback and opinion regarding the same. The feedback is then collaborated with the researches and they are edited in a comprehensive manner to aid the understanding of the subject.

Apart from the editorial board, the designing team has also invested a significant amount of their time in understanding the subject and creating the most relevant covers. They scrutinized every image to scout for the most suitable representation of the subject and create an appropriate cover for the book.

The publishing team has been involved in this book since its early stages. They were actively engaged in every process, be it collecting the data, connecting with the contributors or procuring relevant information. The team has been an ardent support to the editorial, designing and production team. Their endless efforts to recruit the best for this project, has resulted in the accomplishment of this book. They are a veteran in the field of academics and their pool of knowledge is as vast as their experience in printing. Their expertise and guidance has proved useful at every step. Their uncompromising quality standards have made this book an exceptional effort. Their encouragement from time to time has been an inspiration for everyone.

The publisher and the editorial board hope that this book will prove to be a valuable piece of knowledge for researchers, students, practitioners and scholars across the globe.

List of Contributors

Zheng Shen and Frederick E. Beck
Texas A&M University, College Station, TX 77843, USA

Kegang Ling
University of North Dakota, Grand Forks, ND 58202, USA

Emad Waleed Al-Shalabi, Kamy Sepehrnoori and Gary Pope
Department of Petroleum and Geosystems Engineering,The University of Texas at Austin, Austin, TX 78712, USA

Saurabh Mishra, Achinta Bera and Ajay Mandal
Department of Petroleum Engineering, Indian School of Mines, Dhanbad 826004, India

Vahid Alipour Tabrizy
Department of Reserve Replacement, ˚ Asgard PTC, Statoil ASA, 7500 Stjørdal, Norway
Department of Petroleum Engineering, Faculty of Science and Technology, University of Stavanger, 4036 Stavanger, Norway

Titus N. Ofei and Sonny Irawan
Petroleum Engineering Department, Universiti Teknologi PETRONAS, Bandar Seri Iskandar, 31750 Tronoh, Malaysia

William Pao
Mechanical Engineering Department, Universiti Teknologi PETRONAS, Bandar Seri Iskandar, 31750 Tronoh, Malaysia

Yuxue Sun, Fei Teng and Jingyuan Zhao
Northeast Petroleum University, Daqing, Heilongjiang 163318, China

Wei Yu and Kamy Sepehrnoori
Department of Petroleum and Geosystems Engineering, The University of Texas at Austin, Austin, TX 78712, USA

Eshagh Rezaee Nezhad
Department of Chemistry, Faculty of Sciences, Shahid Chamran University, Ahvaz 61357-43337, Iran
Department of Chemistry, Payame Noor University, P.O. BOX 19395-4697, Tehran, Iran

Fariba Heidarizadeh
Department of Chemistry, Faculty of Sciences, Shahid Chamran University, Ahvaz 61357-43337, Iran

Sami Sajjadifar and Zahra Abbasi
Department of Chemistry, Payame Noor University, P.O. BOX 19395-4697, Tehran, Iran

M. Yadav and Sumit Yadav
Department of Applied Chemistry, Indian School of Mines, Dhanbad 826004, India

P.N. Yadav
Department of Physics, Post Graduate College, Ghazipur 233001, India

Albina Mukhametshina and Elena Martynova
Harold Vance Department of Petroleum Engineering, Texas A&M University, 3116 TAMU-407 Richardson Building, College Station, TX, USA
Gubkin Russian State University of Oil and Gas, 65 Leninsky Prospekt,Moscow, Russia

Liu He, Gao Yang and Li Yiliang
RIPED, PetroChina, Beijing 100083, China

Li Guoxin
Exploration & Production Company, PetroChina, Beijing 100083, China

Thiago Feital and José Carlos Pinto
Programa de Engenharia Qu´ımica/COPPE, Universidade Federal do Rio de Janeiro, 68502 Rio de Janeiro, RJ, Brazil

Públio Lima
Escola Polit´ecnica, Universidade Federal do Rio de Janeiro, 68529 Rio de Janeiro, RJ, Brazil

Maurício B. de Souza Jr.
Departamento de Engenharia Qu´ımica/EQ, Universidade Federal do Rio de Janeiro, 68542 Rio de Janeiro, RJ, Brazil

Gilberto Xavier
PETROBRAS Petr´oleo Brasileiro S.A., Rio de Janeiro, RJ, Brazil
Faculdade de Tecnologia de Resende, Universidade do Estado do Rio de Janeiro, Rodovia Presidente Dutra Km 298, Resende, RJ, Brazil

Mario Jorge Lima
PETROBRAS Petr´oleo Brasileiro S.A., Rio de Janeiro, RJ, Brazil

Marcel Joly
PETROBRAS Petr´oleo Brasileiro S.A., Rio de Janeiro, RJ, Brazil
Departamento de Engenharia Qu´ımica, Universidade de S˜ao Paulo, S˜ao Paulo, SP, Brazil

Upendra Singh Yadav and Vikas Mahto
Department of Petroleum Engineering, Indian School of Mines, Dhanbad 826004, India

Adriane B. S. Serapião
DEMAC, IGCE, UNESP, CP 178, 13506-900 Rio Claro, SP, Brazil

Antonio C. Bannwart
DEP, FEM, UNICAMP, CP 6122, 13081-970 Campinas, SP, Brazil

K. K. Salam, A. O. Alade, A. O. Arinkoola and A. Opawale
Petroleum Engineering Unit, Department of Chemical Engineering, Ladoke Akintola University of Technology (LAUTECH), PMB 4000, Ogbomoso, Nigeria

Qin He, Shahab D. Mohaghegh and Vida Gholami
Department of Petroleum and Natural Gas Engineering, West Virginia University, Morgantown, WV 26505, USA

Shayan Tavassoli, Wei Yu and Kamy Sepehrnoori
Department of Petroleum and Geosystems Engineering,The University of Texas at Austin, Austin, TX 78712, USA

Farzam Javadpour
Bureau of Economic Geology, Jackson School of Geosciences,The University of Texas at Austin, Austin, TX 78713, USA

Hossein Rahmati, Mahshid Jafarpour, Saman Azadbakht, Alireza Nouri and Dave Chan
Department of Civil and Environmental Engineering, University of Alberta, Edmonton, AB, Canada T6G 2W2

Hans Vaziri and Yuxing Xiao
BP America Inc., Houston, TX 77079, USA

Mohamed Iqbal Pallipurath
Mechanical Engineering Department, TKM College of Engineering, Kollam, Kerala 691005, India